T0139205

Managing
TRUST
in Cyberspace

Managing
TRUST
in Cyberspace

Edited by
Sabu M. Thampi
Bharat Bhargava
Pradeep K. Atrey

CRC Press is an imprint of the
Taylor & Francis Group, an **informa** business

A CHAPMAN & HALL BOOK

CRC Press
Taylor & Francis Group
6000 Broken Sound Parkway NW, Suite 300
Boca Raton, FL 33487-2742

Printed on acid-free paper
Version Date: 20131029

International Standard Book Number-13: 978-1-4665-6844-0 (Hardback)

Visit the Taylor & Francis Web site at
http://www.taylorandfrancis.com

and the CRC Press Web site at
http://www.crcpress.com

Contents

Preface

Traditional security mechanisms restrict access to authorized users only, in order to protect resources from malicious users. However, in many contexts, we must protect ourselves from those who offer resources, so that the problem is in fact reversed. This improper usage cannot be addressed by traditional security mechanisms, and the issues relate more to trustworthiness than security. Trust is a vital factor in our daily coexistence with other people, who can be unpredictable. Trust helps to reduce the uncertainty caused by this unpredictability to an acceptable level. The notion of trust includes trust management systems. These systems gather information required for creating a trust relationship and dynamically monitor and fine-tune present relationships. Thus, trust management provides a basis for cooperation to develop. The schemes include both centralized and distributed approaches.

The main aim of this book is to provide relevant theoretical frameworks and the latest research findings in the area of trust management. This includes cross-disciplinary examination of fundamental issues underpinning computational trust models. The book thoroughly discusses various trust management processes for dynamic open systems and their applications. Moreover, most of the chapters are written in a tutorial style so that even readers without a specialized knowledge of the subject can easily grasp some of the ideas in this area. There are 21 chapters in this book which discuss trust and security in cloud computing, peer-to-peer (P2P) networks, autonomic networks, multiagent systems, vehicular *ad hoc* networks, digital rights management, e-commerce, e-governance, embedded computing, and a number of other topics. The intended audience of this book mainly consists of graduate students, researchers, academics, and industry practitioners working in areas such as distributed computing and Internet technologies. The book is not written in textbook style. Most of its content is based on the latest research findings. It also discusses future research directions in trust management. It is hoped that this book will influence more individuals to pursue high-quality research on trust management.

We acknowledge and thank many people for our success in completing this task. We convey our appreciation to all contributors, including the authors. In addition, we are deeply indebted to the reviewers. We thank CRC Press Project Coordinator David Fausel and the CRC Press Editorial support staff for the excellent editing and formatting. Many thanks go to Aastha Sharma, Commissioning Editor at Taylor & Francis India, for her help and cooperation.

Dr. Sabu M. Thampi
Indian Institute of Information Technology and Management-Kerala

Dr. Bharat Bhargava
Purdue University

Dr. Pradeep K. Atrey
University of Winnipeg

Contributors

Aakanksha
Department of Computer Science
University of Delhi
New Delhi, India

Ajith Abraham
Machine Intelligence Research
    Labs (MIR Labs)
Auburn, Washington

Muhammad Akhlaq
Information and Computer Science
    Department
King Fahd University of Petroleum
    and Minerals
Dhahran, Saudi Arabia

Farag Azzedin
Information and Computer Science
    Department
King Fahd University of Petroleum
    and Minerals
Dhahran, Saudi Arabia

Manoj Balakrishnan
Indian Institute of Space Science
    and Technology (IIST)
Trivandrum, Kerala, India

Punam Bedi
Department of Computer Science
University of Delhi
New Delhi, India

Sumitra Binu
Department of Computer Science
Christ University
Bangalore, Karnataka, India

Stephen Cai
Auckland University
    of Technology
Auckland, New Zealand

Priya Chandran
Department of Computer Science
    and Engineering
National Institute of Technology
    Calicut,
Kozhikode, Kerala, India

Rahul Chandran
Auckland University
    of Technology
Auckland, New Zealand

Nitin Singh Chauhan
Infosys Labs
Infosys Limited
Hyderabad, Andhra Pradesh,
    India

Brijesh Chejerla
Department of Computer
    Science
Missouri University of Science
    and Technology
Rolla, Missouri

Pethuru Raj Chelliah
Wipro Consulting Services
Bangalore, Karnataka, India

Narayan C. Debnath
Department of Computer Science
Winona State University
Winona, Minnesota

Sabariyah Din
Razak School of Engineering and
    Advanced Technology
Universiti Teknologi Malaysia
Kuala Lumpur, Malaysia

Stephen Faatamai
Auckland University of Technology
Auckland, New Zealand

Eliot Foye
Auckland University of Technology
Auckland, New Zealand

Richa Garg
Infosys Labs
Infosys Limited
Hyderabad, Andhra Pradesh, India

Bhavna Gupta
Department of Computer Science
Keshav Mahavidhyalaya
University of Delhi
Delhi, India

Sanchika Gupta
Indian Institute of Technology
Roorkee, Uttarakhand, India

Mindy Hsieh
Auckland University of Technology
Auckland, New Zealand

Gary Huo
Auckland University of Technology
Auckland, New Zealand

Alex Pappachen James
Department of Electrical and
    Electronic Engineering
School of Engineering
Nazarbayev University
Astana, Kazakhstan

Shyam P. Joy
Department of Computer Science
    and Engineering
National Institute of Technology
    Calicut
Kozhikode, Kerala, India

Harmeet Kaur
Department of Computer Science
Hans Raj College
University of Delhi
Delhi, India

Karuppanan Komathy
Easwari Engineering College
Chennai, Tamil Nadu, India

Vimal Kumar
Department of Computer Science
Missouri University of Science and
    Technology
Rolla, Missouri

Padam Kumar
Indian Institute of Technology
Roorkee, Uttarakhand, India

Edith AiLing Lim
School of Business and Design
Swinburne University
    of Technology
Kuching, Malaysia

William Liu
Auckland University
    of Technology
Auckland, New Zealand

Sanjay Madria
Department of Computer Science
Missouri University of Science
    and Technology
Rolla, Missouri

Sajjad Mahmood
Information and Computer Science
    Department
King Fahd University of Petroleum
    and Minerals
Dhahran, Saudi Arabia

Maslin Masrom
Razak School of Engineering and
    Advanced Technology
Universiti Teknologi Malaysia
Kuala Lumpur, Malaysia

Mukesh Mohania
IBM Research Lab
New Delhi, India

German Montejano
Universidad Nacional de San Luis
Universidad Nacional de La
    Pampa
San Luis, Argentina

Emery Moodley
Auckland University
    of Technology
Auckland, New Zealand

Al-Sakib Khan Pathan
Department of Computer Science
International Islamic University
    Malaysia
Kuala Lumpur, Malaysia

Shivani Prasad
Auckland University of
    Technology
Auckland, New Zealand

Daniel Riesco
Universidad Nacional de San Luis
San Luis, Argentina

Sravan Rondla
Infosys Labs
Infosys Limited
Hyderabad, Andhra Pradesh, India

Ashutosh Saxena
Infosys Labs
Infosys Limited
Hyderabad, Andhra Pradesh,
    India

Sugathan Sherin
School of CS and IT
Indian Institute of Information
    Technology and Management-
    Kerala (IIITM-K)
Technopark Campus
Trivandrum, Kerala, India

Makkuva Shyam Vinay
Indian Institute of Space Science
    and Technology (IIST)
Trivandrum, Kerala, India

Axel Sikora
Department of Electrical
    Engineering and Information
    Technologies
University of Applied Sciences
    Offenburg
Offenburg, Germany

Steven Sivan
Auckland University
    of Technology
Auckland, New Zealand

Tony Thomas
Indian Institute of Information
    Technology and Management-
    Kerala (IIITM-K)
Trivandrum, Kerala, India

Roberto Uzal
Universidad Nacional de San Luis
San Luis, Argentina

Pooja Vashisth
Department of Computer Science
University of Delhi
New Delhi, India

Ravi Sankar Veerubhotla
Infosys Labs
Infosys Limited
Hyderabad, Andhra Pradesh, India

Wei Q. Yan
Auckland University of Technology
Auckland, New Zealand

1

Analyzing Trust and Security in Computing and Communications Systems

Al-Sakib Khan Pathan

CONTENTS

Defining Boundaries of Trust and Security

Whether trust should be considered completely within the perimeter of security in computing and communications systems is a debatable issue. In normal human life, we see these two terms go hand in hand to define the relationships we might have with fellow human beings. Trust among people sets the level of security felt by each person involved in various relationships. If person A does not trust person B, person A may not feel secure in the company of person B. Similarly, these terminologies could also retain the same meanings for our modern computing and communications equipments. However, a notable difference is that whereas trust and security are quite interrelated in human-life scenarios; in technical fields, these are considered as two clearly different issues with distinct boundary lines. In computing and communications fields, trust is a kind of vague term that sets the outline of a task or communication event, based on which the operation can be performed. On the other hand, security is a broad concept that ensures communications go forward in a desired manner, maintaining the core aspects of security intact, that is, privacy, authenticity, authority, integrity, and nonrepudiation. The relation between trust and security could be seen in the way that security includes the concept of trust partially, and trust stays as a wrapper before any secure or insecure communication happens within a network of devices.

As we progress toward a big transformation in the information technology (IT) world with the development of innovative communications and computing technologies, managing trust among participating entities has become one of the major concerns today. The entities may mean different types of technical devices as well as the users or the people who are associated with these devices (or who are taking advantage of these devices for their daily interactions among themselves and with other species). To initiate any communication, it is considered that the initiator already gives a trust value to the entity that is communicated. Based on the acceptable reply, the communication door may be opened or terminated, which means that either that trust is established initially or broken afterward. If trust is maintained by both entities and they enter into any secure communication using some kind of mechanism, we say that both the entities have used trust to achieve secure communication and the security of the communication has ensured a better trust level among the participants. In real life, it may take a long time to establish trust among entities based on some kind of knowledge base about the participating entities. A suspicious action by any entity may threaten long-established trust; again a proven wrong action of an entity may completely tear down long-established trust.

Various Types of Security Models

General Types of Security Models

As with all security designs, some trade-off occurs between user productivity and security measures. The goal of any security design is to provide maximum security with minimum impact on user access and productivity. There are many security measures, such as network data encryption, that do not restrict access and productivity. On the other hand, cumbersome or unnecessarily redundant verification and authorization systems can frustrate users and prevent access to critical network resources. Some trust value should be given preference in such cases. We should understand that network is a tool designed to enhance production. If the security measures that are

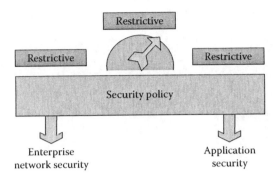

FIGURE 1.1
Network security policies.

put in place become too cumbersome, they will actually affect productivity negatively rather than enhancing it.

Here, we mention that networks are used as productivity tools, and they should be designed in such a way that business dictates security policy. A security policy should not determine how a business operates. As organizations are constantly subject to change, security policies must be systematically updated to reflect new business directions, technological changes, and resource allocations.

Three general types of security models exist: *open, restrictive,* and *closed.* Figure 1.1 shows a conceptual diagram for these. As shown in Figure 1.1, our inclination is often some kind of restrictive for either enterprise or application security. Now, let us see what these models are about.

Open Security Model

Today, the Internet and network technologies have expanded into numerous dimensions. When a simple definition is given such as the *Internet is the network of networks,* it does not explain how complex and dynamic the coverage of networks could be or how layered the communication processes could be. Figure 1.2 shows a diagram that depicts a typical scenario of such openness and diversity of network and communications technologies bound together under the concept of the "Internet" or other concepts that are derived from it. For instance, Internet-based intranets or extranets commonly termed as virtual private networks (VPNs) is a technology for using the Internet or another intermediate network to connect computers to isolated remote computer networks that would otherwise be inaccessible. Such networks could also include wireless network connectivity. Various remote sites could use other kinds of technologies, such as public switched telephone network (PSTN). A PSTN consists of telephone lines, fiber optic cables, microwave transmission links, cellular networks, communications satellites, and undersea telephone cables, all interconnected by switching centers, thus allowing any telephone in the world to communicate with any other. Given this

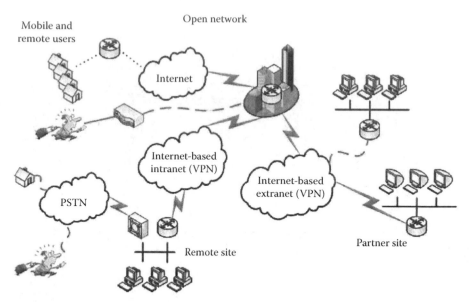

FIGURE 1.2
Open security model for a dynamic open network model.

dynamism in the open network's infrastructure, a common consensus of security objectives is literally impossible to reach. Here, the notion of an open security model as we perceive it comes in. We define an open security model as a setting or set of security policies and principles which could be generally applicable for an open network model as presented in Figure 1.2. The common objective is that all transactions and communications would meet the basic idea of security, that is, authenticity will be ensured, authorization will be maintained, data integrity could be verified, confidentiality will be ensured, and nonrepudiation could be confirmed. However, which portion needs which level of security (or, security parameters) should be a matter of open standards and requirements.

Closed Security Model

In a closed network, a limited number of parties are involved, and often there might be a central network administrator who could be given sufficient power to administer over the network on various policies. A closed network model is shown in Figure 1.3. In such a setting, it is possible to impose strict regulations as to who could participate, who could not, methods of authentication, mechanisms and protocols of communications, validity of a participating node, and so on. Hence, a closed security model is locally applicable for a particular, small portion of a network, and is more reserved in terms of principles than of those in an open network. For instance, in Figure 1.3, any of the PSTN networks can be administered with their own set of security rules.

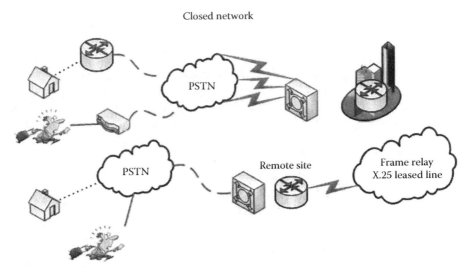

Closed network

FIGURE 1.3
A closed network where closed security model is applicable.

The question is *where does the matter of trust come into play here?* The answer, as understood, is *this should be determined by the decision-making authority.*

Restrictive Security Model

The restrictive security model is basically a concept applicable for a network of any size. The idea behind this is that a restrictive set of rules are applied on the network, which can be meaningful to various kinds of nodes and players in the network. As a general restriction of security policy, each of the participants can be treated as legitimate entities unless proven as a rogue unit in the network.

Reputation-Based Trust and Security

Now let us learn more about trust. Because establishing trust requires long-time observed behavioral analysis of an entity (in most of our perceived domains), a commonly used technique is to give some kind of reputation value [1] to each participating entity in a system. This value increases with reliable contributions of the entity (i.e., a device or equipment or a user or a client) in the system, whereas it decreases with wrong or suspicious behavior. Reputation can be analyzed using various methods: Sometimes peer information is used, sometimes some sort of regional information is kept, at other times global knowledge is used for smaller scale systems for maintaining the value of reputation, and so on. Various works have mentioned using game theory [2], probability model [3], fuzzy logic [4], and other techniques (e.g., distributed reputation management [5], using reputation matrix [6] for reputation analysis).

Like reputation-based trust, reputation-based security uses a similar idea of keeping the reputation of computing and communication devices at an acceptable level, but the difference is that it is channeled toward maintaining protection that is sufficient to meet the aspects of security at the system's operational level. Often trust remains as a wrapper and security is more meaningful for the actual implementation of some steps for communications using the reputation that is gained.

Trust and Security in Prominent Future Technologies

Pervasive and Ubiquitous Computing

Ubiquitous computing or pervasive computing is used interchangeably to label the use of computing technologies and devices in daily human life. Its basic idea is to make technology easily accessible to human beings instead of coercing them to adapt to technological advancements. This field basically stems from the human–computer interaction (HCI) field, which then got support from other fields, and a complex blend of various other areas contributed to its development. If technical devices blend with usual human life, trusting the objects within the environment comes forward as an issue of paramount importance. Such trust among the electronic/ technical devices can be achieved through reputation-based systems as discussed earlier.

On the other hand, security in the case of pervasive computing is often as difficult to define as the reality is; it is possible to give a sense of security in pervasive computing environment [7], but as the devices could be of different platforms and features, it is actually a complicated task to ensure proper security in pervasive communications.

Internet of Things and Future Internet

The Internet of Things (IoT) is the terminology used for uniquely identifiable objects (or things) and their virtual representations in an Internet-like structure. The idea of IoT considers the same ubiquitous environment as that is for ubiquitous or pervasive computing. The basic difference between pervasive computing and IoT is that when the same setting is seen from a conceptual angle, it is termed a pervasive environment, but when seen from an angle of identifiable objects taking part in the system, it is called IoT, an Internet-like network of networks. As IoT and pervasive computing deal with the same environment of connected devices (i.e., different kinds of devices of diverse natures and computing powers), trust and security management are also similar to the methods discussed earlier. This means that a scheme developed for a pervasive computing environment could also be applicable in most of the cases to IoT, when the device-level trust and security are dealt with.

Although IoT has been clearly defined by this time, Future Internet is still an unclear term. Future Internet basically refers to a wide variety of research topics related to the idea of the Internet connecting numerous networking devices around the globe. If simply a relatively faster Internet with new devices and techniques is brought forward at the end, it could end up as just an extension of the current Internet. However, the basic vision behind Future Internet is that it is not the Internet we have seen so far; it may have a new way of working, it may have a new method of connecting devices, and there might even be a complete clean-slate approach of developing it. As the full operational definition is not yet finalized, trust and security in Future Internet are also in the preliminary survey stage and cannot be outlined clearly.

Wireless *Ad Hoc* and Sensor Networks

A wireless *ad hoc* network is a combination of computing nodes that can communicate with each other without the presence of a formal central entity (infrastructureless or semiinfrastructure based) and could be established anywhere and at any time. Each node in an *ad hoc* network can take the role of both a host and a router-like device within the network. There might be different forms of *ad hoc* networks such as mobile *ad hoc* networks (MANET), vehicular *ad hoc* networks (VANET), wireless mesh networks (WMN), wireless sensor networks (WSN), body area networks (BAN), personal area networks (PAN), etc. Though all of these contain some common features of *ad hoc* technology, WSN is a network to mention particularly as this type of network comes with the extra feature that it might have a base station, such as a fixed central entity for processing network packets and all the other sensor nodes in the network could be deployed on an *ad hoc* basis.

The basic characteristics of *ad hoc* networking require trust for initializing the communications process. In most of the cases, some kind of assumed model is needed to introduce the notion of belief and provide a dynamic measure of reliability and trustworthiness in the *ad hoc* scenario [8]. Despite having the extra advantage of some kind of central authority in WSN (where every participating entity is not very distributed like pure *ad hoc* networks), trust-related research works still need concrete achievements to reach a satisfactory level. Regarding security, there are hundreds of works on key management, secure routing, security services, and intrusion detection systems for any kind of *ad hoc* network [9].

Cloud Computing

Cloud computing is a recently coined term in which distributed processing is seen from the perspective of providing "computing service." The basic concept of cloud computing is payment based on the usage of resource and the resources of IT that do not exist at the users' end. This strategy enables the users to do tasks with their resource-constrained computing devices that they could not have done before. To support such types of communications and computing, there are many research issues to consider for resources to be managed efficiently and securely. In actuality, service dynamism, elasticity, and choices offered by this highly scalable technology are very attractive for various enterprises. These opportunities, however, create many challenges on considering trust and security. Cloud computing has opened up a new frontier of challenges by introducing a different type of trust scenario [10] in which technology needs the primary trust of the participating entities to start functioning.

While trust may still remain relevant to the cloud concept, security puts a major difficulty when authentication is taken into consideration, for example. The idea of cloud computing is to keep the identity of service providers in a cloudy state, but authentication, as a part of security, would require correct identification of objects so that the property of nonrepudiation could also be ensured. Nonrepudiation, in plain terms, means the ability to ensure that a party in a contract or a communication event cannot deny the authenticity of its own messages, packets, signature, or any kind of document that it generates. These conflicting principles have kept cloud computing security still a very challenging area of research. One possible solution is to maintain some kind of trusted service provider and using them through a trusted service-providing server or service manager. Another question in this field is that in such setting, how long should an established trust be maintained, or how can it be torn down while retaining the concept of cloud intact? (That is, the clients do not know from exactly which locations they are getting the required services or how their requests have been processed.) In fact, this is an area where trust and security are intertwined. The elements of cloud computing also blend into the notion of Future Internet, leading to the concept of cloud networking. Hence, advancement in the areas of trust and security in cloud computing may also contribute to the same areas in Future Internet.

Next-Generation Converged Networks

Before stating the relation between trust and security, let us recall the basic difference between the two terms *hybrid* and *converged*. While some research areas can be broken into more subareas and topics, there is a trend among researchers to think about combining different technologies. The term *hybrid* is commonly used for such cases. Often people talk about combining the physical structure, or various device platforms, or benefits, or services, or features of different technologies to get a hybrid system. However, the term *network convergence* or more specifically *converged network* goes beyond the notion of network hybridization or a hybrid network. In the telecommunications field, network convergence refers to the provision of telephone, video, and data communication services within a single network. So, while hybrid simply refers to combining apparently disjoint technologies/services, convergence refers to the appropriate mixing and smooth functionality of the joint system. That is why a converged network demands one pipe to deliver all forms of communication services. There is little doubt that the process of network convergence is primarily driven by the development of technology and demand. One of the primary goals of such integration is to deliver better services and lower prices to consumers. Instead of paying for each service or technology separately, users may be able to access a wider range of services and choose among more service providers, which would make the users the main beneficiaries as well as ensure the provider's profit. In fact, convergence allows service providers to adopt new business models, offer innovative services, avail special offers, and enter new markets.

Another definition of network convergence or simply *convergence* could be given as a broad term used to describe emerging telecommunications technologies and network architecture used to migrate multiple communications services into a single network. It means the future networks are also included in this concept. In some literature, a formal definition of converged network is thus: a cross-national perspective, as integration and digitalization. Integration, here, is defined as *a process of transformation measure by the degree to which diverse media such as phone, data broadcast, and information technology infrastructures are combined into a single seamless all purpose network architecture platform* [11]. Digitalization is not so much defined by its physical infrastructure but by the content or the medium. Sometimes *digitalization* means *breaking down signals into bytes consisting of ones and zeros*; it is as simple as that!

To address the issue and understand it from practical perspectives, we may observe that many recent initiatives such as next-generation networks (NGNs) and IP multimedia subsystems (IMSs) have been undertaken to provide seamless architecture for various access technologies, which harmonize with the objectives set by network convergence. Besides other innovative approaches, by integrating different technologies—service delivery platforms (SDPs) and IMS, for example—easy service delivery, execution, and management are now possible. Other issues, such as point-to-point and broadcasting communications, data and multimedia-oriented services, mobile and cellular systems, and fast fiber and mobility, remain challenging tasks and, to say the least, the end-to-end picture is still unclear. As it should be understood by this time, when the full picture is not clear, the same type of uncertainty in defining the boundaries of trust and security remains as is in the cases of other innovative technologies presented before.

Conclusion and Open Issues

In today's world, big companies such as Amazon and eBay (or similar companies) use a kind of rating system that is based on trust or that lays the foundation of trust on products sold using those channels. Security is built on the trust and is mainly used for allowing monetary transactions.

Trust management in reality is a complex mixture of different fields if we consider the computing and communication technologies together. In this chapter, some future technologies are discussed but there are numerous other emerging fields such as near-field communications (NFCs), electronic knowledge management, nanocommunication networks, etc., that will also need the support of trust and security. Researchers and practitioners from various fields such as networking, grid and cloud computing, distributed processing, information systems, HCI, and human behavior modeling could be the contributors, and the combination of different fields under this umbrella will become inevitable in the coming days.

Whatever the advancements we could get on a concrete scale in the coming days, some fundamental questions could still remain such as how long should an established trust be maintained or kept valid in any computing and communications system? If periodic refreshing and reestablishing of trust is needed, what could be the optimal interval for different settings? Will periodic reestablishment of trust affect the established secure channels? Will security be able to act on its own when trust parameters are completely ignored? The major questions will still remain: Will there be clear boundaries between trust and security in future innovative technologies, or should trust be considered within the perimeter of security?

References

1. Shmatikov, V. and Talcott, C., "Reputation-based trust management," *Journal of Computer Security*, 13(1): 167–190, 2005.
2. Seredynski, M. and Bouvry, P., "Evolutionary game theoretical analysis of reputation-based packet forwarding in civilian mobile ad hoc networks," *IEEE International Symposium on Parallel & Distributed Processing (IPDPS)*, May 23–29, IEEE, Rome, Italy, pp. 1–8, 2009.
3. Dong, P., Wang, H., and Zhang, H., "Probability-based trust management model for distributed e-commerce," *IEEE International Conference on Network Infrastructure and Digital Content (IC-NIDC)*, November 6–8, IEEE, Beijing, China, pp. 419–423, 2009.
4. Li, J., Liu, L., and Xu, J., "A P2P e-commerce reputation model based on fuzzy logic," *IEEE 10th International Conference on Computer and Information Technology (CIT)*, June 29–July 1, IEEE, Bradford, UK, pp. 1275–1279, 2010.

5. Bamasak, O. and Zhang, N., "A distributed reputation management scheme for mobile agent-based e-commerce applications," *The 2005 IEEE International Conference on e-Technology, e-Commerce and e-Service (EEE)*, March 29–April 1, IEEE, Hong Kong, pp. 270–275, 2005.
6. Wei, X., Ahmed, T., Chen, M., and Pathan, A.-S. K., "PeerMate: A malicious peer detection algorithm for P2P systems based on MSPCA," *International Conference on Computing, Networking and Communications (IEEE ICNC)*, January 30–February 2, Maui, HI, pp. 815–819, 2012.
7. Pagter, J. I. and Petersen, M. G., "A sense of security in pervasive computing: Is the light on when the refrigerator door is closed?" *Proceedings of the 11th International Conference on Financial Cryptography and 1st International Conference on Usable Security (FC/USEC)*, Heidelberg, Germany. Springer-Verlag, Berlin, 2007, available at http://usablesecurity.org/papers/pagter.pdf
8. Pirzada, A. A. and McDonald, C., "Establishing trust in pure ad-hoc networks," *Proceedings of the 27th Australasian Conference on Computer Science (ACSC)*, Dunedin, New Zealand, 26, pp. 47–54, 2007.
9. Pathan, A.-S. K., *Security of Self-Organizing Networks: MANET, WSN, WMN, VANET*, Boca Raton, FL: Auerbach Publications, 2010.
10. Khan, K. M. and Malluhi, Q., "Establishing trust in cloud computing," *IT Professional*, 12(5): 20–27, 2010.
11. Menon, S., "The role of diverse institutional agendas in the importation of convergence policy into the Indian communications sector," *Info Journal*, 7(4): 30–46, doi:10.1108/14636690510607286, Emerald, 2005.

2

Cloud Computing: Trust Issues, Challenges, and Solutions

Sanchika Gupta, Padam Kumar, and Ajith Abraham

CONTENTS

What Is Trust?

Defining Trust

There are many definitions of trust; however, we have listed down some of the important ones which are as follows:

- It is the percentage in which one party meets the behavior as expected by the other.
- It is the degree in which the first party behaves exactly as it was expected from the second party. If the degree is high, it represents a higher trust on the first party by the second one. In security domain, trust acts as an important parameter to determine the threat model of a system.
- It is represented in the form of a trust model. It can also be referred to as confidence.
- It is generally a binary relationship between two entities. It is established between two entities based on certain common attributes over which the confidence is analyzed and measured.

Figure 2.1 shows that trust is a relationship between two entities.

Trust Modeling

Trust modeling is the technique used for evaluating trust of a system. Trust modeling identifies the issues that can affect trust of a system and helps in identifying points where a low degree of trust can degrade the system usability. It helps to identify measures that can be applied on the system for making it more trustworthy for end users.

We can generalize the concept of trust modeling by taking an example into consideration. Let us take an example of a university transferring final examination papers to a college affiliated to it. The university had established a new scheme in which they transfer the examination papers to colleges

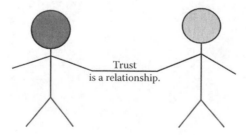

FIGURE 2.1
Trust is a relationship.

through normal e-mail communication system. The paper of a particular subject is sent with proper IDs, which is then collected at the other end and is printed one day before the day of final examination. Two parties communicate with each other through an underlying e-mail delivery system.

The university and colleges have enough security techniques to ensure that the college will take printout of only those exam papers that are sent from the university without any information leakage. The whole system is dependent on the trust over the e-mail delivery system (e-mail service provider). An e-mail communication system that behaves in the expected way as the university thinks (timely delivery of message and no tampering, spoofing, or information leakage) will have higher trust as compared to other e-mail delivery systems that cannot assure expected services. Hence, in this case the attributes over which the trust of an e-mail service provider can be analyzed are as follows:

1. Information leakage
2. Timely delivery of message
3. Spoofing and fraud detection schemes availability
4. Message integrity

If some e-mail delivery system can assure the university on these expected behavioral attributes of service, university will have more trust on the provider, and hence will have more confidence on services provided by it. But the questions are as follows: How is the assurance of expected services fulfilled by the provider? How will the provider make sure that the examination papers will reach the college properly as expected? How will he stand ahead of all other competing mail delivery systems for getting the bid from the university?

This is achieved by understanding the trust requirements for the system. By fulfilling the necessary attributes of the systems in a known and standard way and with the use of advanced technologies and architectural designs, the system will obtain valuable trust of its customer. For example, now in this case if a mail delivery system can prove to the university that it uses a 256-bit secure sockets layer (SSL) connection for encrypted communication of the message and does not store any part of the message in plaintext in its servers, the university can make sure that there will be no information leakage as data are strongly encrypted.

The e-mail system can also show that it appends a message hash and checks it at the other end; hence, it is impossible to tamper the data and hence they have mechanisms for providing message integrity. Also, it is possible to show that they use simple mail transfer protocol (SMTP) authentication and can identify any spoofed communication going from other ends. On-time delivery can be assured by sending a mail failure message in a few minutes after sending the message fails, so that the university can take proper action to deliver the exam papers in a timely manner instead of assuming they have reached safely at the other end.

Trust is defined hierarchically on any system. However, the trust over external entity, which is a medium of e-mail communication, is less, but even having a high amount of trust with external entity does not mean that the whole system is totally secure. For example, it is the responsibility of the university that the correct examination papers are sent with trusted e-mail communication medium by the people responsible for it. As the persons who are responsible for sending correct exam papers are internal to the university, they have more trust level when compared to an entity whose services are availed externally for completing the service. Hence, trust level of a system over various entities varies according to the entity, its type, and on evaluation of their expected behavior.

The next questions that come to mind are the following: Why do we care about trust? What is the role of trust and the trust model that we discuss? Why are we evaluating trust, and of what help will it be for us in determining the security aspects of a system?

Trust evaluation helps in the development of the trust model that represents hierarchical trust over various entities of a system. The trust model identifies various areas of the system and the trust an entity expects from other entities. Evaluating this trust model helps in determining confidence over entities of the system. A trust model represents those major areas of the system, where if trust is lost, the system would be vulnerable to a number of threats. For example, the trust model of the system discussed above will describe the external entity e-mail service provider to provide a higher trust level for the whole system to sustain and behave in the way it is meant to be. If entities have a higher trust over each other and if they abide by it, a system is expected to work in the best possible way.

The example that we discussed is for a quick description of trust and what it means. By entities, we will be referring to providers, users, or computational systems and subsystems based on the context. Our main focus will be considering the aspects of cloud computing and identifying the need for trust in it. What has been done and what is left will also be discussed.

What Is Cloud Computing?

Cloud Computing

In the past, we were using various technologies for computing, which includes distributed computing, grid computing, etc. One of the recent technologies, which provides computing resources on a more scalable, user-friendly, and pay-per-usage model, is available in the market and is known as cloud computing. This new service is known as the next-generation architecture for IT enterprises. Cloud computing services attract their customers because of the enormous advantages they provide. It is a scalable, efficient, and economical solution for storage and computing, in which users store their data in cloud storage servers dispersed across geographies that are managed centrally by cloud service providers. Essentially, it means that "NIST defines cloud computing as a model that provide on demand network access to shared resources that can be configured, allocated, managed, release with minimal efforts and interaction."

Users can employ well-known or specialized software and Web applications, demand for platforms for the execution of their programs and IT infrastructure overall for handling the entire set of IT operations. The concept of cloud computing from IT resources can be hired on rent, which helps in increasing the resource utilization of the IT infrastructure. Some big IT firms have spare resources that remain unutilized. This sharing of resources increases the efficiency of physical infrastructure with the use of virtualization technologies, and hence cloud computing provides a low-cost solution computing and data storage solution to its users. It provides many other facilities to its users and providers including remote and on-demand access, easier management, maintenance, and allocation with scalability of resources on demand.

Current Trends in Cloud Computing

Cloud computing-based services are getting a great amount of success both in the market for individual users and small business organizations, which do not want to spend money in buying and managing physical computing resources. Examples of cloud computing services that we use one day or the other include online data storage facilities, and e-mail service provider facilities where they provide handling of customized e-mail services for small business organizations. Others include the use of online Web applications such as word processing software and online management software, and platforms for running users' own programs, which include Web and remote code execution platforms.

Due to the alarming need of cloud computing services, a large number of highly reputed and well-known firms such as Amazon, Google, and Microsoft have already started to come forward in the market. Well-known clouds by these firms include Amazon EC2, Google Cloud, and Google App Engine and Windows Azure. However, the concept of cloud computing is named and popularized now as a service, but it was known since the past [1]. The use of free services that resemble the way cloud computing behaves was there in the form of free e-mail services, virtual guest operating system services, and online file storage services. But proper management and involvement of virtualization technologies have made this form of computing advantageous, which is the key difference between cloud computing and other computing services. Figure 2.2 shows an abstracted view of the cloud computing environment with the interacting parties.

The concept of cloud computing is to have adequate IT infrastructure, resources, allocation, management, and release on demand on the basis of pay-as-you-go model that resembles the telecommunication market where big companies with excessive amount of operational resources rent them to a

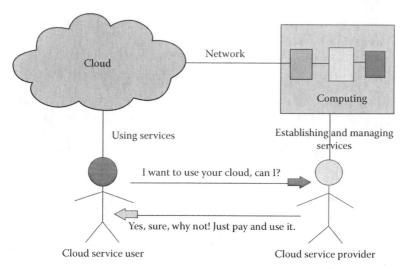

FIGURE 2.2
Cloud computing.

group of startup telecommunication companies that are not able to establish
their own physical operational resources for functioning. The concept of renting resources that are additional and of no use to the telecommunication provider helps them gain more money from their surplus hardware for efficient
utilization. Cloud computing is a combination of concepts of technologies,
including distributed computing, virtualization that increases utilization
of resources by driving physical resources virtually, and high-performance
computing. The new paradigm of renting resources and modeling it on a
pay-as-you-go model is what makes cloud computing lead the market. It is
a new era of computing on the Internet, which makes its future look bright.

Cloud Services

Cloud can provide two types of resources: (1) computing resources and
(2) storage resources. Based on the type of service that cloud computing
provides, it is broadly classified thus: storage cloud and compute cloud. The
service varies from data storage to online platforms for remote code execu-
tions and known software in the form of easily accessible Web applications.

Some of the essential characteristics that a cloud should provide according
to NIST [2] include the following:

1. On-demand self-service
2. Ubiquitous network access
3. Resource pooling
4. Rapid elasticity
5. Measured service

Cloud provides its offerings in three well-known forms of service:

1. *Software as a service (SaaS)*. In this form of service, a cloud user can access applications running over cloud servers by using a Web browser or other customized interface. Clients do not have to worry about management or maintenance of software and do not have control over how the software behave. The user can have some level of privileges to perform configuration changes to his/her view of soft ware or application running over cloud.

2. *Platform as a service (PaaS)*. In this service, a user can deploy his own created applications over the Internet by using the platform of a cloud service provider including programming platforms and libraries, etc. However, users have little control over infrastructure but have enough control over their own software configuration and working.

3. *Infrastructure as a service (IaaS)*. In this service, a user has control over the infrastructure including data storage and network and computing platforms. The user can run the platform they wish and install software accordingly on the rented infrastructure. However, the maintenance, allocation, management, scaling, and release still remain in the control of cloud service providers.

Advantages of Cloud Computing Services

Cloud computing provides to its users the choice of availing services without having to purchase any physical resources, infrastructure, storage, etc. Also, users do not have to worry about establishment and management of services. They can just obtain the service by paying for its use. The solution or services the users purchase can be customized to fit their particular needs. After some time when the users think that the resources are not needed anymore, they can stop paying for the services and the resources they leave can be used by other users in a fresh manner. The important thing is that users just see the service without clearly knowing how the service is provided or how and where the data are stored, etc. The users merely see a virtual copy of the actual physical resource customized by the service provider. Cloud computing is a totally flexible and customized form of computing.

Requirement of Trust in Cloud

Trust Requirement in Cloud

As cloud computing is a service which is packaged and provided to a user, it requires some service-level agreements (SLAs) and understanding between the user and the provider of the service. Today, we use cloud services in

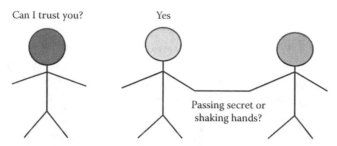

FIGURE 2.3
Need for trust in cloud services.

our day-to-day life. From e-mail services to free storage services are all a part of the cloud computing environment. We can refer to them as free cloud computing services, but they are not much different from the services provided by the cloud to a user on a pay-per-usage model. Figure 2.3 is a self-descriptive example of why we need trust in cloud computing environment as the client has questions regarding trust from the cloud service provider as it can pass client data to third parties.

Trust is required in the cloud computing environment [3] because of the service model it provides. It provides a distributed model of service where a user or the customer of the service has generally a low level of control over the service it takes. However, the cloud service provider can have distributed or centralized control over the service it provides to its customers. As a user places his secret data in a storage cloud, he may have certain questions that need answers from the cloud service provider:

> As a cloud is providing data storage services, a user wants the data to be kept to his reach only even if it is stored in a shared manner on physical disks. Hence customer requires trust over cloud in providing data security for the data stored by them in cloud [4].
>
> It's not only data security user also expects that when the data is accessed from the remote location the data cannot be sniffed and converted to information. User also expects that the data can't be seen by the cloud service provider itself and should be kept in encrypted form. Also a user trusts that the data should not be tampered when it lies inside the cloud storage.
>
> In case of computing as a service a user wants that the application it is using works with user authentication. The application or software that a user is using as a service should be trustable and follow standards and must be compliant to certain guidelines that are governed by trusted third parties.
>
> Also when a customer uses certain platforms (platform as a service) for executing their own software and applications a user expects that the platforms will not open or use any vulnerabilities present in the application for any unwanted usage. The platform should be secured and must be standards compliant. Similar trust requirements are expected by a user using infrastructure as a service.

A stakeholder or a customer using the distributed services provided by cloud may want assurance that the business critical data must be secured and the processes which are involved in securing it that includes process for accessing the data, for storing it and securing it must follow standards that are well governed by a trusted third party. This assurance is for increasing the trust that the cloud service provider is keeping the business critical data safe and away from reach of insiders and from other users availing the services from the cloud service provider.

In a cloud computing service, either for data storage or computing, a user expects certain trust-based attributes to be fulfilled by the cloud service provider, which are required for the proper functioning of the cloud computing services [5]. Establishing trust on a cloud is a major task that needs to be addressed. For the use of the services provided by the cloud, it is required that both the hard and the soft trusts of their customers are fulfilled. This means that they should provide hard trust features such as encryption of storage, security of online transactions, but at the same time, also provide soft trust parameters including user-friendliness, theoretical trust, etc. It is required that persistent trust should be provided by the cloud service providers to their users. By persistent trust, we mean trust that is permanent and can be proven to the other entity with the implementation of technological mechanisms [6]. For example, any implementation that can help the user increase his trust on the provider will generate a persistent trust for the cloud service user.

Trust and Cloud: Current Trends

If we consider the current trends and customers' experience in using online cloud services, it can be observed in the surveys that users are insecure about the services provided. If you consider online mail services and free storage accounts provided by well-known cloud service providers, users are unsure if their data are safe or not. Many think that the data that are stored in such clouds are generally compromised and used for other means. This is one of the reasons why people generally have started encrypting data at their end and stored them for the purpose of sharing with others using free storage providers.

This lack of trust affects the soft trust of the user over services. This previously had happened with the usage of online transaction services when the lack of trust causes problems to many of the users when using the services. The lack of trust is because of the unavailability of a persistent trust mechanism system that can assure the customer that his information will not get revealed to any third party or to the service provider with which the transaction is done. The technological deployments of trust such as encryption of the data (credit card information) and standards-compliant transaction processing system which are approved by trusted third party have changed the trust level of the user drastically. Similar deployments are necessary in cloud computing to assure users of placing their data and using cloud computing.

Solutions for Ensuring Trust in Cloud Computing

Various researches are going on in the field of providing trust solutions in the cloud computing environment. This section gives a brief description about the past work with a critical review on what researchers did and how it can help ensure better trust in the cloud computing environment. This critical description of the taxonomy of cloud trust solutions with current techniques in the area of ensuring trust in cloud computing will help industry professionals and researchers to get a better understanding of current issues and needs for trust in cloud computing. Figure 2.4 presents the taxonomy of the proposed solutions for ensuring trust in the cloud computing environment. This taxonomy classifies the proposed techniques based on the kind of approach applied for ensuring trust. The approaches overlap but provide a clear description of which techniques could be implemented under their category to make them different from other approaches, and how this helps in having a better trust environment. The approaches that form the basis of this taxonomy with the techniques that come under them are explained next.

Trust Evaluation Based

The solutions under this approach are those that use trust evaluation components that evaluate trust on the basis of predefined features. After evaluation of trust by individual components, the final trust value is calculated from their combination. These solutions have the advantage that trust is calculated by individual trust components, and hence one can be more confident on the values received. But they are complex and need correct definition of features,

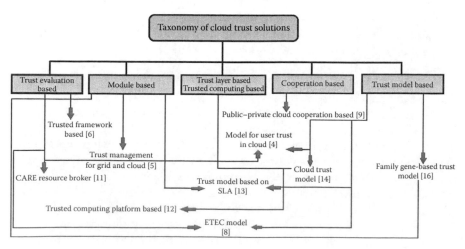

FIGURE 2.4
Taxonomy of cloud trust solution.

which form the basis of trust because if some of the high-value features are missed, one may not receive a correct overall trust value to rely on. The techniques under this category are as follows:

1. *A model for user trust in cloud computing.* Ahmad Rashidi [7] focused on the fact that trust is one of the major concerns to be handled in cloud environment. He has taken the statistics and data from reports such as Gartner, Cloud Security Alliance, and others to identify the questions that are related to trust in cloud computing in the mind of end users.

   The main concerns of users' trust in the cloud environment are listed as follows: data location, investigation, data segregation, availability, long-term viability, regulatory compliance, backup and recovery, and privileged user access.

   These provide a model of a user's trust in cloud computing. For evaluating the model, the users performed a user survey with a questionnaire and have done the statistical analysis of their results with SPSS 19 and fit was measured with analysis of moment structures or AMOS 18. The model was examined for maximum likelihood. Research results show that backup and recovery provide the strongest impact on user's trust in cloud computing with a recorded value of 0.91 from the AMOS.

   Analysis of the results illustrates that data segregation and investigation have a weak impact on user's trust on cloud computing. His findings include determination of eight parameters that can be looked upon for ensuring better trust in cloud by the user. In addition, the model proposed with the findings can help the industry to look upon the concerns that mostly affect the trust and find appropriate solutions for ensuring them.

2. *Trust management system for grid and cloud resources.* Manuel et al. [8] proposed a new trust model that is based on CARE resource broker. The technique proposed by the authors focuses on the evaluation of trust on grid and cloud systems based on three evaluator components that include (1) security-level evaluator, (2) feedback evaluator, and (3) reputation trust evaluator.

   There are various measures that are utilized by each of these trust evaluators. The security-level trust evaluator uses these parameters for evaluating the trust: authentication type, authorization type, mechanism used for self-security, multiple authentications, and authorization mechanisms. The degree of trust is based on the grade that is provided to each of the security parameters based on the strength of their implementation in the system.

   There are three steps involved in feedback evaluation: (1) feedback collection, (2) feedback verification, and (3) feedback update. Finally, the reputation evaluator uses the grid and cloud resource parameters such as computing power and network capabilities to calculate the trust value. At last, all the values obtained from the three trust evaluators are then used to calculate the final summation that becomes the final trust value of the cloud or the grid system.

The final trust value can then be used for accessing the cloud services and is applicable to heterogeneous cloud computing. However, the concept is not implemented and is in progress, but an initial prototype is already tested by the authors by simulation.

3. *Modeling and evaluation of trust in cloud computing environments.* Qiang et al. [9] illustrated the fact that trust evaluation models are unavailable in the cloud computing environment. The authors proposed an extensible trust evaluation model named extensible trust evaluation model for cloud computing environments (ETEC) and also described various trust measures including trust, trust degree, trust relation, trust service, trust model, trust chain, recommendation trust, experience or knowledge, direct trust, and time-based forgetting function.

After analysis, the authors also proposed the ETEC model that works on recommendation trust and direct trust. ETEC algorithm with its component algorithms including direct trust, recommendation trust, and dynamic trust is explained. ETEC is actually a time-variant and space-variant method for evaluating direct trust and recommendation trust.

Module-Based Solution

Module-based solutions are those in which the system is divided into specific modules with predefined tasks and responsibilities to ensure trust in the cloud environment. The solutions are based on modules that perform a specific operation for ensuring security, and these techniques are easy to manage. However, the task of dividing the system into modules and assigning responsibilities that are properly synchronized require real efforts for their implementation. The techniques under this category are explained as follows:

1. *SLA-based trust model for cloud computing.* Alhamad et al. [10] proposed a trust model based on SLA. SLA agents, cloud consumer module, and cloud service directory are the components of this trust model. Various tasks that are performed by the SLA agents include grouping the consumers into classes based on the needs, designing of SLA metrics, negotiating with cloud service providers, selecting service provider based on nonfunctional requirements such as quality of service, and monitoring the consumer activities.

The responsibility of cloud consumer module is in requesting the services externally in the cloud environment. The information of cloud service provider who is meeting functional requirements for getting chosen by consumers is present in the cloud service directory. The requirements that user needs to view using the cloud service directory include database provider, hardware provider, and application provider. Service providers also advertise the services provided to cloud users through the cloud service directory.

The model looks quite novel for ensuring trust in the cloud computing environment, but the authors have not implemented the model till now. Hence, the effectiveness of the solution is yet to be determined and can be judged only after its implementation.

Layer-Based and Trusted Computing-Based Solutions

Techniques in this approach are those in which the system is divided into multiple trust layers that have specific trust responsibilities. In trusted computing-based solutions, a layer of software stack generally referred to as trusted computing platform is used for ensuring trust on certain operations that can vary from authentication to authorization, etc. Layer-based techniques are very much similar to trust evaluation-based approaches but with a small difference that the architecture is based on trust layers that are different in their specific roles and responsibilities from trust evaluation components. Trusted computing-based techniques are however more specific for their applicability to heterogeneous environments. The techniques under this approach are described as follows:

1. *A cloud trust model in a security aware cloud.* Sato et al. [11] identified a new issue of social security trust. They approached the social security problem by dividing it into three subareas including multiple stakeholder problems, open-space security problem, and mission critical data-handling problem. They have addressed multiple stakeholder problems, which happen because of the interaction of third parties. The parties that are generally involved in cloud communications include cloud service provider, client, the stakeholders, and rivals of business.

   An SLA between the cloud service provider and the client is provided by the client that contains the administrative and operational information. The SLA plays a specific role in defining the policies between the three interacting parties to ensure a greater trust on each other. Open-space security problem tries to address the issue of the loss of control over a user's data when the data are transferred to a cloud service provider.

   The authors proposed solving this problem by data encryption, which requires a key management infrastructure. Mission critical data-handling problems look for the issues that are of concern when mission critical data need to be stored on cloud servers. They have proposed that if mission critical data are stored in a private cloud, the problem can be easily addressed but with a drawback of increased cost of setting up the private cloud. The authors proposed and developed a trust model called "cloud trust model." Trust layers such as internal trust layer and contracted trust layer have been added as additional layer in the cloud environment.

The work of internal trust layer is to handle ID and key management operation. Mission critical data can also be stored in the internal trust layer for better trust and security needs. Cloud service providers also provide trust to its clients in the form of three documents, which are service policy/service practice, ID policy/ID practice statement, and the contract. The parties can ensure a higher level of trust based on their needs, and hence can implement a more trustable cloud infrastructure. However, the authors have only discussed the issues and the work exists as proposals without implementation.

2. *Cloud computing system based on trusted computing platform.* Zhidong et al. [12] discussed the various trust aspects in cloud computing. Concretely, they have talked about the trusted computing platform in cloud computing. The authors have also proposed a method to improve the security and dependability of cloud computing environment.

The application of trusted computing platform is used for ensuring authentication, confidentiality, and integrity in the cloud computing environment. The authors have provided this model as trusted platform software stack as software middleware in the cloud environment. The proposed model however provides identify management and authentication and it is applicable to heterogeneous cloud.

3. *Trust cloud: A framework for accountability and trust in cloud computing.* Ko et al. [13] focused on how important the requirement of trust is in cloud computing and have discussed key issues and challenges in achieving a trusted cloud using detective controls. They have also proposed a trusted framework that addresses the issue of accountability using policy-based approaches.

The authors referred to one of the researches that Fujitsu Research Institute conducted in 2010 and reported that 88% of the cloud users are worried about who can access their data on physical servers. Also, a report by the European Network and Information Security Agency states that loss of governance is a major risk to cloud services nowadays. They also include findings from cloud security alliance report V1.0 which states that increasing accountability and auditability will increase the trust of the user on the cloud services and decrease the probability of the enlisted threats.

S. Pearson and A. Benameur [14] cited in their research about the important components that affect the trust of the cloud, which include security, privacy, accountability, and auditability. They also suggest that the use of detective and preventive controls for ensuring trust in cloud will be the best combination for ensuring trust in cloud in both physiological and technical manners. The authors list down some of the complexities that are introduced because of elasticity in cloud. These include challenges introduced by virtualization such as tracking of virtual to physical mapping and vice versa, scale scope and size of logging, and live and dynamic systems.

For ensuring better trust in cloud computing environment, they tried solving the problem by including five abstraction layers in their trust cloud framework: (1) trust cloud accounting abstraction layer, (2) system layer, (3) data layer, (4) workflow layer, and (5) policy law and regulation layer. They have described how accountability and logging can be achieved with the help of the defined layers. They have also described that file-centric logging can be performed at the operating system end or in the file system or on the cloud's internal network. Data-centric logging can also be performed by consistency logger or provenance logger. With workflow layer, they have focused on audit trail logs found in software services in cloud. Their focus is on issues related to governance in cloud, automated continuous auditing, patch management auditing, and accountability of services.

The research also includes some of the related work done, including the work done by Cloud Security Alliance (CSA), HP Labs, Max Planck Institute, etc. The authors are however working on trust cloud framework that is capable of providing users with a view of accountability in cloud and the research is in progress.

Cooperation-Based Solutions

Certain trust solutions are based on cooperation between operational entities. One of the examples is the cooperation and trust between public and private cloud in a hybrid cloud environment that can help both the entities develop a higher overall trust. The approach is good as cooperation increases trust because the information existing with other trusted entities can help increase the overall trust in the cloud environment. This information sharing helps increase the level of trust with a drawback that the trust on the entities that are involved in sharing the information should be ensured by the communicating parties. The detailed description of solution in this category is as follows:

1. *Establishing trust in cloud computing.* Khaled and Khan [15] illustrated the issue of how a cloud service provider can gain its customer's need of trust when a third party is processing important data of the user in various countries. According to the authors, using new emerging technologies can help cloud service provider obtain hard trust from its effective users.

They have identified two challenges while describing hybrid cloud architecture with the example of a SoftCom Image pro software application running on the private cloud end of the hybrid cloud architecture. The two challenges include diminishing control and lack of transparency. The authors explained how the assurances to each other from the public cloud and the software running on private cloud can help provide better trust to the end user.

The authors have provided an overview of some of the emerging technologies that can be used for ensuring trust, which include the following:

- *Remote access control.* This is a case in which a user can have remote access facilities and greater jurisdiction over their data.
- *Reflection.* Through this, the cloud service provider informs the user about its strengths and weaknesses and how security policies are addressed. This will help the user choose the trusted service and also ensure the trust from his end if the degree of trust required by the user is not provided by the service provider.
- *Certification.* A trusted third party provides trusted certification to the cloud service provider by judging the services according to global standards and based on the trust assurance measures that are implemented by them.
- *Private enclaves.* In this technology, a set of secure enclaves can be made for users demanding increased trust with enhanced logging, auditing, incident response, and intrusion detection capabilities having global standards.

The techniques as referred by the authors are quite impressive and can be implemented for having better trust in the cloud environment.

Trust Model-Based Solutions

There are various overlapping solutions in this category. The solutions under trust model are those which create a trust model for trust evaluation based on features, layers, or modules. These solutions follow well-defined mathematical and analytical models for calculating trust in the cloud environment. Trust model-based solutions are more flexible in including other models and approaches for trust evaluation and such solutions are known since the past. The solutions under this category are explained as follows:

1. *Trust management in cloud computing.* Firdhous and Hassan [16] focused on various existing trust models for distributed systems and cloud computing environment. They have discussed the cloud computing paradigm with the explanation of services provided

by it at various layers of computing. The authors have focused on defining trust in terms of human behavior and how it is taken into consideration by computer science researchers to ensure trust to end users.

The authors have identified some of the common factors of trust such as "it plays a role only when the environment is ascertain and risky, and is built up by prior knowledge and experience." They also identified previous researches in which the characteristics of trust are divided into groups. One of the examples is of McKnight and Chervany who divided 16 characteristics of trust into five groups: (1) competence, (2) predictability, (3) benevolence, (4) integrity, and (5) others.

The authors also refer to some other classification of trust where the authors classified trust on the basis of four dimensions: (1) subjective versus objective trust, (2) transaction-based versus opinion-based trust, (3) complete information versus localized information, and (4) rank-based versus threshold-based trust. They also describe various trust models that are developed for distributed systems including cuboid trust, Eigen trust, Bayesian network-based trust management (BNBTM), AntRep, Semantic Web, global trust, peer trust, PATROL-F, trust evolution, time-based dynamic trust model (TDTM), trust ant colony system (TACS), trust model for mobile agent systems based on reputation (TRUMMAR), comprehensive reputation-based trust model (PATROL), META-TACS, context-aware trust- and role-based access control for composite Web services (CATRAC), and Bayesian network-based trust model.

The authors have critically analyzed the existing research work and have come up with the findings that most of the models that are proposed remain short of implementation and very few of them have been simulated to prove the concept. The authors have critically commented the research work that was done in the past according to their capability and applicability in the cloud environment and focused on the fact that there are no complete solutions for trust management that exist in cloud, hence there is a need for solutions that can be implemented and that stay on a strong base.

2. *Family gene-based cloud trust model.* Tie Fang et al. [17] proposed a nontraditional trust management module and a security framework for ensuring trust in cloud. Some of the security measures based on trust are also provided by the authors. A family gene technology-based trust model is also proposed by Tie Fang et al. [18] on authentication, authorization management, and access control mechanisms. They have utilized the concepts of biologic gene technique and provided the cloud trust model based on family gene model also known as CTFG model. They have utilized formal

definitions from the family gene technology and applied them to the domain of trust and security in cloud. They have also shown experimental results that prove it to be a better approach for establishing trust in cloud.

3. *Trusted anonymous execution: A model to raise trust in cloud.* Zhexuan et al. [19] provided their focus into the security issue where unrestricted access to user data from remotely installed software can cause security risk to SaaS. They have taken this issue into consideration and provided an approach where software can be separated from the data.

The authors proposed a mechanism where there are four parties, namely, resource provider, software provider, data provider, and coordinator. The responsibility of resource provider is to implement data and software; it should also provide the platform for the execution of software. In the proposed scheme, the data providers are those who have control and ownership of data and software providers have control and ownership of software that run on data. The last party in the team, called the coordinator, is responsible for connecting software, data, and resource provider together and providing interface between data and applications. While any of the software or data providers submit resources to the resource provider, it would be encrypted and then stored by the resource provider.

A coordinator helps the data provider in identifying the software to run over the data on the resource provider. The reference identity generated during execution is stored by the data provider and after execution a data provider can download and perform other operations on the results obtained. Software and resource provider charge the data provider for use of services. The operational logs created earlier are used by the software provider to understand which software were operated on the data without knowing the identity or content of data provider.

This is a new approach toward the development of a more trustworthy cloud. However, the authors may have to look into how they can assure the data provider that the software providers are not running any algorithm in the background that can read the contents of the data provided to them for execution.

Challenges in Deploying Trust-Based Solutions in Cloud

We list the questions regarding trust that cloud users have in their minds and how a cloud service provider can answer these with challenges [20] that needs to be tackled.

Trust: User Perspective

The user generally calculates the belief or trust or confidence on the services a provider provides on the basis of the following soft trust parameters:

What is the brand of the service you are having? You can answer that it is the overall online reputation of the brand whose cloud services the user is using.

Recommendation is the second parameter a service is assessed with. If any service is recommended by a trusted third party, a user finds it easier to generate a trust and get used to a service with a higher level of confidence.

The third is with use of the service they want to generate trust with. Many users use the free or paid cloud service and give them a trial. They look for some time for any breaches that they can detect and from their own knowledge rank the service and create a trust with the service they are using.

Many users create trust relationship in a contractual manner with the service provider.

Other reasons may also exist, but we want to have a clear description of a cloud user thinking about assessing a cloud service with no technological aspects involved. These characteristics are generally found with customers who use free cloud services and users with customers who have low risk with the data stored on a cloud service provider's end. However, they play the role of assessing the services with a small business organization thinking of using the service with some other technological or hard trust aspects taken into consideration. The hard trust concerns for assessing any cloud service with respect to a cloud service user are as follows:

How are the data getting stored in the service provider's end and what are the mechanisms that are going to be used to access it? Are they standards-compliant and approved by trusted third parties.

Are the data stored in a meaningful form or as garbage value (encrypted data to others)? How strong is the scheme used? Can cloud providers have privileges to obtain data as meaningful information?

How data integrity is assured when data are stored at cloud service provider's end?

Is the network access to storage systems secure and encrypted? How is the authentication to access done and how is it secured?

What are the availability aspects of using the storage service of cloud?

Figure 2.5 depicts how a cloud user can assess the trust from a cloud service provider by getting answers in both technical and psychological form from cloud service provider's end.

FIGURE 2.5
Assessing trust in cloud.

In the case of using the computing services of cloud, the user is concerned about asking the following questions to the cloud service provider:

Are the software standards-compliant and secure to be used? Will their usage cause any vulnerability exploitation to their end systems?

Is the platform provided by cloud service providers secure for the software to run? Is the platform transparent and not do something for his interest in case of platform, as a service?

How much can I believe on the infrastructure provided by the cloud service provider?

A cloud service provider must be ready to answer these questions for providing a trusted service to its users. A healthy trust relationship between the cloud service provider and the cloud service user can be built if these questions can be answered in both technological and nontechnological manners to the stakeholders.

Generally, cloud users assess a cloud service security on the basis of data security mechanisms available, transparency measures in both execution and storage in computing and storage cloud services, and availability and reliability of the services. A cloud service user also looks for good authentication mechanisms including two-factor authentications with proper mechanism for insuring the auditing and logging.

However, the answer to these questions may help the user in establishing trust (both security and availability) on cloud services:

Are the services standards-compliant, and do they follow the government regulations in which they run?

Are the cloud services globally accepted?

Will they provide audit logging monitoring and reporting?

Do they provide access to customer data, and if yes, in which situations?

Can the data be seen at cloud service provider's end or not. How are the data stored, encrypted, or unencrypted? Is the integrity of data maintained?

If a user deletes his data, then what is the surety that cloud service provider will permanently delete it and not use for his own means?

What happens in the event of a crash or failure? Are there any backup or recovery mechanisms that keep the data safe?

How will the data get protected in case of network-based attacks? What preventive or defense measures are in place to thwart such attacks? What is the network model? How does data flow in encrypted or unencrypted format.

As cloud service providers use various well-known hypervisors to provide virtualized sharing of physical resources, how then will the user handle the well-known hypervisor vulnerabilities existing and exploited. Does the user have any such defense measures?

How will the user ensure security of his data in an environment where others are also viewing the same resources virtually? The assurance of security in a multitenancy environment is of critical importance.

Does the cloud service provider have a threat model for the service he provides, and hence can identify threats to the system before they are exploited? How secure are the security policies, and how are they implemented?

How will access control be provided to the user, and how will access control parameters (user name and passwords or any two-factor authentication technique) be kept safe in the environment?

Does the cloud service provider provide software and Web applications that are secured and up to date with patched known vulnerabilities?

Does the cloud service provider update the applications used with patches released so that they will not cause harm to the cloud user, and if yes, then how frequently?

Will the cloud service provider ensure any investigatory services through his audit logs and monitoring in case any breach to user data is reported?

The answers to these questions will surely help the cloud service user and provider in creating a healthy trust relationship between them.

Challenges in Establishing Trust on Cloud Computing

It is not the case that the cloud service provider does not know about the aspects that they need to look into for providing trust to their users and can hence increase the growth of their business. Many researchers have addressed the issue and they know these well. The deployment and implementation of such schemes is delayed or not taken into consideration because of some of the challenges that need to be tackled for providing trust to cloud computing environment user, and these are as follows:

Many cloud service providers are ready to establish a higher and better trust-based mechanism in terms of technological aspects also referred to as hard trust but are scared of the degradation in the performance of cloud services. As the cloud is a pay-per-use model, the resources are limited and shared, and if the cloud resources are utilized for ensuring the conditions that a user expects for trust, then he may need to pay more for the services he is availing and with a degraded performance in some cases. Some solutions may also make the system user unfriendly and may be complex to use. The cloud service provider needs to identify and answer that how such a change for increasing trust level among his customers will be increased with minimal changes and degradation to the service based on various measures.

The second question that comes to mind is how such measures will be provided in a standards-compliant way. How should the services be provided by vendors in a transparent and globalized manner so as to gain the confidence of the users?

As there is no complete model for ensuring the trust that can be implemented in cloud, and if such solution exists, then which solution or trust model is to be used so as to get maximum benefit in assuring trust to users?

There are many more challenges that include difficulties in implementation in deploying the models in cloud because of its complexity. These and many more challenges need to be addressed for deploying a trust cloud environment for users. Figure 2.6 is a representation of how the cloud service provider thinks while implementing and assuring it to the cloud users.

FIGURE 2.6
Challenges in establishing trust.

Ensuring Trust in Cloud Environment

To provide trust in cloud computing, many technological implementations are used and deployed [21]. To ensure data integrity and security, data encryption and message digest schemes can be used for the data stored on cloud servers. Better authentication and access control systems are also required. There are certain techniques that can be used by the cloud service provider, which allow them to take encrypted data from a cloud service user and process it. With so many of these problems, other implementations and measures can be utilized that will allow the cloud service providers to ensure data security that will eventually increase user trust over cloud computing services. Many can be taken from cloud user's end to ensure security of its data at service provider's end. These are as follows:

A cloud service provider can provide a user privileges to dynamically check the status of the cloud services availed by him. For example, he may login remotely to the cloud server and can provide restriction which a cloud service provider can see and which it cannot. By having such access to its data through the resources of cloud service provider, he will remain confident that his data are secure at cloud service provider's end. In the case of computing as a service, a cloud user may fix rules that will delete his personnel data when a cloud service provider or any cloud service user other than he logs in to his account. He may also set alerts that will get raise an alarm when unauthorized access happens with proper auditing and logging mechanisms of such access. This kind of technology can be provided as a single customized service to a cloud service user for increasing his trust level on a cloud service provider with larger benefits and little increased cost for managing and maintaining such customer-controlled facility. In this technique, even if the data of the owner lie at various different locations, he can have full control over who can see the data and which part of it.

When a cloud provider provides transparency of the services it uses for accessing and storing the data with other technological aspects of how the data are stored in its repository, a cloud service user can determine the weak and vulnerable areas that can pose threats to its data or services in cloud. A deeper analysis of the security provided by a cloud service provider will help the customer identify the weak areas over which they need to establish security on their own. Such transparency will help customers provide better security at their end and have an increased trust on the services they use from a cloud service provider. For example, if it is known that the data can be seen as plaintext to an employee of cloud service provider, the customer will then store his data in encrypted formats instead of plaintext.

Third-party certification techniques can help increase user trust over cloud services. As every cloud service provider has their own security mechanisms that cannot be accessed globally in general and may or may not be compliant to a given standard a cloud service user may find it difficult to access the security, and hence trust on the cloud service provider is doubtful. However, if techniques are implemented in which a trusted party can assess the security mechanisms provided by a cloud service provider and can rank them on the basis of confidence they are providing in the security services, a cloud user may gain the same confidence that will help him in selecting the cloud service provider for meeting his expectations for a particular service. In such a mechanism, there is need for a trusted third party that can transparently judge the security of a cloud service provider for the various services it provides and a set of well-known security parameters over which security of the system can be judged. Many such techniques in which a third party gives rating in the form of trust cards are available nowadays. Trust measures in these techniques generally collect a large amount of behavioral aspects of the system with its reputation and historical evidences of its working to define the trust a stakeholder can expect from a cloud service provider.

For users that demand higher trust levels with the available infrastructure of a cloud service provider, a private enclave can be created that will be externally separated from other cloud user's data and have added security measures and control mechanisms for intrusion detection incidents and their responses. Such private enclaves will have a higher security defense measures that include firewalls, network traffic filters, and intrusion detection systems, and may have better auditing, event log management, and incident response components. They will have standards-compliant implementation of security measures that will be different from normal security measures. These types of services can be helpful in increasing trust by cloud service users who demand higher level of trust and also do not want to provide such mechanisms from their end.

Figure 2.7 describes how a cloud service provider can ensure trust to its users by answering the cloud user's question with the help of soft trust and hard trust measures.

FIGURE 2.7
Establishing trust measures.

Conclusions

Trust is a relationship between two entities. Trust is commonly known as the confidence of the second party that the other party will work according to its expectations. Trust modeling concepts are used to present trust levels that exist between various entities in a system that will eventually help system designers identify weak trust areas. This information helps them incorpo rate better practices to include hard or soft trust as required to make the system more trustworthy. Trust modeling is used for identifying and demonstrating trust over real-time systems. This chapter discusses trust issues and challenges in a cloud computing environment. Cloud computing is a new paradigm of network access-based computing that can be accessed on a pay-per-usage model by customers varying from individual users to small-business organizations.

The services are provided in the form of cloud storage services and cloud computing services. The software, platform, and infrastructure as a service are provided by cloud service providers nowadays. Cloud distributes physical resources into virtual counterparts with the help of virtualization technologies that are rented to users on a pay-per-usage model, which eventually increases efficient resource utilization. However, as cloud is growing at a good pace a large population of users have questions regarding the trust they can have on cloud computing. There is need for establishment of trust in cloud computing as clients using rented resources may not want to incur any risk from the storage service or applications they use on the cloud service. As user data are stored in a distributed manner and the data can be confidential, users want to be assured by the cloud service provider that the data will remain available and secure and their integrity maintained. Also, the cloud user would want to know if the cloud applications or platforms they use do not expose their vulnerabilities to third parties as also not exploit or compromise their end systems. They also would want an assurance that their data will not be shared with a third party and that the data will not get deleted completely in case the service agreement ends.

It is a requirement that these and many other questions will be answered with technical and nontechnical aspects because an increase in trust will affect the population of users who may want use them. To solve these and many others questions, added implementation measures such as intrusion detection systems, storage encryption, data anonymity, and others be taken to help increase the trust but at the same time utilize cloud resources and hence increase maintenance and operational cost. But the requirement is an immediate need and the challenges for ensuring trust need to be tackled in a positive way that results in a more trusted cloud environment. This chapter lists some of the known solutions that can be easily implemented on cloud including the concept of third-party service certification and private enclaves. Cloud computing environment requires trust as one of the major areas of improvement for its better growth. An increased trust will attract increasing number of users to the services with greater confidence level.

References

1. Buyya, R., Chee Shin, Y., and Venugopal, S. "Market-oriented cloud computing: Vision, hype, and reality for delivering IT services as computing utilities," *10th IEEE International Conference on High Performance Computing and Communications (HPCC'08)*, pp. 5–13, 2008.
2. Brown, E. "NIST issues cloud computing guidelines for managing security and privacy," *National Institute of Standards and Technology Special Publication 800-144*, U.S. Department of Commerce: Washington, DC, 2012.
3. Pearson, S. and Yee, G. "Privacy, security and trust in cloud computing," London: Springer, 1–306, 2012.
4. Nick Coleman, M. B. "Cloud security who do you trust? IBM," IBM Global Services: Armonk, NY, 2010.
5. Subashini, S. and Kavitha, V. "A survey on security issues in service delivery models of cloud computing," *Journal of Network and Computer Applications*, 34(1): 1–11, 2011.
6. Krutz, R. L. and Vines, R. D. *Cloud Security: A Comprehensive Guide to Secure Cloud Computing*, Wiley: Indianapolis, IN, 2010.
7. Ahmad Rashidi, N. M. "A model for user trust in cloud computing," *International Journal on Cloud Computing: Services and Architecture*, 2(2): 1–8, 2012.
8. Manuel, P. D., Thamarai Selvi, S., and Barr, M. I. A. E. "Trust management system for grid and cloud resources," *First International Conference on Advanced Computing*, December 13–15, Chennai, India, IEEE, pp. 176–181, 2009.
9. Qiang, G., Dawei, S., Guiran, C., Lina, S., and Xingwei, W. "Modeling and evaluation of trust in cloud computing environments," *3rd International Conference on Advanced Computer Control (ICACC)*, January 18–20, Harbin, China, IEEE, pp. 112–116, 2011.
10. Alhamad, M., Dillon, T., and Chang, E. "SLA-based trust model for cloud computing," *13th International Conference on Network-Based Information Systems*, September 14–16, Takayama, Japan, IEEE, pp. 321–324, 2010.
11. Sato, H., Kanai, A., and Tanimoto, S. "A cloud trust model in a security aware cloud," *10th IEEE/IPSJ International Symposium on Applications and the Internet (SAINT)*, July 19–23, Seoul, Republic of Korea, IEEE, pp. 121–124, 2010.
12. Zhidong, S., Li, L., Fei, Y., and XiaoPing, W. "Cloud computing system based on trusted computing platform," *International Conference on Intelligent Computation Technology and Automation (ICICTA)*, May 11–12, Changsha, China, IEEE, pp. 942–945, 2010.
13. Ko, R. K. L., Jagadpramana, P., Mowbray, M., Pearson, S., Kirchberg, M., Qianhui, L., and Lee, B. "Trust cloud: A framework for accountability and trust in cloud computing," *IEEE World Congress on Services (SERVICES)*, IEEE, pp. 584–588, 2011.

14. Pearson, S. and Benameur, A. "Privacy, security and trust issues arising from cloud computing," *IEEE Second International Conference on Cloud Computing Technology and Science (CloudCom)*, November 30–December 3, Indianapolis, IN, IEEE, pp. 693–702, 2010.
15. Khaled Q. M. and Khan, M. "Establishing trust in cloud computing," *IT Professional*, 12(5): 20–27, 2012.
16. Firdhous, M., Ghazali, O., and Hassan, S. "Trust management in cloud computing: A critical review," *International Journal on Advances in ICT for Emerging Regions*, 4(2): 24–36, 2011.
17. Tie Fang, W., Bao Sheng, Y., Yun Wen, L., and Yi, Y. "Family gene based cloud trust model," *International Conference on Educational and Network Technology (ICENT)*, June 25–27, Qinhuangdao, China, IEEE, pp. 540–544, 2010.
18. Tie Fang, W., Bao Sheng, Y., Yun Wen, L., and Lishang, Z. "Study on enhancing performance of cloud trust model with family gene technology," *3rd IEEE International Conference on Computer Science and Information Technology*, July 9–11, Chengdu, China, IEEE, pp. 122–126, 2010.
19. Song, Z., Molina, J., and Strong, C. "Trusted anonymous execution: A model to raise trust in cloud," *9th International Conference on Grid and Cooperative Computing (GCC)*, IEEE, pp. 133–138, 2010.
20. Protogenist Info Systems, Technology Research, "Trust challenges of Cloud computing," 2012. http://blog.protogenist.com/?p = 1068 (Accessed on February 23, 2013).
21. Khan, K. M. and Malluhi, Q. "Establishing trust in Cloud computing," *IEEE IT Professional*, 12(5): 20–27, 2010.

3

A Survey of Trust and Trust Management in Cloud Computing

Vimal Kumar, Brijesh Chejerla, Sanjay Madria, and Mukesh Mohania

CONTENTS

Introduction

Over the years, the need for better computational capability, better storage facility, additional infrastructure, and the possibility of ease of access to various platforms and applications has driven cloud computing paradigm to where it is today. Cloud computing may mean different things to different people, and accordingly many definitions for cloud computing have been given. Jansen and Grance [1] try to concisely and formally define cloud computing in its entirety as given below:

> Cloud is a large pool of easily usable and accessible virtualized resources (such as hardware, development platforms, and/or services). These resources can be dynamically reconfigured to adjust to a variable load (scale), allowing also for an optimum resource utilization. This pool of resources is typically exploited by a pay-per-use model in which guarantees are offered by the infrastructure provider by means of customized service level of agreements (SLAs).

There is a rapidly growing trend of adopting cloud computing among companies irrespective of their size, individual users, and researchers. This has led to the proliferation of a number of cloud service providers (CSPs) in every sphere of cloud computing, ranging from storage to software, platform, and infrastructure. Some examples of such CSPs are Amazon Web Services, Rackspace, Google App Engine, Microsoft Cloud Services, CSC, Salesforce, GoGrid, etc. A large list of such CSPs is given in Ref. [2]. Many of these CSPs provide similar services and compete with each other for consumers. This is good from a consumer's perspective; however, the consumer needs a systematic method of choosing the best CSP, based on its capabilities and the consumer's own application-specific needs. This is where the concept of trust plays a role. Trust can be used as means of making an informed decision when it comes to assessing a cloud and choosing one for the given application. Trust management however is not trivial in cloud computing. Clouds generally are highly distributed systems, often crossing geographical and political boundaries. Laws regarding data, privacy of data, and privacy of data owners differ vastly when political boundaries are crossed, which results in conflicting perspectives in case of a contention. The proprietary nature of the clouds lends to their being nontransparent to outsiders. This makes evaluation of a cloud difficult from a trust management perspective. Lack of standards pertaining to security and performance in cloud computing also makes trust management a challenge.

Clouds are deployed under three different models, namely, *private*, *public*, and *hybrid* clouds. The issues concerning trust vary across different deployment models. The idea of cloud computing can be attributed to large organizations using their existing infrastructure and creating a layer of abstraction and virtualization over it. This helped the organizations increase the usage and throughput of their hardware and gain on demand elastic provisioning of resources. This was the beginning of the private cloud.

Trust management in a *private cloud* is not an issue as the organization does not rely on a third-party cloud provider. Issues pertinent to cloud computing such as perimeter security, data security, hardware security, availability, reliability, and privacy cannot be held as a cloud service provider's responsibility because the private cloud is on the premises of the organization that controls it. Generally, none of the aspects that determine the trust of a CSP are applicable to the private cloud scenario, as the control over processing, storage, and maintenance of data rests with the internal IT department. Trust issue may arise only when a CSP is requested to assist with the maintenance and management of the data by the IT department. Again, the CSP only provides the solutions and does not have direct access to the data or the hardware. The only scope for a possible trust parameter that can be used for consideration in determining the trustworthiness of a CSP is to determine how effective the solutions provided are against known attacks. This again is information no organization will divulge and is therefore hard to use in determining trustworthiness in terms of the security solutions employed by cloud providers.

Public cloud is what most people mean when they use the term *cloud computing*. Services in a public cloud are used on a *pay-per-use* basis. One of the major benefits of using a public cloud is the ability to scale out and scale in elastically. Clients can request additional services when they have higher load than normal and relinquish these services when the additional load is no more there. This elasticity of services saves money. However, the cost associated with elasticity comes in the form of security concerns and loss of control over one's data. Thus, from the perspective of a consumer who is comparing various CSPs, security and privacy of data, amount of user control on the data, and performance benefits must play a role in determining the trust in each CSP. The definitions and parameters mentioned in the rest of this text are principally modeled under the premise of a consumer using the public cloud model.

The *hybrid cloud* deployment model is a straightforward solution to the limited scale out problem of the private cloud model and limited security in the public cloud model. There are two scenarios in which a hybrid cloud model can be used. First, at peak load times when an organization needs more resources than the private cloud can provide, services from a public cloud can be used on *pay-per-use* basis to assist the private cloud. Second, organizations classify their data and application in terms of security. Data and application which need greater security are deployed in the private cloud and other data and applications can be placed in the public cloud to make use of the benefits a public cloud provides. This scenario is called a hybrid cloud where the public and private clouds work together to achieve a balance of security and performance. Trust issues that correspond to the public cloud model translate to the hybrid cloud model as well. This is because, even though a private cloud is involved, the scale out requires the use of a public cloud and thus the issues of security, data management, confidentiality, and privacy come into the picture. We will discuss in length how each of these and other parameters of QoS affect trust assessment and management in a cloud computing scenario.

This chapter is divided into two distinct parts. In the first part, we discuss the notion of trust in cloud computing; in the second part, we discuss how trust is managed in cloud computing. We start by discussing trust in a general sense and how it is applicable to cloud computing as well. We then talk about the different types of trust we see in the literature on cloud computing and survey various papers dealing with each of those types of trust. This is followed by a section on trust management system (TMS). We define four distinct components of a TMS and discuss the functions and requirements of each of the component in detail. Alongside, we also survey papers relevant to each of the component.

Trust

A General View of Trust

Trust can imply different things to different people because it presupposes some risk, but is not explicit about the nature of the risk [3]. Over time, different approaches have evolved depending upon the trust context. These contexts are all relative. Trust starts from how people evaluate the behavior of others based on several factors from/of different dimensions. In social interactions, the implicit risk involved in trusting another person or entity is the basis for the development of trust management schemes. When this trust management permeates into inter-enterprise relationships or between an enterprise's and an individual's transactions, concepts such as auditing and transparency are established to maintain a universally accepted level of trust among people and organizations. Jøsang et al. [3] say that trust provides a decision support base for choosing online services, and it is derived in some way from the past experiences of the two parties and others. As observed by the paper, the usage of trust also inspires the service providers to maintain high quality of service in order to enhance reputation which in turn results in an increase in business. Hence, trust not only provides a basis for selection of services for customers but is also economically beneficial for the providers of such services. The definition of trust becomes increasingly blurred with increase in the use of distributed systems in day-to-day transactions. Apart from the ability of an organization to provide service to an individual/group of individuals with acceptable reliability, critical issues such as security also start defining the level of trustworthiness of the provider.

Trustworthiness emerges from the amount of dependence a party seeks from another party and the amount of reliability (probabilistic) one party provides depending upon the outcome of a transaction, which is governed by risk. Based on this, trust can broadly be classified as *reliability trust* and *decision trust*.

Reliability trust is one in which the trusted entity takes upon itself the onus of providing the service requester a specified amount of reliable service. To formally define it: "it is the subjective probability by which an individual/entity expects that another individual/entity perform a given action on which its welfare depends" [3].

Decision trust is the type of trust where one party relies on the services of another party based on its assessment of the level of risk involved and the kind of risk involved in the transaction. To formally define it: "trust is the extent to which one party is willing to depend on something or somebody in a given situation with a feeling of relative security, even though negative consequences are possible."

Applicability to Cloud Computing

As is the case with any concept that has been commercialized, the structure of cloud computing inherently has inter-enterprise and enterprise–individual transactions that are built on the analysis of the behavior of concerned parties based on trust, as given in Ref. [3]. Both reliability and decision trust are applicable to the cloud-computing paradigm where the consumer uses several parameters to establish an understanding between him and the CSP to provide him with an agreed level of service. This has led to trust being used as a means to differentiate between services provided by different service providers.

Types of Trust in Cloud Computing

As stated before, trust may mean different things to different people and entities and this makes defining a singular view of trust difficult. In cloud computing this difficulty is compounded by the vastness, heterogeneity, and diversity of technologies that are involved. To streamline the understanding of the various trust management techniques existing in the literature for cloud computing, we categorize them by the kind of trust they deal. A classification of various kinds of trust has been given by Grandison and Sloman [4] and refined by Jøsang et al. [3]. In this section we discuss the most relevant ones in a scenario. Because trust management in cloud computing is a relatively new area of research, along with papers treating some form of trust and trust management specifically in cloud computing, we also consider related works treating trust management in the key enabling technologies of cloud-like service-oriented architecture (SOA) and Web services. It should be noted that the survey of papers here is not exhaustive but representative; we have chosen papers which represent each category and provided brief summaries for better understanding of the classification.

Access Trust

Access trust is that which a party places in a second party when the second party is accessing the services and resources provided by the first party. Access control and access policies have traditionally been used to exercise this trust. In a cloud computing scenario, however, maintaining access trust becomes much harder because of the presence of a number of parties, each having contrasting objectives. Song et al. [5] demonstrate this problem. The paper discusses the relationships between the parties involved in a cloud computing application. These parties are the software provider, the data provider, the coordinator and the resource provider. The software and the data provider upload their software or data on the resource provided by the resource provider. The coordinator facilitates the service discovery and the resource provider provides the platform for the execution of the software on the data. The trust issue in the architecture arises when a data provider uses software from the software provider, or a third party wishes to use the software and the data. In such cases there needs to be a guarantee that the software being used is not malicious and is only doing what it says (i.e., in the case of a private dataset, the software does not make copies of the data behind the scene). A trust issue also arises when the software is provided to the data providers that they do not make unauthorized copies of the software. The paper proposes a solution based on logging and isolation. Data from the data provider and applications which work on that data from the software provider are kept encrypted and accessing the keys triggers a logging mechanism. A service discovery mechanism is used by the coordinator to find appropriate software for a dataset and vice versa. The resource provider provides an isolated environment for the applications to run. The software provider and the data provider are unaware of the identities of each other and only know that the execution is in place through an execution reference id. This model, though helps in accountability and anonymous execution, does not provide any guarantee that the software will not make copies of the data. However, to ensure that a copy of the software is only used for the dataset it was intended for and is not copied without authorization, the authors propose a trust binding of software and the data. This is done by the resource provider using trusted platform module (TPM). Because the data are kept encrypted by a key, each specific instance of the software is modified to work only for a particular key. Thus, any instance will need to be modified for a new dataset and unauthorized modification would not be allowed by the TPM. Sundareswaran et al. [6] propose a decentralized information accountability framework to keep track of the actual usage of the users' data in the cloud. The paper provides a treatment of access trust in clouds by leveraging the programmable capability of Java JAR files to enclose the logging mechanisms together with users' data and policy. To improve the user control over the data, they provide distributed auditing mechanisms. In a cloud computing scenario because data goes through resources (hardware and software) owned by different entities, conventional data handling cannot guarantee accountability and data privacy. The authors propose cloud information accountability (CIA) to overcome these issues. The CIA has two major components, the logger and the log harmonizer. The logger's main tasks include automatically logging access to data items that it contains, encrypting the log record, and periodically sending them to the log harmonizer. The log harmonizer is responsible for auditing. The architecture has two JARs; the outer JAR and the inner JAR. The outer JAR contains the access policies and the inner JAR contains the encrypted actual data in several JAR files. The access policy determines access to the inner JARs by methods such as times access and location-based access. This mechanism can survive various attacks such as the copying attack where attacker copies the JAR files, the disassembling attack where the JAR file is attempted to be disassembled, the man in the middle attack, and the compromised Java virtual machine (JVM) attack. Ko et al. [7] look at cloud accountability from another perspective. The paper discuses key issues and challenges in achieving a trusted cloud through the use of detective controls and present the TrustCloud framework to deal with access trust. The paper discusses challenges introduced by virtualization, logging from an operating systems perspective versus logging from a file-centric systems perspective, the scale, scope, and size of logging, and data provenance. Their TrustCloud framework has different abstraction layers; the system layer which takes care of the operating systems, the file systems, and the cloud's internal network; the data layer, which supports the data abstraction and facilitates data-centric logging through a provenance logger and a consistency logger; the workflow layer which focuses on the audit trails and audit-related data found in the software service in the cloud. In the framework, there is a policy, law, and regulation layer which spans all the three layers. How the data are stored, who accesses them, where the data are stored is covered by these policies. In all, this paper does a good job of mentioning all the issues hitherto not mentioned previously in cloud computing literature.

Provision Trust

Provision trust is the trust of a party in the services or resources provided by a service provider. This class of trust is most pertinent to the trust we are looking to establish in a TMS for cloud computing. Spanoudakis and LoPresti in their paper [8] deal with provision trust in an SOA. The authors note that while computing trust, one needs to take into account subjective as well as objective information. Subjective information is defined as the recommendations and opinions of other users and objective information is the information about the behavior and the quality of the service acquired during the run time. In the conceptual model presented in the paper, trust in a service depends upon evidence collected by monitors. Monitors consist of rules and assumptions on which TMPs are to be collected as evidence. Evidence is of two types, subjective and objective. Each evidence, however, is associated with a context and is only valid in that context. Trust is thus calculated by a third party and distributed in trust cards. A trust card becomes a trust certificate when the third party assumes legal responsibility for it. A mobile agent-based technique for provision trust assessment in Web services has been discussed in Ref. [9]. The authors provide their model in the paper which consists of the Web service agent framework (WSAF) server. The WSAF server acts as an intermediary between the service providers and the clients. The agents reside on the WSAF and proxy the service for the customer. It exposes the same interface to the client as the service provider but also monitors the service. The paper provides a feasible mobile agent model; however, it lacks completeness because it only considers SLA compliance. The customers of the service and the service providers provide the WSAF server with their policies and SLAs in the form of XML files in a predefined schema with predefined attributes. When a user asks for a service the WSAF server matches the best possible service provider with the policy XML provided by the user and exposes the service provider's interface to the user. The model given by Habib et al. [10] provides with a means to help consumers identify reliable and trustworthy cloud providers. The type of trust the authors deal with in the paper is provision trust. They propose multifaceted TMS architecture for a cloud computing marketplace. They look at different attributes of cloud trust such as security and performance compliance, which are assessed by multiple sources of information and come up with a trust management scheme to help choose a trustworthy cloud provider. They consider quantitative and qualitative information to factor into a TMS. In order to do this, they use *certain trust* and *certain logic* as the basis for trust metric. In certain trust, one models the trustworthiness of an entity based on opinions that express one's belief that a certain proposition is true. Each opinion is modeled as a tuple of values, $o = (t, c, f)$, where t denotes the average rating, c the uncertainty associated with the average rating, and f the initial expectation assigned to the truth of the statement. The trust management scheme has a *registration manager* where cloud providers register to be able to act as sellers in the cloud marketplace. Next is the *consensus assessments initiative questionnaire engine* which allows cloud providers to fill in the CAI questionnaire by providing an intuitive graphical interface through the RM. Then there is the *trust manager* (TM) which allows cloud users to specify their requirements and opinions when accessing the trust score of the cloud providers. This is done by providing them with a Web-based front end. It is followed by what the authors call a *trust semantics engine* (TSE) which models the configuration of propositional logic terms (PLTs) that are considered to be the trusted behavior of a cloud provider in terms of a specific attribute. Then they have a *trust computation engine* (TCE) which consists of operators such as AND, OR, and NOT, used in PLTs to compute the corresponding trust values. Finally, they have a trust update engine (TUE) which allows collecting opinions from various sources and roots. Noor et al. [11] propose the *Trust as a Service* (TaaS) framework to improve ways of trust management in cloud environments. They introduce an adaptive credibility model that distinguishes between credible trust feedbacks and malicious feedbacks by considering cloud service consumers' capability and majority consensus of their feed-backs. Their framework, which focuses on provision trust, is constructed using the SOA to deliver TaaS. In particular, their framework uses Web services to span several distributed TMS nodes that help several custom-ers give their feedbacks. The framework uses the recommendation and feedback system to establish trust. In other words, they use a subjective logic framework. The trust feedback collection and assessment works by providing the user with a primary identity, the CSP with an identity, a set of trust feedbacks, and the aggregated trust feedbacks weighted by cred-ibility. Each trust feedback is normalized and represented in a numerical form between 0 and 1, where 0 represents negative feedback, 1 represents positive, and 0.5 represents neutral feedback. The reference point is cal-culated using the majority consensus, where the age of the customer in terms of the number of days he/she actively used the cloud service comes into consideration to determine the credibility of the feedback assessment. The authors largely base their experimental results on online feedback obtained by epinions [12].

Identity Trust

Defined as certification of trustees in Ref. [4], identification trust is the confidence a party is able to place in the credentials submitted by another party. Identity management systems deal with this class of trust in cloud computing. Sato et al. [13] use a combination of the types of trust. They aim to handle identity and infrastructure trust discussed in the next sub-section. The paper discusses a hierarchy of trust, at the core of which is internal trust. Internal trust is analogous to TPM in case of hardware; it is a platform in which you have a guarantee of trusted operations. The authors place identity management and key management features in inter-nal trust so that secure and trusted services could be built up on them. Anything outside the internal trust circle is untrusted. The authors then define contracted trust as trust which comes into effect as a result of the binding through a contract or SLA which would identify a general level of assurance regarding each service. An untrusted entity can become trusted by entering into a contract and trustworthiness is calculated using the level of assurance specified in the contract. The paper however fails to pro-vide a concrete model of how the internal trust and the contracted trust interact with each other and only provides the definitions of the various components.

Infrastructure Trust

Grandison and Sloman [4] define infrastructure trust as the guarantee that the infrastructure indeed works as advertised. For computer hardware TPM has been proposed to be used for providing this guarantee. A number of initial papers on trust management in cloud computing mention this class of trust. Li et al. [14] discuss their trust model called multitenancy trusted computing environment model (MTCEM) which handles infrastructure trust. The authors observe that in a cloud computing model the objectives of the CSP and the cloud user are in contrast to each other. While the cloud user wants the highest quality of service at the lowest cost, the CSP wants to get the most money out of their resources. Models such as Wang et al. [15] thus assume the CSP to be an adversary of the user. The paper further says that with this adversary model it is not wise to allow the CSP to handle all the security. MTCEM separates and divides the security duty between the cloud users and service providers. It calls for and draws the boundaries for a shared responsibility of security between the CSP and the user. Security and hence trust in MTCEM is based on platform attestation and transitive trust provided by TPM. MTCEM, however, is mainly only for infrastructure as a service (IaaS), where the responsibility of the trusted infrastructure up till the VMM (first level) is of CSP and from the Guest OS to the application (second level) is of user. The chain of trust is started from the core root of trust measurement (CRTM) of the TPM, which is always trusted. Each subsequent component of the system is attested by a trusted component until the whole system is attested and trusted. The user or a trusted third party (TTP) on the user's behalf can then evaluate the chain of trust provided by the CSP and use it further on the user's part of the security responsibility. Fu et al. [16] talk about the security of the software deployed by the users on the cloud. Trust in this case is the assurance that user's software has not been tampered with by adversaries, who may either try to directly run malicious software on the cloud or may plant malicious code in the legal software. We classify it as a form of infrastructure trust because the premise is that the user does not trust the cloud infrastructure to be sufficiently secure to protect his software. The architecture provided in the paper creates a trusted running environment for the user software by embedding watermark in it and verifying the watermark before execution. The architecture though is specifically for software written in Java and running on JVM. The user, before deploying the software, embeds the watermark in the software using any suitable watermarking algorithm. The prototype implementation has a pool of 11 algorithms to choose from. The watermark embedder accepts a jar file and a watermark and outputs a watermarked jar file. The JVM on which this software will be running is also modified and enhanced with a watermark recognizer module. The watermark recognizer module is specially tailored to extract the above watermark. This JVM is then installed on the machines on which the user software is supposed to run. The watermark recognizer module in the JVM fires randomly during the run time to check the validity of the user software continuously and stops execution if the validation fails. The implementation of the algorithm shows that there is only a small size and execution time overload (Table 3.1).

TABLE 3.1

Classification of Papers on Trust in Cloud Computing according to the Type of Trust They Deal With

Model	Access Trust	Provision Trust	Identity Trust	Infrastructure Trust
Song et al.	X			
Li et al. [MTCEM]				X
Spanoudakis et al.		X		
Fu et al.				X
Sato et al.			X	X
Maximilien et al.		X		
Habib et al.		X		
Sundareswaran et al.	X			
Ko et al.	X			
Noor et al.		X		

Managing Trust in Cloud Computing

Adopting a public cloud essentially means migrating one's data to data centers outside one's premises. This migration of data means relinquishing the control of data to the CSP and trusting the cloud systems for the security of the data and applications which operate on the data. A survey by Fujitsu Research Institute [11] found that 88% of the potential cloud customers want to know what goes on in the backend physical server. Consumers are worried about where their data is stored and who has access to the data. This is the motivation behind various TMSs that are being developed to provide the user with the means to mitigate the risks involved in unknown ventures. Trust management is a field of active research; the terminologies pertaining to trust management though are ambiguously defined. On more than one occasion in literature, researchers have used the terms *trust, trust assessment,* and *trust management* interchangeably. In this section, we attempt to eliminate this ambiguity between these terms and try to create the overall picture of a complete TMS in cloud computing. The concept of Web service is very closely related to cloud computing and trust management in Web service has been extensively studied. We make use of the existing Web service research to define and elaborate on the required components of a comprehensive TMS in cloud computing. In SOA, TMS is the part of the system which answers the following questions.

1. A number of service providers exist in the market. What should be done to find out those that are of high quality but less advertised.

2. What parameters are most suited for calculating the trust and reputation of the service provider? We call these trust measure parameters (TMPs). A related question is how are these TMPs collected?

3. How are these TMPs used to calculate a tangible quantity which we call trust? Trust assessment, as part of the TMS, deals with this question, which is answered by a solid theoretical model of trust calculation.

4. How can users of the system provide feedback about their experiences to other users and use it to further refine their own trust in future?

The answers to these questions provide us with the components of a TMS. A TMS consists of the following components, namely, *service discovery, TMP selection and measurement, trust assessment,* and *trust evolution* as shown in Figure 3.1. In the rest of this section, we go through each of these components in detail and elaborate the functions that each of these components need to perform and survey papers that mention specific techniques for such functions.

Service Discovery

The World Wide Web Consortium (W3C) [17] defines service discovery as follows:

> the act of locating a machine-processable description of a web service-related resource that may have been previously unknown and that meets certain functional criteria. It involves matching a set of functional and other criteria with a set of resource descriptions. The goal is to find an appropriate Web service-related resource.

Although this definition has been created for Web service discovery, we can, without loss of generality, replace *Web* in the definition with *cloud*. The first step toward adopting a cloud-based solution is to find out the solutions/services available in the market which can cater to the needs of your applications and satisfy your requirements initially. In the current scenario this step is mostly manual; however, with rapid increase in the number of CSPs, soon it will become necessary to find a well-defined method to perform service discovery. This will help consumers save money by discovering not very

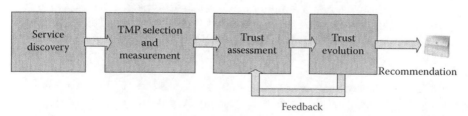

FIGURE 3.1
Components of a trust management system.

well known but competent CSPs automatically. Universal description and discovery interface (UDDI) and Web service description language (WSDL) have been used for service discovery on the Web service. The challenges in cloud service discovery though are compounded because services are offered at different layers [infrastructure as a service (IaaS), platform as a service (PaaS), and software as a service (SaaS)] and involve not just software but hardware resources too. Traditional SOA service discovery methods cannot penetrate across the layers and hence are rendered ineffective. Thus, service discovery methods specifically geared towards cloud computing are needed. Service composition along with service discovery is also an integral part of this component of the TMS. Wei and Blake [18] point out that while standardization efforts in Web services have not borne much fruit yet, in cloud computing the services inherently are centralized owing to the nature of the cloud. The authors further elaborate that a federation of cloud can bring uniformity and open doors for service discovery that would benefit end users to create and compose services analogous to Web services. Service discovery and composition is an interesting area of research in cloud computing domain. Earlier, solutions based on ontology [19,20] and attributes [21] have been proposed. Cloudle [19,20] is an ontology-based cloud service search engine which also takes into account the semantic relationships between different cloud computing concepts. Cloudle first defines some cloud computing concepts which each object will fall into. Each object (objects can be service providers) has well-defined properties and the properties have well-defined values. It then performs three different kinds of similarity search, concept similarity, object property similarity, and data type property similarity, and returns the results to the user based on the similarity score. Goscinski and Brock [21] introduce resources via Web services (RVWS) framework for state Web services for cloud computing. The paper recognizes that the state of resources in cloud computing is dynamic, and thus, stateless Web services cannot be used in such a scenario. In RVWS, the Web services consist of two kinds of attributes; state attributes and characteristic attributes. Characteristic attributes are attributes which do not change with time such as the total hard disk on a machine and the core count. while state attributes are dynamic attributes such as CPU utilization and disk in use. The publishers or the resource providers in this case publish their resources to a dynamic broker. The dynamic broker exposes a Web interface for the consumers and makes use of the RVWS framework to select appropriate resources for them.

TMP Selection and Measurement

After the service discovery phase is over, we have a handful of CSPs we know who are capable of handling the task we have. A customer will need to choose one out of these CSPs. The next step thus in trust management is to figure out the parameters on which the CSPs will be judged and compared. We labeled these parameters as TMPs and the process of selecting suitable TMPs and measuring their values is called TMP selection and measurement.

TMP Selection

TMP selection is a very application- and context-specific component of the TMS. The purpose of this component is to qualitatively try to find parameters that can be used to evaluate trust in a cloud service from the end users perspective. A good TMP would be easily measurable from the service provider's side and would be contextually related to the service the end user wants to use on the cloud. The selected parameters collectively should also be able to appropriately define one of the types of trust discussed earlier. Jøsang et al. [3] and Spanoudakis et al. [8] categorize TMPs in the objectivity–subjectivity dimension. Objective measures are determined by assessing the behavior and the quality of the service through a formal method during run time. On the other hand, subjective measures are the ratings given to the parameter by the users based on their experience of the service. Alhamad et al. [22] have defined SLA metrics for infrastructure, platform, software, and storage as a service in cloud computing. Suitable metrics or a combination thereof from the lists provided by them can be used as TMPs in trust management. CloudCommons [23] has come up with a service measurement index (SMI), which categorizes different measurement indices into a hierarchy with different attributes. These attributes again have subindices which can also be used as TMPs. In order to be able to actually measure all these parameters; transparency from the CPS's viewpoint is highly essential. Big players in the cloud market such as Amazon, Microsoft, and Google do not yet provide the user with full transparency and capabilities for tracking and auditing of file access history. Currently, users can only monitor the performance of the virtual hardware provided to them and the system event logs of the cloud service.

We define trust to be composed of three major categories of TMPs orthogonal to each other, which are *privacy and security, performance,* and *accountability*. Trust can then be viewed as a vector in three-dimensional space and the value of trust can be taken in the vector form or as the magnitude of the vector or in any other granularity level as deemed fit by the environment. As with the CloudCommons SMI, each category can be dissected in further detail and a TMS may choose to go into sufficient detail for the implementation, including only details which are required and leaving out those not necessary. In what follows, we discuss these categories focusing on the issues which have been seen most often in cloud computing literature and round up the section with a discussion on their measurement.

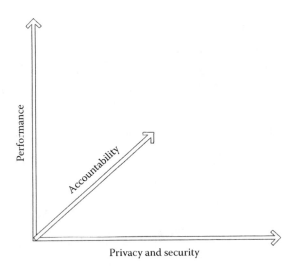

Privacy and security

Privacy and Security

Privacy is the adherence of a service provider, by legal or nonlegal norms, for the protection of confidential information of a client. Privacy plays a major role in defining the trustworthiness of a CSP as it lays down the assurance which a CSP gives to a client about the confidentiality of the client's content. The parameters that encompass privacy are consent, restrictiveness, and transparency of the client. They are used in laying down service-level agreements (SLAs). A service provider is required to comply with the norms stated in the SLA. Another factor that determines the amount of privacy of the data placed on a cloud provider's infrastructure is the segregation of virtual machines (VMs) on the physical machine. The virtualization concept helps us as it allows users to share the logical space on a physical machine. At the same time, it also provides a shared platform for multiple customers and their VMs. The isolation and separation of those VM activities is critical to guaranteeing the confidentiality and integrity of the operations, without interference from other customers. A CSP who can provide this with much confidence and conform to it will be placed very high on the trust assessment architecture that is developed.

Security issues in cloud computing include loss of governance, data, application, isolation failure, compliance risks, malicious insider in the cloud, data leakage, data protection risks, etc. The technical report in Ref. [24], talks about the risks involved in the cloud's economies of scale and flexibility from a security viewpoint. The massive concentrations of resources and data present a more attractive target to attackers, but cloud-based defenses can be more robust, scalable, and cost-effective. These security issues can be categorized under the following heads:

Perimeter Security Perimeter security refers to the physical security and protection of the cloud infrastructure from outside intruders. Firewalls, intrusion detection systems (IDSs), intrusion prevention system (IPS), demilitarized zones (DMZs), log inspection, system file integrity checks, etc. are the measures usually taken for perimeter security. CSA, in its consensus assessment initiative, also lists facility security as one of the parameters which can be grouped with perimeter security.

Hardware Security   Possible malicious insiders can have access to the data/ metadata of another cloud user using many side channel attacks. The level of trustworthiness also depends upon the ability of the CSP to control/prevent users from performing such unwanted activities. Ristanpart et al. [25] talk about several security issues related to virtualization in third-party clouds. Security researchers including Sailer et al. [24] have looked at methods for establishing trust in the hypervisors using the trusted computing platform module (TPM) to attest to the integrity of its software, guaranteeing it, for example, to be free from malicious or accidental modification of its core operating system/data. This is in addition to mechanisms for establishing trust within VMs themselves as they are transferred between hypervisors and physical servers. Trustworthiness can also be improved by making the hardware more resilient in the face of attacks or failures, as was discussed by Wang et al. [15].

Software Security   The cloud computing paradigm and associated technology of virtualization enhance the attack vector giving rise to new sources of threat along with traditional ones. While encryption, authentication, and digital signatures have long been used to provide confidentiality and data integrity in software, these measures are often inadequate in the cloud. Various approaches to software security have been proposed by researchers, including watermarking the software and using a chain of trust from a trusted hardware such as TPM's CRTM. Assurance of the hypervisors' ability to isolate and establish trust for guest or hosted VMs is critical, as this forms the root node for multitenant machine computing and thus could prove to be a single point of failure, because the hypervisor can potentially modify or intercept all guest OS processing. This access can be a serious potential hindrance to the security of a CSP.

Data Security   Because users place sensitive and private data in cloud, data security becomes an important issue. It becomes a major issue when users do not have control on how and where their data is placed on the cloud. Data security in cloud can be broken down into data confidentiality, data integrity, and data isolation. Data confidentiality is the protection of data from eavesdroppers and attackers who may try to access data in an unauthorized manner. Encryption and access control are often used to implement data confidentiality. Data on cloud is also dynamically updated by users, which includes modification, deletion, reordering, etc. Verification of the integrity of data in such cases becomes a challenging problem. However, the challenge of data security in cloud computing is compounded due to virtualization of resources and multitenancy. In a multitenant environment more than one user may have access to the same physical resource. In such a case it is important to have proper measures to isolate data from different users working on the same physical machine.

Performance

The three important components of performance of a service in cloud computing are availability, response time, and hardware throughput. A number of other quality of service parameters can be part of the overall performance of a service but we consider those that occur most often in the literature and discuss them in detail.

Availability Availability of the cloud system is of high importance to the customer. The on-demand, elastic, scalable, and customizable nature of the cloud, as well as the extremely high network traffic at the CSP affects availability. Availability can be affected by inside issues such as load balancing and outside issues such as denial of service attacks. Glenn Archer, Australia's First Assistant Secretary of Policy and Planning at the Department of Finance and Deregulation said that he *used to think that the issue of availability in the cloud was a significant advantage* but the recent outages at Amazon EC2 and Microsoft Azure show that even the biggest names in the business struggle at it. He also added that it is really hard to ensure such high availability as claimed by the CSPs. This clearly shows that a CSP's word (as written in the SLA) cannot be considered to be entirely true when it comes to defined metrics. We take our understanding of the definition of availability from the one defined in the EC2 SLA, which states that availability is the "Annual Uptime Percentage," where, uptime is the period of external connectivity barring the 5 minute periods of maintenance downtime during a service year. The baseline for availability can be gathered from the CSPs SLA. Most CSPs provide an availability guarantee in their SLA. It should be kept in mind that the onus of detecting an SLA violation, which in this case is availability dropping below a previously agreed-upon number, is on the consumer. This however demands continuous monitoring by the user.

Response Time Response time is usually measured as the time difference between sending a request to a cloud and the subsequent reception of those results. Response time consists of the connect time, round trip delay, and the processing time for the request. CloudSleuth (http://www.cloudsleuth.net) [26] and Alistair [26,27] attempt to measure response time. Whereas Ref. [26] calculates response time as the time taken by sending a request for an image and downloading a set number of images, Ref. [27] calculates it as an http request response time. Refs. [26,27] show that for the same request, response time of different clouds is different and thus they can be ranked accordingly. By analyzing the results of both Refs. [26,27] we can see that the results are mostly comparable. The small difference between the results can be attributed to the difference in methodologies and the time of observation. This observation enables us to use this metric for evaluation of various CSPs.

Hardware Throughput   Performance of the hardware of a cloud is one of the most critical and the most difficult to measure among the metrics. This owes partly to the difference in terminologies and quantum of cloud used by the various CSPs and partly because of the variance in the requirements of the cloud customer. Lack of standards in defining the quantum of resources makes comparing them difficult. For example, Amazon EC2 calls each unit of resource as EC2 compute unit (ECU), vCloud calls each unit of resource as virtual processing unit (VPU), and Microsoft Azure calls each unit as compute instance, and all these units have different configurations. Hardware performance is also closely related to cost because it depends on the quality of resources and resources in a cloud are offered as services. Thus, instead of the performance of the hardware we should measure the hardware throughput, which is hardware performance per unit cost. With more money you can buy better resources but the question is will that translate into better performance? And, is the change in performance due to change in cost uniformly across all CSPs? The three important components of hardware performance of a system are its CPU performance, disk IO performance, and the memory performance. Different applications have different resource needs and are affected by different performance parameters. For example, adding more CPU resources cannot help a memory-intensive application. This also means that clouds should be evaluated in accordance to the applications the user is supposed to be running because a generic evaluation would not provide accurate results. Hence, applications should be divided into three different categories; CPU intensive, disk intensive, and memory intensive applications. When comparing the performance of the hardware, the comparison should be performed within the categories and not across.

Accountability

Accountability supports the other two categories of TMPs. It is the extent to which the expectation from a CSP to provide the services agreed upon hitherto is fulfilled. This expectation can be in terms of *privacy* and *security* or in terms of *performance*. It can be deemed as the responsibility of the CSP to conform to all the norms mentioned in the SLAs or adherence to generally accepted standards, processes, and procedures. In a general SOA, accountability from hardware or software or a platform-dependent perspective will be related to the performance/health of the service being provided. When a client moves his/her data to the cloud they lose access to the operation of the service. CSPs then make claims about the quality of their service, the performance of their hardware, and the security which they provide integrated with the offered services. Accountability in this case is the level to which such claims are fulfilled and in some cases verified by the client.

The transparency accorded by accountability enhances the trust which can be placed in an entity (which is the CSP in this case). A CSP which mentions in the SLA about the level of health that is provided and delivers it eventually regardless of any matter will be rated high in the trust assessment architecture. To gauge the level of accountability of a CSP, a number of factors can be looked into. We discuss three most important factors later.

Logging Logging in cloud is affected by virtualization, privacy, and access control. As we know, virtualization allows higher resource utilization but it also brings with it the issue of a CSP having to track the different virtual servers on one of many physical servers. This means that there could be a possibility of tracking multiple operating systems, which could get complex and would have to be dealt with appropriately. This brings us to the concept of logging. Depending on the logging information from the perspective of an operating system or a file-centric system, policies have to be defined appropriately in a well-defined manner that clearly tell us which information to log in to and which to not. This is because, in an elastic system such as the cloud, logging information that is not pertinent could violate the prior agreements that a service provider makes with a client. This could be seen as a violation of privacy and access control. Therefore, the policies have to be defined, which specify the areas that loggers are assigned to log. As an example, in Ref. [6] accountability is taken as keeping track of the users accessing a client's data and making sure only eligible users are given the access. To improve user control over the data, a distributed auditing mechanism based on logging is described, which also takes into account the privacy concerns of the client.

Auditing An audit enables the auditor and the party on whose behalf the auditor works to ensure that the CSP is following the processes and standards and has the security controls in place which were agreed to in the SLA. From the viewpoint of information assurance, the concept of cloud computing is similar to IT outsourcing. Cloud computing, however, has features such as multitenancy, scale in, and scale out, which introduce a number of complications for which no prior security controls exist. Many standards pertaining to security and privacy of processes, data, and infrastructure exist for information security which can also be extended to cloud computing. The CloudCommons SMI defines a metric called auditability under the accountability category which is the ability of a client to verify that the service provider is adhering to the standards, procedures, and policies that have been agreed upon in the SLA. Clearly, a CSP following widely known and accepted standards and more open to audits by a client can be trusted more than others.

Compliance In practice, regulatory compliance is the established mechanism to ascertain a level of trust in information security. Compliance means establishing policies, procedures, and controls toward fulfilling the requirements of a standard provided by an authority. There exists legislation that covers many areas from governance and ownership to security and privacy, for example, Health Insurance Portability and Accountability Act (HIPAA) for health data, Payment Card Industry Data Security Standards (PCI DSS) for payment cards, and Sarbanes–Oxley Act (SOX) for disaster recovery. Cloud computing however complicates regulatory compliance situation. Data in a cloud is replicated at various places which may be geographically and physically far apart. This potentially introduces a very tricky situation in which the same data may need to be compliant to different regulations, which, in extreme cases, may be contrasting. The customer thus needs to make sure that either the data is localized with respect to the laws or there is no such conflict of laws. CSP's transparency and flexibility in this regard should be taken into account when evaluating trust. A CSP may also use resources from or subcontract the work to other CSPs, for example, an SaaS provider using an IaaS provider's resources. In this case it needs to be insured that all such providers in the chain be compliant. The onus of regulatory compliance as stated by Arshad [28] lies on the customer and hence, in trust computation, a CSP that is transparent to the customer regarding how his data will be handled will be considered to be more accountable. Some solutions to the regulatory compliance problem have been proposed in literature, for example, regulatory compliant cloud computing (RC3) [28]. RC3 proposes to first classify a customer's data into sensitive and nonsensitive data. Sensitive data are then processed and stored in regulated zones and nonsensitive data can be stored elsewhere. Regulated zones can be inside the client's perimeter (private cloud) or a trusted compliant public cloud. A number of other papers propose storing only encrypted data on the cloud and working on this encrypted data to solve the compliance problem. Encrypting data before storing is not a big issue; however, aggregating, querying, and applying other operations on cryptographic data are not easy and incurs a large overhead.

TMP Measurement

We now talk about some of the existing ways to measure the TMPs outlined above and discuss their merits and demerits. In some cases, one or more of the metrics can be gathered from the cloud provider's SLAs, white papers, and other published artifacts. For example, Amazon's EC2 in its SLA states 99.95% availability; which in terms of hours, comes out to be about 4 hours 23 minutes of downtime in a year, apart from scheduled maintenance. However, an SLA does not guarantee that things will not go wrong at run time. For instance, in March 2011 EC2 suffered an outage and according to Amazon it lasted for almost 11 hours. Thus, this single outage accounted for far more than the downtime Amazon guarantees in a year. According to a Sage Research report [29], on average, yearly unscheduled downtimes are about 44% of total downtimes. Such unscheduled downtimes may translate to huge monetary losses to the clients. According to the same report the average cost of network downtime is $2169 per minute. Also, in Ref. [1] the definitions of the metrics are based on business objectives and the business objectives of the cloud customer and the CSP are obviously different. Although published artifacts provide us a baseline, they cannot be entirely relied upon by a client and a more accurate and meaningful measurement of the metrics is required. A number of Web sites and tools such as CloudSleuth [26], CloudHarmony [30], CloudStone [31], and Cloud CMP [32] also exist, which measure these metrics. CloudHarmony [30], CloudStone [31], and Cloud CMP [32] are tools which consist of a set of benchmarks that measure the performance of clouds by deploying test applications in the cloud. These tools however suffer from two drawbacks. First, they do not take into account other forms of measurements such as SLAs and user feedback and hence are unidimensional; second, these tools do not provide continuous monitoring. CloudSleuth is a more distributed tool than the rest of them. It consists of 30 backbone nodes all over the world out of which 18 are in the United States and 12 are outside. CloudSleuth runs a transaction on each of the CSP it wants to monitor periodically. It measures two metrics, response time and availability. Each of the backbone nodes runs the transaction on the clouds in its region. The transaction consists of downloading a page with 40 images, which simulates a Web site browsing experience. The response time is calculated as the average time taken for the images to download. Availability is measured as the percentage of successful transactions over a period of time. It can be argued that CloudSleuth's response time test is very generic and may differ vastly from each individual customer's requirements and load as it provides an aggregate view that is not what is desired to perform a holistic assessment of the trust from the viewpoint of availability. Papers such as Ref. [11] rely on feedbacks from the cloud consumer about the services he/she has used to calculate the consumer's trust in the service. For their experimental evaluation, they use feedback data from epinions [12]. Again use of feedback alone makes this method of evaluating a cloud unidimensional because it only considers a consumer's viewpoint. A consumer's viewpoint is most important in evaluating a cloud. However, it should not be the sole criterion, because consumer feedback may depend on a number of factors unrelated to the cloud's performance and thus may be flawed. Moreover, as observed with cloud comparison Web sites and tools, feedbacks too do not offer continuous monitoring of the cloud. To test the three performance components, we can choose any existing benchmarks keeping in mind that because we want to run the benchmark tests periodically during the runtime, we want them to be short and unintrusive. We need to test if CPU performance test suites such as Geekbench [33] and UnixBench [34] can be used. Dbench [35] can be used to test disk IO and Geekbench and UnixBench can again be used to measure memory performance. The performance measurements can then be aggregated and averaged over the cost of the machine to obtain the performance throughput. A more subjective measure of the performance can also be collected from the user feedback. To have a fair comparison, most existing measurements are done on machines running a single VM. This though is not the case with the day-to-day functioning of the cloud where physical machines are multitenanted and a user's VM has to compete for resources on a machine, thus affecting the overall performance. Hence, the existing measures do not accurately reflect the runtime performance of the cloud. When measuring the performance, external affects caused by multitenancy should also be accounted for.

Trust Assessment

The trust assessment component of the TMS uses the measured values of the TMPs selected and collected in the previous component and using a mathematical model calculates a final value, which we call trust. Trust assessment techniques can be classified based on how the trust is modeled mathematically.

Ratings-Based Modeling

In ratings-based modeling, customers rate the services provided by a service provider on a scale of 1 to n, typically n being 5 or 10. The overall trust in a service is computed by taking the average of all the customer ratings. Amazon and eBay use such a rating system for the services and products provided by them. This modeling is very simple but has great scope for errors and attacks. The recently proposed SMI for CloudCommons from Cloud Service Measurement Initiative Consortium (CSMIC) is one such model. In the CloudCommons marketplace, consumers can buy cloud services provided by a multitude of CSPs. SMI defines a set of business relevant key performance indicators (KPIs) which, as mentioned before, are *accountability, agility, assurance, financial, performance, security and privacy,* and *usability.* Users assign weight to these KPIs and evaluate a service based on the weight-assigned KPIs. These evaluations are then submitted to the CloudCommons, which collects all such data to differentiate between services with similar weight distribution. In addition, CloudCommons also provides a star-based rating much like Amazon, which may act as an initial filter when you are looking for certain services.

Bayesian Modeling

Bayesian modeling of trust encodes probabilistic relationships among the TMPs [36] and is based upon the Bayes' rule which states that [37]:

$$P(h|e) = P(e|h).P(h)/P(e),$$

where:

$P(h|e)$ is the probability of the hypothesis given the evidence
$P(e|h)$ is the probability of the evidence given the hypothesis
$P(h)$ is the prior probability of hypothesis
$P(e)$ is the prior probability of the evidence

Dependence of trust on the various TMPs is modeled in a relationship graph, which is a directed acyclic graph (DAG). The TMPs may then depend upon each other or other factors. The output of a Bayesian network is the degree of trust, which can be placed in a service provider given the prior and posterior probabilities. Hien Trang Nguyen et al. [38] note that Bayesian networks are ideal in service-oriented/cloud architecture because they can be used to calculate the probability of a hypothesis under various conditions, for example, with various TMPs. A Bayesian network-based trust model was proposed by Wang and Vassileva [37]. Although the model has been proposed for trust in a peer-to-peer user agent-based model, it can be easily extended to SOA and hence to a cloud computing. The trust is evaluated based on some TMPs that are selected beforehand, such as the quality of file and download speed. Each agent develops a naive Bayesian network for each service provider with which it has interacted. The root of the Bayesian network is a node *T*, which has two values *satisfying* or *unsatisfying* that represents the agent's reflection of the preceding transaction. The agent maintains a conditional probability table for each of the leaf nodes of the network (the TMPs). After each transaction, the agent calculates the conditional probability $P(T = satisfying|TMPs\ are\ satisfied)$. Multicontext, personalized trust that captures more than just binary assessment of a transaction was allowed in Hien Trang Nguyen work [38], which extended this work. In contrast to Ref. [37], in Ref. [38] a user-defined interest level is used for each TMP to calculate *T*. This ensures that trust can be calculated based on the context- and application-specific needs.

Belief Modeling

While most trust models work with only two factors, trust and distrust, belief models such as Jøsang's [39,40] also account for uncertainty which may arise in some circumstances. Jøsang's belief model, the metric of trust propagation is called opinion, which is denoted by the quadruple (b, d, u, a) where b is belief, d is disbelief, u is uncertainty, and a is atomicity, where $b + d + u = 1$. Past experiences of a customer are quantified in opinions and these opinions can then be combined using a Bayesian network or Dempster rule to produce a final trust score. This model is used for trust-based Web service selection in Ping Wang et al.'s work [41]. The paper assumes a TTP which acts as a trust repository. When the consumers interact with a service provider they generate an opinion of it, which takes the form of a quadruple (b, d, u, a). The paper then defines operations on this quadruple which can be used to generate a recommendation based on the opinion provided by a consumer and the opinion the TTP has of the consumer. A consensus operation is also defined to aggregate the opinions provided by multiple consumers. The reputation of a service provider is established based on the opinions submitted by the consumers. A similar approach is followed in Ref. [42] for Web service selection using a TTP.

Fuzzy Modeling

Ratings from individuals in the form of feedback vary from person to person and tend to be of an uncertain nature. To work with this uncertain nature of data, researchers [43,44] have proposed fuzzy models of trust. In fuzzy models of trust, the raters rate the quality of service in linguistic parameters, such as poor quality, good quality, and high quality. A membership function then maps the parameters in the range of 0–1. Fuzzy logic then provides rules for reasoning with such a set of data. Fuzzy models are easier to grasp by the end user because the user deals with linguistic parameters such as *high importance, low importance, good quality,* and *poor quality* rather than numbers. In the model proposed in Ref. [43] users evaluate services on some prefixed parameters in such fuzzy terms. The paper then defines fuzzy queries, which can be run using these parameters to again produce output in fuzzy linguistic terms which are more end user-friendly. A trust model for IaaS clouds based on neuro-fuzzy theory is introduced in Mohammed Alhamad et al.'s works [45]. The paper takes into account four TMPs, namely, *scalability, availability, security,* and *usability.* Each of the TMPs, can take three values, *low, medium,* or *high.* The Sugeno fuzzy inference model is used and trained with the data collected from a survey. The fuzzy inference model provides fuzzy output based on the input data which is then defuzzified to produce a reputation score.

Trust Evolution

Early trust models would statically calculate a value of trust based on the given TMPs and that trust was the basis of all the decisions in the future. However, now we understand that trust needs to evolve continuously and adjust dynamically, taking into account current and past performance experiences by the user as well as others who are in the same situation. This is done by providing feedback to the trust model in the form of trust calculated by the user (direct trust) and the trust referred (referred trust) by others. A few questions then arise. How does trust calculated at time t affect trust calculated at time $t + 1, t + 2, \ldots, t + n$ and beyond, and when does it become useless. Referred trust in the form of recommendations should be used carefully because ambiguity always surrounds the authenticity and correctness of the referred trust. How much can the trust value received from other sources be trusted and how much impact the referred trust should make in trust computation. These issues are tackled by the trust evolution component of the TMS. Once the trust assessment phase is over and a well-defined method of calculating trust is in place, the trust evolution component kicks in and continuously tries to refine trust with each new experience with the help of other users as well.

The first issue in trust evolution is how to quantify the effect of trust calculated at time t over trust at time $t + x$. Most papers we have reviewed thus far have some form of feedback based on built-in past experiences. In Ref. [46] a very simple approach is taken. Trust is defined as the probability σ that a user task will be serviced within the response time τ specified by the customer. Initial trust score T_0 is calculated based on the mean arrival rate, service time, and the number of virtual servers. The service time of every subsequent task served by the server will either be greater or less than the required response time. A positive response is used to improve the trust by adding to the trust value (T_n) the fraction by which the service time was better than the required response time and subtracting the fraction in case of negative response. Trust at time t thus in this case depends only on direct experience and the trust value at time $t - 1$. An identity management scheme to allow only a trusted set of users to access the data stored in the cloud is given in Ref. [6]. The evolution of trust in this scheme is based upon the access logs. During auditing, if the logs remain true to the identities, then the trust value of the CSP rises and the user provides with better ratings and recommendations, which helps new users assess the CSP's trustworthiness and the cycle of trust evolution continues.

The simplest of methods to quantify referred trust is to use transitivity of trust. This method is followed in Ref. [42]. If x user has a trust T_x in a CSP c and another user y has trust t_y in user x, then user y's trust in CSP c is calculated as $t_y * T_x$. Transitive trust however fails to take into account the context in which the trust is used. Moreover, long chains of trust can be easily poisoned by attackers. Talal and Quan [11] introduce the concept of the age of the user to quantify the effect of referred trust on trust computation. As previously discussed, Noor et al. determines the trust in a CSP based on the ratings and recommendations provided by the user. Age of a user is defined by the number of good recommendations provided by the user. Good recommendations are those which are closer to the mean of recommendations provided by a group of users. The recommendation provided by a user is thus based on the age of the user and how close to the mean the user's recommendation is. Refs. [37,38] discuss the use of Bayesian network to probabilistically build a relationship among the TMPs and thereafter build a decision-based system, which will enable users to determine the effect of one parameter on the overall trust value. Recommendations in this case are context specific and each user's recommendation is weighted according to how trustworthy the user is considered in the system. The user's trustworthiness is calculated from his past recommendations. In Ref. [38], a user's own feedback is also considered as a recommendation and the source is considered to be fully trustworthy and the same process is used to factor this in as feedback.

Conclusions

The cloud computing paradigm has gained in prominence in recent years and with a number of cloud services and cloud providers available, trust in cloud computing has become an attractive field of research. The literature however is scattered with a number of definitions and solutions to the problem of trust and trust management in cloud computing. We have tried to bring together all this scattered research under a common heading and have tried to discuss how all of them are related to and different from each other in this text. We started with a generic definition of trust and then broke that definition down into four types, which are most related to cloud computing: *access trust, provision trust, identity trust,* and *infrastructure trust,* and provided the definition of each of these different types of trusts relevant to cloud computing. Several papers exist that explain about trust in cloud computing, all from different perspectives. We classified some of those papers into the above four different types of trust. This classification can be used for future papers on or other papers dealing with trust in cloud computing that we may have missed. This will provide a commonality to the papers that talk about trust in cloud computing, but each touches a different aspect. We also differentiated between the often-confused terms of trust, trust management, and trust assessment. We broke down the TMS into four components: *service discovery, TMP selection and measurement, trust assessment,* and *trust evolution.* Management of trust begins with the *service discovery* component. Service discovery is the least-researched component of trust management in cloud computing. We discussed how service discovery fits in and will impact trust management in future and reviewed the few early solutions which exist today. Service discovery leads to the *TMP selection and measurement* component where the system or the customer decides which TMPs are relevant to their application and should be used in trust computation. We categorize the TMPs into three classes, *privacy and security, performance,* and *accountability.* We discussed each of the TMP in depth and ended the section with a discussion on measurement of the TMPs, using the TMPs selected. After the TMPs have been decided upon and the TMP measurement techniques are in place, a formal method of computing trust from these TMPs is required. This is accomplished in the *trust assessment* component. We talked about four major kinds of techniques used in trust computation: *ratings-based modeling, Bayesian modeling, belief modeling,* and *fuzzy modeling.* We again reviewed papers which have used these techniques in trust assessment in cloud computing and related disciplines. A TMS is incomplete without a discussion on how the system will adapt dynamically and take into account feedbacks and recommendations. We end the paper with a discussion on the *trust evolution* component, which does this for us. In the section on trust evolution, we discussed papers which try to answer how feedbacks and recommendations can be used with a trust assessment engine.

References

1. Jansen, W. and Grance, T., "DRAFT guidelines on security and privacy in public cloud computing," *NIST Technical Report NIST SP 800-144*, 2011.
2. CloudXL, "Cloud computing and software as a service providers." http://www.cloudxl.com. Retrieved on January 8, 2013.
3. Jøsang, A., Ismail, R., and Boyd, C., "A survey of trust and reputation systems for online service provision," *Decision Support System Journal*, 43(2): 618–644, 2007.
4. Grandison, T. and Sloman, M., "A survey of trust in Internet applications," *Communications Surveys & Tutorials*, 3(4): 2–16, 2000.
5. Song, Z., Molina, J., and Strong, C., "Trusted anonymous execution: A model to raise trust in Cloud," *Proceedings of the 2010 Ninth International Conference on Grid and Cloud Computing*, November 1–5, IEEE, Jiangsu, China, 2010.
6. Sundareswaran, S., Squicciarini, A., Lin, D., and Huang, Shuo, "Promoting distributed accountability in the Cloud," *2011 IEEE International Conference on Cloud Computing (CLOUD)*, July 4–9, Washington, DC, pp. 113–120, 2011.
7. Ko, R. K. L., Jagadpramana, P., Mowbray, M., Pearson, S., Kirchberg, M., Liang, Q., and Lee, B. S. "Trust Cloud: A framework for accountability and trust in Cloud computing," *IEEE World Congress on Services (SERVICES)*, July 4–9, Washington, DC, pp. 584–588, 2011.
8. Spanoudakis, G. and LoPresti, S. "Web service trust: Towards a dynamic assessment framework," *Proceedings of the 4th International Conference on Availability, Reliability and Security*, March 16–19, IEEE, Fukuoka, Japan, 2009.
9. Maximilien, E. M. and Muninder, P. S., "Toward autonomic web services trust and selection," *Proceedings of the International Conference on Service Oriented Computing (ICSOC)*, November 15–19, ACM, New York, 2004.
10. Habib, S. M., Ries, S., and Mühlhäuser, M., "Towards a trust management system for Cloud computing," *IEEE 10th International Conference on Trust, Security and Privacy in Computing and Communications*, November 16–18, Changsha, China, 2011.
11. Noor, T. H. and Sheng, Q. Z., "Trust as a service: A framework for trust management in cloud environments," *The 12th International Conference on Web and Information Systems (WISE)*, October 11–14, Sydney, NSW, 2011.
12. Epinions, http://www.epinions.com
13. Sato, H., Kanai, A., and Tanimoto, S., "A Cloud trust model in a security aware Cloud," *Proceedings of the 10th IEEE/IPSJ International Symposium on Applications and the Internet*, July 19–23, IEEE Computer Society, Seoul, Republic of Korea, 2010.

14. Li, X. Y., Zhou, L. T., Shi, Y., and Guo, Y., "A trusted computing environment model in cloud architecture," *Proceedings of the 9th International Conference on Machine Learning and Cybernetics*, July 11–14, Qingdao, China, 2010.
15. Wang, C., Wang, Q., Ren, K., and Lou, W., "Ensuring data storage security in cloud computing," *17th International Workshop on Quality of Service*, July 13–15, IEEE, Charleston, SC, 2009.
16. Fu, J., Wang, C., Yu, Z., Wang, J., and Sun, J.-G., "A watermark-aware trusted running environment for software clouds," *5th Annual ChinaGrid Conference*, July 16–18, IEEE, Guangzhou, China, 2010.
17. The World Wide Web Consortium (W3C), http://www.w3.org
18. Wei, Y. and Blake, M. B., "Service-oriented computing and cloud computing: Challenges and opportunities," *IEEE Internet Computing*, 14(6): 72–75, 2010.
19. Jaeyong, K. and Kwang Mong, S., "Ontology and search engine for Cloud computing system," *International Conference on System Science and Engineering (ICSSE)*, June 8–10, Macao, pp. 276–281, 2011.
20. Jaeyong, K. and Kwang Mong, S., "Cloudle: An ontology-enhanced Cloud service search engine," *Web Information Systems Engineering—WISE 2010 Workshops*, Hong Kong, China, December 12–14, pp. 416–427, Springer, Berlin, 2010.
21. Andrzej, G. and Michael, B., "Toward dynamic and attribute based publication, discovery and selection for cloud computing," *Future Generation Computer Systems*, 26(7): 947–970, 2010.
22. Alhamad, M., Dillon, T., and Chang, E. "Conceptual SLA framework for cloud computing," *4th IEEE Conference on Digital Ecosystems and Technologies*, April 13–16, IEEE, Dubai, 2010.
23. Service Measurement Index Version 1.0, http://www.cloudcommons.com/documents/10508/186d5f13-f40e-47ad-b9a6-4f246cf7e34f
24. Daniele, C. and Giles, H., "Cloud computing: Benefit, risks and recommendation for information security," *ENISA Technical Report*, ENISA, Crete, Greece.
25. Ristanpart, T., Tromer, E., Sacham, H., and Savage, S. "Hey, you, get off of my Cloud: Exploring information leakage in third-party compute Clouds," *Proceedings of the 16th ACM Conference on Computer and Communications Security*, November 9–13, Chicago, IL, 2009.
26. CloudSleuth, http://www.cloudsleuth.net
27. Alistair, C., "Cloud performance from the end user perspective," BitCurrent WhitePaper BT_CPEU_0411, March 31, 2011, http://www.bitcurrent.com.
28. Arshad, N., "Web-application architecture for regulatory compliant Cloud computing," *ISSA Journal*, 10(2): 28–33, 2012.
29. IP *Service* provider downtime study: Analysis of downtime. Causes, costs and containment strategies, Sage Research, Prepared for CISCO SPLOB, August 17, 2001.
30. CloudHarmony, http://www.cloudharmony.com
31. Ang, L., Xiaowei, Y., Kandula, S., and Zhang, M., "CloudCmp: Comparing public cloud providers," *Proceedings of the 10th ACM SIGCOMM Conference on Internet Measurement*, ACM, Melbourne, VIC, November 1–3, pp. 1–14, 2010.
32. Sobel, W., Subramanyam, S., Sucharitakul, A., Nguyen, J., Wong, H., Klepchukov, A., Patil, S., Fox, A., and Patterson, D. "Cloudstone: Multi-platform, multi-language benchmark and measurement tools for web 2.0.,"*1st Workshop on Cloud Computing and Its Applications*, Chicago, IL, October 22–23, 2008.
33. Primate Labs Geekbench 3, http://www.primatelabs.ca/geekbench/.

34. Freecode UnixBench, http://freecode.com/projects/unixbench
35. Dbench, http://dbench.samba.org/.
36. David, H., "A tutorial on learning with Bayesian networks," *Technical Report MSR-TR-95-06*, Microsoft Research, 1996.
37. Wang, Y. and Vassileva, J., "Bayesian network-based trust model," *Proceedings of the 2003 IEEE/WIC International Conference on Web Intelligence*, October 13–17, IEEE Computer Society, Halifax, NS, 2003.
38. Nguyen, H. T., Zhao, W., and Yang, J., "A trust and reputation model based on Bayesian network for web services," *Proceedings of the 2010 IEEE International Conference on Web Services (ICWS)*, July 5–10, Miami, FL, pp. 251–258, 2010.
39. Jøsang, A., "Trust based decision making for electronic transactions," *Proceedings of the 4th Nordic Workshop on Secure Computer Systems (NORDSEC)*, November 1–2, Stockholm University, Kista, Sweden, 1999.
40. Ali, A. S. and Rana, O. F., "A belief-based trust model for dynamic service selection," *Economic Models and Algorithms for Distributed Systems*, Birkhäuser, Basel, pp. 9–23, 2010.
41. Wang, P., Chao, K. M., Lo, C. C., and Farmer, R., "An evidence-based scheme for web service selection," *Information Technology & Management*, 12(2): 161–172, 2011.
42. Shamila, E. S. and Ramachandran, V. "Evaluating trust for web services access," *IEEE International Conference on Computational Intelligence and Computing Research (ICCIC)*, December 28–29, Coimbatore, India, pp. 1–4, 2010.
43. Nepal, S., Sherchan, W., Hunklinger, J., and Bouguettaya, A., "A fuzzy trust management framework for service web," *Proceedings of the 2010 IEEE International Conference on Web Services*, July 5–10, IEEE Computer Society, Miami, FL, 2010.
44. Griffiths, N., Chao, K.-M., and Younas, M., "Fuzzy trust for peer-to-peer systems," *Proceedings of the 26th IEEE International Conference on Distributed Computing Systems Workshops*, July 4–7, IEEE Computer Society, Lisbon, Portugal, 2006.
45. Alhamad, M., Dillon, T., and Chang, E., "A trust-evaluation metric for Cloud applications," *International Journal of Machine Learning and Computing*, 1(4): 416–421, 2011.
46. Firdhous, M., Ghazali, O., and Hassan, S. "A trust computing mechanism for cloud computing," *Proceedings of ITU Kaleidoscope 2011: The Fully Networked Human?—Innovations for Future Networks and Services (K-2011)*, December 12–14, Cape Town, South Africa, pp. 1–7, 2011.

4

Trust Models for Data Integrity and Data Shredding in Cloud

Ashutosh Saxena, Nitin Singh Chauhan, and Sravan Rondla

CONTENTS

Introduction

Information technology (IT)-enabled services over the Web are rapidly expanding with the increased adoption of cloud computing services. The cloud services model has a clear advantage over traditional IT data centers as it utilizes shared resources and helps individual enterprises reduce cost. With the launch of many cloud-enabled services and increased use of smart mobile gadgets, cloud has entered our daily lives as well. However, cloud services-enabling technologies, cloud service characteristics, and cloud operational models bring in certain risks and challenges along with the advantages [1,2]. There is genuine concern on losing control over data, which is being seen as a major drawback of the otherwise well-liked and talked-about cloud movement.

Cloud customer companies are also uncertain about secure data management practices followed by cloud service providers (CSPs). Threat exposure has increased in the cloud's multitenant data centers as sources of attack are not only restricted to coming from outside but compromise could also occur from tenants as well in the same environment. The chances of threat realization such as unauthorized access, data loss, and unavailability of services are being escalated for containment due to the characteristics of cloud service operations.

One facet of cloud computing is storage as a service (StaaS). This essentially means that the owners of the data relegate their data to cloud storage offered by a third party. The cloud owner may charge a fee from the data owner depending on the time and space occupied by his data. StaaS has been envisaged as a popular solution to the growing concern of rising storage costs of IT enterprises and individual users. With the rapid rate at which data are being produced, it has become very costly for users to update their hardware frequently. Apart from the economic advantage of low storage costs, individuals and enterprises benefit from significant reduction in data maintenance costs, increased availability, and resilience to data losses by using cloud storage.

In the StaaS model, the cloud moves the owner's data to large data centers at remote locations where the user does not have any physical control over his data. However, such unique features pose many new challenges and risks to the security and trust of the owner's data which need to be extensively investigated. Verifying the integrity of the data stored in the cloud environment gains significance in the StaaS model [3].

Another concern in the StaaS model is the possibility of data misuse by the CSP after the end of the storage contract period. There is a probability of data being used for malicious or unintended purposes, if it is not verifiably shredded. The owner of the data may like to have a proof that his data are securely and effectively shredded in the service provider's storage after the service period. Technically, it is challenging to ensure that no data are left over with the CSP after the service contract period ends. In this chapter, the trust issues related to data shredding in remote cloud environment and a possible solution for providing comprehensive verifiable proof for data shredding in cloud are presented.

Cloud Storage: Present Scenario

Cloud storage is the latest buzzword in IT circles. In simple terms, it means storing data in a remote cloud. Some of the enterprises that offer cloud storage services are Amazon's Simple Storage Service (Amazon S3) [4], Nirvanix's CloudNAS [5], EMC's Atmos, etc. Consumers of cloud storages range from individual users to blue chip enterprises. For instance, *The New York Times* uses Amazon's S3 to deliver its online articles. Cloud storage enterprises provide an interface in the form of application program interfaces (APIs) or custom-made software through which storage space can be accessed by the consumers.

Cloud storage is also used to provide online data backup services. This comes with all the features of a traditional backup product such as scheduling backup, reporting, and file recovery, with the only difference that data are now being stored in remote cloud storage. Cloud backup services include Carbonite Online Backup, EMC's Mozy, EVault (a Seagate venture), Iron Mountain Digital, and Symantec Online Backup.

Trust and Security Issues in Cloud

"On-demand" and "pay-per-use" are typical phrases that buzz around the definition of the cloud. In recent years, cloud adoption has increased significantly, as it brings cost saving to IT savvy business companies and their service providers. Trust, security, and privacy are still a concern for enterprises that consider migration of data and applications to public cloud. Control boundaries are slightly fuzzy and shift toward a cloud provider. Cloud consumer companies have to trust the cloud provider for secure data and application management practices.

In such a scenario, the cloud customer desires greater assurance and transparency from service providers concerning handling of data. Further, it is challenging for cloud customers to monitor and evaluate the security control applied on their data by the CSP.

Trust can be gained by applying a security-oriented or a nonsecurity-oriented mechanism. In a cloud context, one way of gaining trust is by applying security mechanisms such as access control, authentication, and encryption by the cloud provider on the customer's data. Another aspect of gaining trust is by forming appropriate service-level agreements (SLAs) and contracts between the CSPs and their customers. Further, reputation and brand image are associated with trust, which impacts negatively if there is a breach of trust or privacy.

In the following subsections, we will discuss some of the security- and privacy-related issues that impact the trust factor of cloud customer.

Confidentiality Issues

Data confidentiality in cloud is more critical compared to that in traditional enterprise computing because an enterprise has no control over storage and transmission of data once it is in the cloud. A cloud customer depends on the cloud provider for applying appropriate levels of access control mechanisms for protecting their data. Encryption is one of the obvious options to ensure confidentiality. It is also critical to know the algorithms and encryption key sizes applicable, even if customer gets an assurance of applying encryption on the data stored. Some of the algorithms are cryptographically weak and can increase the risk for potential breaches. Key management practice is another concern when encryption is used to protect confidentiality. If the cloud customers desire to manage the keys of the encrypted data, they need to have additional resource and capability to manage such a system. It is challenging for CSP also to manage keys of multiple customers securely in a multitenancy environment of cloud.

Integrity Issues

It is critical to ensure that data do not get corrupted or modified once they are stored in the cloud environment. Integrity violation can not only bring losses to valuable asset but also damage the market reputation of the enterprise and have legal implications. There are many ways of violation of integrity in a cloud environment. The reasons for violation of integrity could be unintentional or malicious. Unintentional issues are mostly technical in nature, such as media controller failures, metadata corruption, and tape failures. Data corruption can occur due to data duplication and data-transfer activities in the cloud environment as well. In a cloud data center, integrity of the data depends on the security management practices followed by the CSP. Further, in a multitenant environment, cloud storage may keep on changing and finding the physical location of an individual customer's data could be very difficult.

Issues of Availability

Cloud service provides various resources online, and the service delivery is highly dependent on the available network. There are various threats to the availability of data stored in the cloud environment because of multiple reasons.

An attack on the cloud network or public networks is one of the major threats. Large cloud data centers and their networks are attractive as a single point of attack because they host large amounts of data from multiple cloud users and enterprises. The occurrences of attacks in the cloud environment have also increased because of enhanced communication and network requirements due to the distributed nature of the cloud.

CSPs may lay claim for high availability and service up-time, but there were incidents [6] of outage with many major CSPs earlier. The severity of outage impact varies vis-à-vis criticality and outage situations. Multiple factors drive the availability of applications in the cloud including application architecture, cloud data center architecture, network systems, fault tolerance architecture, reliability of software and hardware used in cloud services, and exposure to human errors and malicious intent. Further, there is concern that cloud customers will be in serious trouble if the provider suddenly shuts down its business. Thus, the business availability of the CSPs is also a serious issue.

The cloud customer needs to be certain and aware about the level and type of service offered by the cloud provider. Though many providers offer backup service for cloud storage by default, these services may not be mandatory and may be offered for an additional price. All such details must be explicitly and clearly articulated in the SLAs. Though SLAs and contracts can attempt to bring some confidence in the cloud customer, a technical solution can provide enhanced trust to the customer about the availability of resources on the cloud.

Privacy Issues

Personal information should be managed securely throughout the life cycle of the data. Privacy requirements establish the organization's accountability with regard to collection, use, disclosure, storage, and destruction of personal data (or personally identified information, PII). Privacy is one of the key concerns for organizations that are willing to utilize cloud services for storing and processing their customer data as physical boundaries do not matter anymore.

Across the globe, there are many regulatory, compliance, and legal requirements for data privacy, depending on industries and jurisdictions. Meeting such requirements and establishing accountability in the cloud is a complex issue, as cloud data may be distributed across multiple jurisdictions.

Organizations also have concern about the ability to provide access to all personal information of an individual once data are stored in the cloud. To meet privacy requirements, a high level of transparency is desired from the CSP with respect to storage location, data transfers, data retention, and data destruction. It is difficult for the cloud customer to ensure that the provider has destroyed the PII data at the end of the retention period and had not retained any copies.

Compliance, Audit, and Legal Issues

Audit and compliance requirements become critical in any outsourcing relationship and the same is applicable in the case of the cloud as well. Numerous federal laws, state laws, and related regulations such as Health Insurance Portability and Accountability Act (HIPAA) and the Gramm–Leach–Bliley Act (GLBA) require companies to ensure protection of security and privacy measures with respect to the data of an individual. Adherence to such laws and regulations helps in building trust for end users of the services. There are standards such as Payment Card Industry Data Security Standards (PCI DSSs) and ISO 27001 imposes an obligation on the service provider to adopt a certain level of security control. Cloud customers in a multitenant cloud environment could have different types of compliance and legal requirements. It becomes challenging for CSPs to meet all requirements individually. There are also issues with providing details of security controls and audit-related information to individual customers and their third-party auditors (TPAs), because shared resources of the cloud contain information about other customers as well. Many laws restrict the transfer of data out of the country. Depending on the country and its applicable laws, the cloud provider needs to control data flow, storage, and backup activities.

Data Ownership

Enterprises and users may enjoy the benefit of placing large-size data on a cloud storage server that range from e-mails, pictures, corporate data, customer relationship management (CRM) data, documents, etc. However, a cloud customer may get into a challenging situation when data need to be exported from one cloud service to another due to unavailability of proper export mechanisms. This situation reduces the cloud customer's control over his own data and the data remain confined to the current CSP.

The remainder of this chapter discusses two major concerns relating to security and trust of the data stored in the cloud storage server. One is to establish an effective proof of data integrity to assure the owner of the data about the correctness and completeness of the data. The other is to highlight the necessity of effective data shredding after completion of the SLA.

Data Integrity Proofs

The increasing use of computers and smartphones has resulted in the spread of IT into the daily lives of the people. This ever-increasing use of IT has led to a boom in data generation and search for additional data storage devices. Storing this ever-increasing data and effectively managing them is a challenge to data owners. In such scenarios, data owners find it lucrative to outsource their data storage requirements to third parties. These third parties take the form of distributed systems, peer-to-peer systems, and more recently cloud storages. Outsourcing data storage enables data owners to off-load their data management and maintenance duties. This is helpful to large, medium, and individual users.

However, along with the advantages of data outsourcing, there are certain concerns that need to be carefully addressed to safeguard the interests of data owners. When the owner's data are outsourced to a cloud storage server, the owner of the data effectively loses control over his data. He is required to access his data through the interface provided by the cloud storage. Some concerns that arise in this changed scenario are the privacy, availability, and integrity of the data.

In this section, we deal specifically with the issue of data integrity when storing the owner's data in the cloud server. The owner of the data after storage may sign an agreement, called an SLA in business parlance. The SLA may stipulate that the cloud owner should not make any modifications to the data without the explicit consent of the data owner. However, in practical situations, there will be need for the data owner to have a mechanism through which he can be assured that the data present in the cloud have not been modified or deleted without his consent. We call such assurance mechanisms as data integrity proofs (DIPs).

Integrity Issues with Remote Data Storage

Enterprise and individuals generate large amounts of data in the due course of their business and find it difficult to manage their ever-increasing data storage needs. It is challenging for small- and medium-sized companies to keep updating their hardware in accordance with increasing data. Hence, they find it effective to store their data with third parties to reap operational and economic benefits. Cloud computing provides an excellent opportunity to such entities by offering flexibility to use storage services on demand according to the changing requirement of enterprises.

However, moving of data to third-party storage essentially means that the data owner needs to trust the third party for protection of data with respect to availability, confidentiality, and integrity. It is also assumed that the CSP will provide the data intact as and when requested by the data owner. While cost, maintenance, and flexibility are a few of the great advantages of cloud computing, security, privacy, and reliability are matters of concern for the cloud customer.

This section throws light on the problem of assuring the owner of the data that the data stored by him in the cloud server is not modified or deleted without his explicit permission. We call such assurances as DIPs. It discusses a few popular DIPs that exist today and presents a comparison of performance of various schemes.

Security and Performance Concerns

The primary purpose of a DIP is to provide the owner of the data with secure proof of the integrity of his data stored in the cloud storage. This proof should assure him that the data stored in the cloud server are necessarily the same, both in content and form. The proof should be able to provide quick and reliable information about any unauthorized additions, deletions, and modifications on his data. Any DIP that is developed will be judged primarily on the effectiveness with which it provides the owner of the data proof of the integrity of his data.

However, the developed DIP schemes should also take into consideration certain performance parameters regarding the resources of the owner, the computation of the cloud server, and the network bandwidth that will be consumed for generating the required DIPs. Some of the basic computational parameters include client storage, client computation, additional server storage, and network bandwidth. Figure 4.1 shows various performance parameters on which the DIPs will be compared.

One of the important objectives of a DIP is to give an integrity assurance to the owner of the data while keeping computation and storage overhead at a minimum on both sides—data owner and cloud server. A good DIP scheme should also have very less communication overhead between the data owner and the cloud server. This becomes very important while considering large amounts of data stored on the cloud.

Also, there is an increasing trend these days in the form of ubiquitous presence of mobile devices and the wide variety of functions for which they are

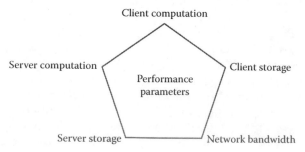

FIGURE 4.1
Performance parameters of a DIP.

used. Most of these functions are data generating (such as photography and video shooting). However, the amount of data generated by these devices is often greater than their storage capacity. Cloud storage offers an effective solution to the storage necessities of such clients in these cases.

It should be noted that these devices come with unique advantages as well as challenges. While these devices offer user mobility and access to information on the move, they are limited by their computation power and storage capacity. Hence, while designing effective and secure DIPs for such clients, their unique performance requirements should be taken care of.

Popular Data Integrity Proofs

The problem of securing a DIP has been well studied and understood by the research community. The problem has been referred to as proof of retrievability [3], provable data possession (PDP) [7], etc.

In the forthcoming paragraphs, we briefly introduce the reader to some well-known DIPs that have been proposed by various authors. All the below-discussed schemes follow the overall architecture as shown in Figure 4.2. The owner of the data repeatedly asks the cloud server for DIPs, and the cloud server replies with the required proof.

Keyed-Hash Approach

A simple DIP can be made using a keyed-hash approach. In this scheme, the owner of the data maintains a secret keyed-hash function $h_k(F)$. Using this function, the owner, before archiving his data file at the cloud, prepares a cryptographic hash of F using the secret function $h_k(F)$. The owner of the data stores this hash as well as the secret key k within itself and archives the file F with the cloud.

To check for the integrity of the data, the owner of the data releases the secret key k to the cloud server and asks it to compute a keyed-hash value using the secret key k. If the value returned by the server is different from the value already stored by the owner, the owner of the data can be sure that his data have been modified by the cloud server. Thus by storing multiple hash values for different keys, the archive can query the cloud server multiple number of times to check about the integrity of his data.

Though this scheme is very simple and easy to implement, it has some inherent drawbacks. The scheme consumes high resources for its

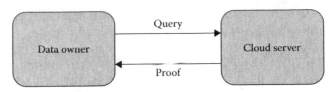

FIGURE 4.2
A typical architecture for a DIP.

implementation. The owner of the data is required to store multiple keys according to the number of times he wants to query the cloud server. The scheme also involves computing the hash value, which may be computationally cumbersome to small/bit/spare clients such as personal digital assistants (PDAs). Also, at the cloud server side, each stage of the query requires it to process the entire file F. This can be computationally troublesome for an archive even for a simple operation such as hashing. The cloud server is also made to read the entire file F, which will be a significant overhead to the cloud server whose intended purpose is only few occasional readings per file [3].

The Sentinel-Based Approach

Ari Juels and Burton S. Kaliski Jr. [3] have proposed a scheme called "proof of retrievability for large files." In this scheme, unlike in the keyed-hash approach scheme, only a single key can be used irrespective of the size of the file or the number of files whose integrity the data owner wants to verify. Also, unlike in the keyed-hash scheme that required the archive to process the entire file for verification every time, the archive in this scheme needs to access only a small portion of the file. This small portion is in fact independent of the length of the data file.

In this scheme, special blocks (called sentinels) are inserted randomly among the other data blocks of a file (Figure 4.3). The total data file with the inserted sentinels is further encrypted so as to make the sentinels indistinguishable from the original data blocks and is then archived on the cloud server. During verification of data integrity, the data owner asks the cloud server to return a few sentinels by mentioning their location.

On receiving the sentinels, the owner decrypts them and verifies whether they are the same as the original sentinel values inserted into the file. If the cloud server has modified the data file, then there can be a reasonable possibility that it might have tampered with some of the sentinels. Hence, there will be a good probability that this unauthorized modification will be detected. This scheme is best suited for storing encrypted files.

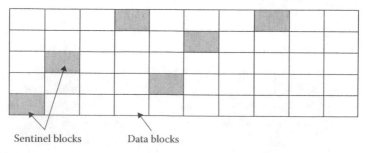

FIGURE 4.3
A data file with data blocks and sentinel blocks.

Provable Data Possession

PDP at untrustworthy stores is a scheme developed by Giuseppe Ateniese et al. This scheme is a public key-based technique that allows for public verifiability of data stored at the cloud server, wherein the challenges can be made by not just the data owner but any public user and the proof can be verified. This interaction for proof can be continued for an unlimited number of times and the verifier can get a fresh proof each time [7].

This scheme allows the verifier who has stored his data at an untrustworthy archive to check if the original data are intact with the archive and have not been modified or deleted without retrieving it. The model generates probabilistic proof of possession by sampling random sets of data blocks from the archive. This sampling of random blocks drastically reduces the I/O costs. The client (verifier) maintains a constant amount of metadata to verify the proof. The network communication is also minimized with a small, constant amount of data that minimize the network communication. This scheme can be used for obtaining proof of possession remotely by clients for large data stored in widely distributed systems.

The verifier stores a small $o(1)$ amount of metadata to verify the archive's proof. The storage overhead at the server is also minimal. It stores a small amount of metadata and the data file. This model allows the archive to generate a proof of data possession by accessing a small amount of the file rather than the complete file as is required with certain protocols. This scheme uses $o(1)$ bandwidth.

This scheme uses homomorphic verifiable tags and because of this property, tags computed for multiple file blocks can be combined into a single value. The client precomputes tags for each block of a file and then stores the file and its tags in a server. Figure 4.4 shows a schematic view of the PDP scheme.

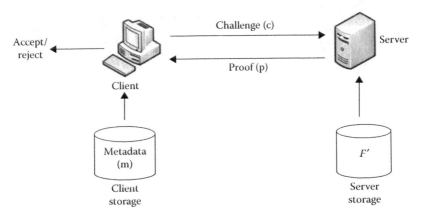

FIGURE 4.4
Schematic view of PDP.

At a later time, the client can verify that the server possesses the file by generating a random challenge against a randomly selected set of file blocks. Using the queried blocks and their corresponding tags, the server generates a proof of possession. The client is thus convinced of data possession, without actually having to retrieve file blocks.

This PDP scheme provides data format independence, which is a relevant feature in practical deployments of data as the data stored at the archive can be in many different formats, and the scheme employed should be able to generate a proof of possession for these data irrespective of the data format. In addition, there is no restriction on the number of times the archive can be queried for generating proof of possession. A variant of this PDP scheme offers public verifiability of the data that can be employed for online storage of data books, astronomical/medical/legal repositories, archives, etc.

The limitation of this PDP scheme is that it applies only to static cases, that is, the data do not change on the server once they are stored. Hence, it can be ideally suited for storing digitalized library books at an outsourced server, where a book once uploaded does not need any future modifications.

There are also schemes for dynamic PDP [8] and secure and efficient PDP [9], which contain provisions for obtaining proof of integrity of dynamically changing data. Table 4.1 shows a comparison of various schemes in terms of performance parameters.

Third-Party Data Auditing

All the above-mentioned scenarios are cases in which the data owner takes upon himself the task of preparing the initial metadata, storing it on the cloud server, query generation, and proof verification. There are also schemes in which the preparation, challenging, and verification of DIPs have been relegated to TPAs [10]. Figure 4.5 shows the typical architecture in such schemes.

TABLE 4.1

A Comparison of Performance Parameters of Various Schemes Presented

Scheme	Server Computation	Client Computation	Network Bandwidth	Client Storage	Comments
POR [3]	$o(1)$	$o(n)$	$o(1)$	$o(1)$	Limited queries
PDP [7]	$o(1)$	$o(1)$	$o(1)$	$o(1)$	Unlimited queries
DPDP [8]	$o(\log(n))$	$o(\log(n))$	$o(\log(n))$	$o(1)$	Unlimited updates
S&E PDP (STATIC) [9]	$o(1)$	$o(1)$	$o(1)$	$o(1)$	Limited updates
S&E (DYNAMIC) [9]	$o(1)$	$o(1)$	$o(1)$	$o(1)$	Limited updates

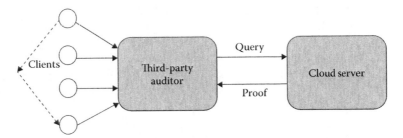

FIGURE 4.5
Third-party auditing for generating DIPs.

In such schemes, the data owners (clients) delegate their responsibility to an entity called TPA to challenge/verify the integrity of the data. The TPA on behalf of the data owners conducts all integrity proofs and sends reports to the data owners about the integrity of their data.

Such schemes offer many advantages compared to the schemes described earlier. They are helpful to small clients such as owners of mobile devices and PDAs, who have limited computational resources. The owner of the data might not be having knowledge of the technical know-how to use DIPs. Using Open-source software, better security audits can be performed by trusted third-party intermediaries.

Though such schemes work well for small clients who by themselves do not have the expertise, these have some limitations too. In this new kind of scenario, a data owner who was earlier required to pay only to the cloud server should now also pay to the TPA. Also, the earlier question of knowing the integrity of the cloud server with respect to its data storage capacity will now consist of knowing the integrity of the cloud server as well as the TPA [11].

Challenges and Future Developments

StaaS is becoming popular, and cloud customers see a clear advantage in outsourcing data to external agents. On the one hand, financial and operational benefits lure enterprise and individual users to use cloud storage; on the other hand, security concerns make them apprehensive about its usage. SLA provisions alone cannot help the customer in gaining confidence on protection of data, but customer-controlled mechanisms to prove the integrity and availability of data are the need of the hour. DIPs are one of the mechanisms for providing such customer-controlled solutions and proofs for assuring the data owner about the integrity of his data.

In the earlier sections, we have introduced the users to various performance and security parameters that need to be accommodated into any DIPs for effective assurance to the users about the correctness of their data.

We have also briefly explained to the reader about the various schemes that exist for providing DIP. It can be seen from the earlier discussion that all the schemes vary in their performance on various aspects. The owner of the data should therefore choose a scheme depending on the resources available to him and other performance parameters. Though there is considerable research on the ways to deal with the problem of DIPs, much needs to be done to improve their efficiency with minimum possible costs.

The following concerns should be taken into consideration for developing any effective proofs of data integrity:

- Client computation: The computation at the client end should be as minimal as possible, both for preparing the metadata and query formulation and evaluation subsequently.

- Server computation: The main work of the storage server will be to provide on-demand access of data to the users. Hence, any DIPs should not consume much of the server's resources.

- Client storage: The additional storage that accrues at the client's end should be kept as minimal as possible.

- Server storage: The additional data storage that accrues at the server owning to the employment of DIPs should be as minimal as possible.

- Network bandwidth: Any DIPs that will be developed should keep the communication overhead between the cloud storage server and the data owner to the minimum. This assumes greater significance when the cloud server charges the users of data on the basis of the data exchanged between the cloud server and the data owner.

Also, there is an increased use of mobile devices, smart phones, PDAs, etc., which have limited computational resources as compared to normal computers. Hence, there is a need to develop DIPs specifically meant for such small clients while taking into consideration their unique requirements.

Data Shredding

As already stated, clouds are a large pool of easily usable and accessible virtualized resources such as hardware, development platforms, and/or services. These resources can be dynamically reconfigured to adjust to a variable load (scale), allowing also for an optimum resource utilization. This pool of resources is typically exploited by a pay-per-use model in which guarantees are offered by the infrastructure provider by means of customized SLAs. StaaS is generally seen as a good alternative for a small- or medium-sized business that lacks the capital budget and/or technical personnel to implement and maintain its own storage infrastructure. However, it has virtually no control over the data once it is stored and the only safety is the SLA provided by the cloud storage facility provider, but what happens once the SLA is over?

Let us consider a scenario, where a cloud customer (end user/enterprise) wants to store sensitive data in the cloud with a cloud storage provider/ vendor for a particular period of time. Now at the end of particular time or a leased period, the cloud customer wants to withdraw the data and dissoci- ate from the service provider. The provider as per the SLA will return the data to the customer or delete it from storage as per the requirements. The question that arises is, if the provider says that he has deleted the data from his infrastructure, what proof does the cloud customer have that the data have completely left the provider's storage infrastructure? The cloud cus- tomer will have to ensure comprehensive destruction of that data from the storage, especially sensitive data that may be under regulatory compliance requirements. Even though a primary privacy protection is ensured by the encryption of the data before it is stored and even after assurances from the provider that the data have been completely erased/rendered useless from their disks, there is no methodology currently to verify it. It will create issues if the provider's data management infrastructure is not compatible with the customer's requirements for destruction (e.g., the provider is able to delete the latest version, but is unable to delete data from archived storage) and some form of leftover data succumbs at last to data leakage. One example of such attack can be the recovery of supposedly erased data from magnetic media or random-access memory.

Challenges in Data Shredding

Data shredding and its validation thereof are not new in the area of digital electronics. Also, various tools focused on data shredding and validation exist in the market. Even though a multitude of technologies exist on data shredding [12,13], among the available variety of software tools, solutions, and systems implementing data shredding techniques, most have some major drawbacks when it comes to a cloud/StaaS scenario. Cloud provides almost infinite virtual space. So, unlike a standalone system, where the lim- ited space needs to be reused, the possibility of such a scenario occurring in cloud is very less. Thus, conceiving a data shredding method for cloud that effectively is a collection of independent machines working in a distrib- uted manner might sound redundant at the outset. But the following are the reasons as to why such a data shredding mechanism needs to be conceived independently for an StaaS scenario.

Administrator Privilege for Shredding Activities

Typical data shredding algorithms have been designed keeping standalone systems in mind, where it is assumed that the person who shreds the data is the sole owner/administrator of the hard disk and thus has the rights to make bit-level manipulations and shredding. However, in the cloud environment, the owner of the data disk and the client who shreds data are different people. Therefore, a need arises where the owner of the data space has to provide explicit permissions for the client to access the bit-level storage mechanisms that bring with itself a huge set of security concerns. Thus, solutions have to be devised for tackling this problem where the provider will have permissions to preset the amount of data the client can shred, etc. Traditional data shredding approaches do not have scope for this.

High Capital Outlay

In addition, the existing products for data shredding are limited from the perspective of the client or vendor or both. Data shredding products require high initial capital outlay as development cost and thus present high risk. Special hardware is required to run these standalone applications, resulting in overhead cost that in turn beats the purpose of reducing cost by the pay-per-use model. It presents worries such as scheduling the shredding operation, disaster recovery and business continuity, or physical access to data shredding services by unauthorized persons.

Continued Maintenance

Regular updates or bug fixes become cumbersome in case of nonstandalone utility as they are tightly integrated with other processes. Also for vendors, a nonstandalone product presents piracy concerns, deployment concerns such as creation of installers, and integration issues as it needs to integrate with running processes. Versioning issues are also present in the case of nonstandalone data shredding product; a new version may not be compatible with the existing OS, hardware, etc. A new release also needs to be tested for integration with all the services it caters. Thus, the above factors demonstrate that data shredding in an StaaS setup must be looked upon from a different angle. The proposed mechanism provides network-based access and management of data shredding service managed from central locations rather than at each customer's site. This enables the customers to access the shredding service remotely via Web application delivery.

Opportunities for Effective Data Shredding—A Possible Methodology

Cloud customers require technical mechanisms to destroy the data, without the help of the provider or any third party. The customer should also be able to get confirmation about the data destruction using a verifiability module. Also, this can be provided as an independent service in the cloud and not as an addendum by the services provider, due to the level of inherent distrust and doubts that come along with a cloud storage setup.

In this section, we present the opportunity and strategy for data shredding in a distributed environment. We also explore the option if the storage infrastructure could be tweaked in such a manner that it necessitates the capability of making a hard erasure with proof, at the end of a storage requirement. Effective and efficient data shredding methodologies cover different phases throughout the data life cycle. Typical phases of efficient data shredding should start from establishing effective SLAs and end by verifying data destruction (after the SLA expires). In the coming subsections, we discuss one possible methodology, its various phases, and typical activities in each phase to achieve effective data shredding.

Establishing Effective SLA

The cloud customer or his company would sign an SLA with the provider, whereby the StaaS provider agrees to rent storage space on a cost-per-gigabyte-stored and cost-per-data-transfer basis and the company's data would be transferred at the specified time over the storage provider's proprietary wide area network (WAN) or the Internet [14]. Under the same preamble, the user may or may not be periodically accessing/updating the data through the multitude of application interfaces that he uses. For example, Web services. Though it is out of the scope of this discussion, it is understood that the SLA must be scrutinized for regulatory checks such as backup services, encryption, and compression algorithms involved for data protection during the storage time. The user must ensure that in the SLA the provider gives the details of the servers, starting from respective geographic locations delving into minute details such as the IP addresses, nodes, and memory blocks, on which their data will be simultaneously distributed. Generally, in a cloud certain number of copies of the same data are created, regularly updated, and stored at various geographical locations to ensure that a mishap in one of the geographical locations does not affect the continuous data access process that any of the client's applications might be doing. In addition, a check must be done on the exposure window time mentioned in the SLA. Others typically need to have clauses in the SLA that include details about archival frequency, archival locations, and the exposure window in which a change made to one node is updated in the rest of the nodes of the cluster, etc.

Exposure Window

An exposure window is the time that is needed to complete an update cycle across various copies of the data stored in various nodes in the clusters. For example, the user application that accesses the data through an interface, namely, a Web service, needs to update one row in one table in the database. So the application sends an update command, SQL or otherwise, to one of the databases to which it is connected. Suppose the cloud vendor is simultaneously storing the client's data in four different servers, namely, D1, D2, D3, and D4. Assume that at the give time, the data being shown to the user application are being retrieved from D1. Now the exposure window is the time, as per the SLA, that is required for these data to be updated in all the other three databases, namely, D2, D3, and D4 across the world, that is, the time during which the data are exposed/not updated. Similarly, post SLA, when the provider does a data erasure, an exposure window is the time required for the erasure done on one node to be reflected in all the respective copies.

End of SLA Period

Suppose that at the end of a particular time period, the user is done with the necessity of storing the data in the cloud. There is a possibility that the user has found a better vendor who is providing the storage facility at a much cheaper rate. At this stage, the user wants to pull out and ensure that the data are no longer with the vendor. Data remain with the existing cloud storage vendor, even if they are in the encrypted state; it is rendered useless so that no privacy leakage occurs even if the data go into the wrong hands.

Methodology for Data Destruction

In the normal scenario, data are deleted by using the operating system command. Further, to confirm it delete process may prompt for confirmation. Usually, normal commands of data deletion delete the link to the data but may not completely erase it along with their traces from the system. So a single delete operation may not destroy the data completely from the cloud systems. The mechanisms that can perform a round of operation at the bit level can help in destroying data more effectively.

In one such possible mechanism, application takes the most significant bit of all the bytes and replaces it with the opposite Boolean value, that is, a 0 will be replaced with 1 and 1 will be replaced with 0. This essentially will render the byte useless. During this destruction process, some specific details will be stored in a static storage media before every destruction operation. This will include a log of the current values of the bit being modified, the relative address of the bit, and the time during which the modification happens. This destruction application is invoked on the database on which the user application is connected, namely, D1, until each and every tuple in the data is systematically and comprehensively rendered useless. However, while doing this we assume that the data are not being stored contiguously, and hence there is a necessity for an algorithm which would respawn the pointer to the address where the next sequential data reside. Further, while modifying a bit value it is required that old and new bit values, current time, and relative address at which the change was made, are stored in a stable storage such as database, so that they can be used for data verification process.

Verifier

Once the destruction cycle is complete, it is necessary that the provider does the verification, preferably in the presence of the user to ensure that the bits have been modified or the data have been destructed in the other copies of the database, namely, the database servers provided by the cloud data storage provider, where all the data are copied, on which any data update made on the first database, D1, will be updated after the exposure time mentioned in the database. For this purpose, the verifier application will next connect to the other databases, D2, D3, etc. Here, it will pick the bits of data from the same address location as was modified in D1. A comparison of this bit value in the secondary databases with the equivalent bit value in D1 on which the destructor application had worked is done next. In conducting the verification, we assume that all the modified bits and its corresponding address values were stored as key–value pairs during the destructor program run and the exposure time mentioned in the SLA, which is the time taken by the provider to update other copies of the update that is made in one database. For getting the right results, the verification/comparison is done only after the exposure time has elapsed.

Conclusion

In the overall context of data storage in the cloud, the question of how to ensure the integrity of retention and dispersal mechanisms of data gains prominence. As data archiving and storing on the cloud gains more prominence, there will be an increased need for efficient and secure DIPs that can handle the varied needs of users. It should also be noted here that cloud customers store data for purposes ranging from simple data backups to online databases. In such a scenario, the DIPs that are being developed should be sufficiently robust to assure proof of integrity to all these users.

Ensuring the complete destruction of data, postusage period attains crucial importance for users storing sensitive or confidential information on the cloud server. Apart from the server stating that it has destructed the data completely, the users might require a proof of data destruction. Such a proof needs to ensure the owner of the data that any redundant copies of his data stored in multiple storage servers is completely destroyed before the end of the SLA period. In developing such data destruction solutions, one needs to carefully consider performance and efficiency parameters. Such solutions need to be developed which consume minimum computational resources of the data owner as well as the cloud server.

We conclude this chapter by bringing to the notice of the reader the importance of effective, efficient, and secure data integrity as well as data shredding proofs for users of cloud storage.

In short, this chapter conveys the various facets of DIPs and data shredding mechanisms along with the challenges of the existing mechanisms and scope for future developments.

Exercise

1. Take a 4 MB data file. Compute its hash value using a few random keys. Store the file in a remote computer. Ask the computer operator to modify the file. Compute the hash value of the new file. Check if both of them match.

2. Take a 100 MB data file. Choose the block size as 512 KB. Identify a few blocks among them as sentinels. Store these sentinels in your local machine. Deploy the original file in a hadoop file system. Ask the administrator of hadoop system to randomly modify a portion of the file. Now query the hadoop system for the sentinel blocks and check whether the modification is being detected or not.

Disclaimer

The views expressed here are the authors' personal views and may not represent those of Infosys. All names of the Web sites, vendors, and applications mentioned in this chapter are intended for informative purposes only with no malicious intent. Readers may check directly for the latest updates.

References

1. Cachin, C., Keidar, I., and Shraer, A., "Trusting the cloud," *SIGACT News*, 40(2): 81–86, 2009.
2. Bisong, A. and Rahman, S. M., "An overview of the security concerns in enterprise cloud computing," *CoRR*, vol. abs/1101.5613, 2011.
3. Juels, A. and Kaliski, B. S., Jr., "Pors: Proofs of retrievability for large files," *CCS'07: Proceedings of the 14th ACM Conference on Computer and Communications Security*. New York: ACM, October 29–November 2, pp. 584–597, 2007.

4. Amazon S3, http://aws.amazon.com/s3/.
5. Nirvanix cloud complete portfolio: Public cloud storage, http://www.nirvanix .com/products-services/cloudcomplete-public-cloud-storage/index.aspx
6. TechTarget Inc., http://searchcloudcomputing.techtarget.com/feature/ Cloud-computing-outages-What-can-we-learn.
7. Ateniese, G., Burns, R., Curtmola, R., Herring, J., Kissner, L., Peterson, Z., and Song, D., "Provable data possession at untrusted stores," *CCS'07: Proceedings of the 14th ACM Conference on Computer and Communications Security*, New York: ACM, October 29–November 2, pp. 598–609, 2007.
8. Erway, C., Küpçü, A., Papamanthou, C., and Tamassia, R., "Dynamic provable data possession," *CCS'09: Proceedings of the 16th ACM Conference on Computer and Communications Security*, New York: ACM, November 2–13, pp. 213–222, 2009.
9. Ateniese, G., Di Pietro, R., Mancini, L. V., and Tsudik, G., "Scalable and efficient provable data possession," *SecureComm'08: Proceedings of the 4th International Conference on Security and Privacy in Communication Networks*, New York, September 22–25, pp. 1–10, 2008.
10. Wang, C., Wang, Q., Ren, K., and Lou, W. "Privacy-preserving public auditing for data storage security in cloud computing," *INFOCOM, Proceedings IEEE*, IEEE, USA March 14–19, pp. 1–9, 2010.
11. Xu, J., "Auditing the auditor: Secure delegation of auditing operation over cloud storage," *IACR Cryptology ePrint Archive*, 304, 2011.
12. Gregory, K., "Method and system for shredding data within a data storage subsystem," Pat No: US 7308543B? (March 2005).
13. Peter, G., "Secure deletion of data from magnetic and solid-state memory," *Proceedings of the 6th USENIX Security Symposium*, July 22–25, USENIX Association, Berkeley, CA, pp. 77–89, 1996.
14. "What is storage as a service (SaaS)?—Definition from Whatis.com," February, 2010. http://searchstorage.techtarget.com/sDefinition/0,,sid5_gci1264119,00 .html (Accessed: June, 2010).

5

Key Management Solutions for Database as a Service: A Selective Survey

Sumitra Binu and Pethuru Raj Chelliah

CONTENTS

Introduction

In today's scenario, efficient data processing is a fundamental and vital issue for almost every scientific, academic, or business organization. To tackle this issue, organizations end up installing and managing database systems to satisfy different processing needs. In case of adopting a traditional solution, the organization needs to purchase the necessary hardware, deploy database products, establish network connectivity, and hire professional people who run the system. But this solution is getting impractical and expensive as the database systems and problems become larger and complicated (El-Khoury et al., 2009). Again, traditional solution entails different costs from the perspective of the investments involved. These concerns are handled efficiently to a great extent by the fast developing technology that goes by the name, "cloud computing."

Due to the surging popularity of the new delivery model, there are conscious efforts by IT experts and architects for pragmatic strategy, plan, and roadmap formulation for systematically modernizing and moving all kinds of IT resources including hardware and software infrastructure, software platforms, middleware and databases, and applications to cloud environments for exploring newer avenues of fresh revenues. The trend is absolutely clear here. That is, everything is being prepared and presented as a service for the human race so as to be publicly discovered, accessed, and used either for free or for a small fee. This service model rings in a litany of transformations on how businesses approach, execute, and meet varying customer expectations in the days ahead. In that direction, databases and database management software solutions are being readied for the cloud space. All kinds of customer data, confidential data, and corporate data are being stored in cloud-based databases in order to be provided as data services in an on-demand basis. Database as a service (DaaS) that is an example for software as a service (SaaS) is one of the service delivery models adopted by cloud service providers.

The decision to outsource databases is strategic in many organizations due to dramatic increase in the size of the databases and increasing costs of internally managing large volumes of information. These databases are hosted by a third party (Hacigumus et al., 2002b), which then provides a "service" to clients who can seamlessly access the data. In addition, organizations that process large volumes of data face the need for database backup, database restoration, and database reorganization to reclaim space or to restore preferable arrangement of data. Migration from one database version to another, without impacting solution availability, is an art still in its infancy (Gupta et al., 1996). Parts of a database solution, if not the entire solution, usually become unavailable during version change. An organization that provides database service takes up the responsibility of database backup, administration, migration from one database version to the another without impacting on availability, etc. Users who wish to access data will now be able to do so with the hardware and software of the cloud service provider instead of their own organization's computing infrastructure. The organization would not be impacted by outages due to hardware, software, and networking changes or failures at the site of the database service provider. The organization would purchase data management as a service by using the ready system maintained by the service provider for its database needs. Data owners can now concentrate on their core competencies while expecting outsourced databases to be managed by experts using the latest solutions at innovative costs. This approach leads to an increase in productivity as well as cost saving.

Challenges in Outsourcing Databases to the Cloud

Along with its numerous advantages, outsourcing databases poses several research questions related to security such as data confidentiality, privacy, authentication, data integrity, and availability. When the data get moved to third-party cloud infrastructures (public cloud), there are some serious issues such as lack of controllability, visibility, security, and auditability. The cloud service provider needs to enforce security at the physical, network, data, and application levels.

The most important security issue from the perspective of outsourced databases is data privacy especially in the presence of sensitive information. Data owners are physically releasing their sensitive data to external servers that are not under their control, and hence confidentiality and integrity may be at risk. Besides protecting such data from attackers and unauthorized users, privacy of data from the external servers also needs to be ensured.

Taking into consideration the security risks involved in availing DaaS, many solutions ensuring privacy of data are proposed based on the category of issues. Issues related to the privacy of data can be categorized into two: (1) privacy of data during transmission and (2) privacy of data at rest or stored data. The first issue, privacy during network transmission, has been studied widely on the Internet and is ensured by two protocols: secure sockets layer (SSL) and transport layer security (TLS) (Karlton et al., 1997). The second issue, privacy of stored data in relational databases, is less studied and is of greater relevance to DaaS model. If DaaS is to be successful, and customer data are to remain on the site of the database service provider, then the service provider needs to find a way to preserve the privacy of user data. A security mechanism needs to be in place that prevents unauthorized users from understanding the sensitive information stored in the database. The literature reveals that encryption is the best technique to tackle this issue. Along with encrypting data using a key that ensures its confidentiality, there should also be a fool-proof mechanism to handle the issues related to key management. For data to be secured in their storage, encryption and decryption keys can be leveraged, but the question is how to keep those cryptographic keys safely and securely? Keys cannot be with the cloud provider as the data are in danger from cloud operators and administrators. The owner can keep the data and if there is any loss of keys, then bringing back the encrypted data to readable text becomes an issue. Thus, key management in the context of powerful emergence, massive adoption, and adaption of the cloud paradigm attains special significance. Students, scholars, and scientists are striving hard to unearth an impenetrable and unbreakable security mechanism for the key management problem. The following paragraphs examine the relevance of key management and the available key management solutions while providing DaaS.

Key Management Role

Database encryption is seen as a solution to prevent exposure of sensitive information even in situations where the database server is compromised. Without encryption, data are not secured and encryption calls for the use of cryptographic keys. An encryption algorithm takes input data (e.g., a field, a row, or a page in the database) and performs some transformation on it using a cryptographic key to produce ciphered text (Sandro and David, 2003). There is no easy way to recover the original message from the ciphered text other than by knowing the right key (Schneier, 1996). However, distributing the keys to valid members is a complex problem. Thus, the security of data also requires proper processing of keys without which data would eventually be compromised and without proper key management data could be lost completely. Cryptographic key management (CKM) plays an important role in storage security and information processing applications (Kyawt and Khin, 2010). The proper management of cryptographic keys is essential for the effective use of cryptography for security. All the proposed solutions to database encryption and key management use the key derivation principle (Atallah et al., 2005) to reduce the number of delivered keys. Another issue sought to be addressed by multiple solutions is the problem of key revocation. Distribution of database encryption keys creates a problem when user's rights are revoked as key revocation requires full database reencryption.

The literature presents us with several different approaches to key management. Key management solutions for outsourced databases can be classified into three categories.

Owner-Side Policy Enforcement Solutions

The user sends the query to the owner, who transforms the query to the appropriate representation on the server. The transformed query is executed on the server side and the results are sent encrypted to the owner. The owner filters out those tuples not satisfying the user's assigned rights, decrypts the results, and sends them to the user in the form of plaintext. Here, the access control policy is enforced by the owner itself.

User-Side Policy Enforcement Solutions

In these solutions, the database is encrypted with symmetric keys that are delivered to the user who can manage the whole querying process, encrypting the query so that it can be evaluated over the encrypted database; then the result is retrieved by decrypting the response. These solutions relocate the access control to the user side, alleviating the owner's charge in the querying process.

Shared Policy Enforcement Solutions

In this category of solutions, access policy enforcement is shared among the data owner, the user, the service provider, and any other participant. Majority of these solutions follow a two-level encryption approach: (1) before outsourcing the data, done by the owner and (2) done by the service provider.

Contribution of This Work

This chapter presents a survey of key management solutions for outsourced databases, identified from surveying the literature pertaining to this area. The key management solutions are categorized into three classes: (1) owner-side policy enforcement solutions, (2) user-side policy enforcement solutions, and (3) shared policy enforcement solutions. We analyze the advantages and the limitations of the various solutions.

Owner-Side Policy Enforcement Solutions

The data will be encrypted by the data owner, who also handles out decryption keys, allowing only authorized persons to access the plaintext. This solution does not permit access to the database through *ad hoc* queries. More flexible techniques based on storing additional indexing information along with encrypted database have also been proposed in the literature (Agrawal et al., 2004; Hacigumus et al., 2002). These indexes are employed by the database management systems (DBMSs) to enable posing queries over the encrypted data without revealing either the query parameters or the data results. The mechanism for querying an encrypted database is as follows: (1) the user sends the query to the owner who maintains the metadata needed to translate it to the appropriate representative on the server; (2) the transformed query is executed on the encrypted database at the server side; (3) once executed, the results are sent encrypted to the owner who decrypts them and filters out those tuples not satisfying the user's assigned rights; and (4) the results are sent to the user in the form of plaintext.

The following two approaches implement owner-side policy enforcement solutions for providing DaaS.

NetDB2

NetDB2 (Hacigumus et al., 2002a) provides database services including tools for application development, creating and loading tables, and performing queries and transactions to users over the Internet. In the system, data and all the products are located on the server side. A user connects to the system through the Internet and performs the database queries and other relevant tasks over the data through a Web browser or an application programming interface such as java database connectivity (JDBC) (Hamilton and Cattell). The goal is to absorb complexity and workload on the server side and keep the client side light weight. The basic NetDB2 system is implemented as three-tier architecture, namely, the presentation layer, application layer, and data management layer.

Encryption Approaches

There are two dimensions to encryption support in databases. One is the granularity of data to be encrypted or decrypted. The field, the row, and the page are the alternatives. The second dimension is software- versus hardware-level implementation of encryption algorithms. The authors in their work have done software-level encryption, hardware-level encryption, and combinations of different encryption approaches.

Software-Level Encryption

For NetDB2 system, encryption done using Blowfish algorithm gives better performance than RSA algorithm (Rivest et al., 1978). Blowfish is a 64-bit block cipher, which means that data are encrypted and decrypted in 64-bit chunks (Schneier, 1996). In this approach, the creator of the encrypted data supplies the key, and the database provides the encryption function.

Hardware-Level Encryption

The NetDB2 system uses an encryption/decryption edit routine called "editproc" for the tables. An edit routine is invoked for a whole row of the database table, whenever the row is accessed by the DBMS. When a read/ write request arrives for a row in one of the tables, the edit routine invokes DES (1977) encryption/decryption algorithm, which is implemented in the hardware, for the whole row.

Advantages. The authors in their work have addressed and evaluated the most critical issue for the successful encryption in databases, namely, performance. To achieve performance, they have analyzed different solution alternatives.

Limitations. The result of a query will contain false hits that must be removed in a postprocessing step after decrypting the tuples returned by the query. This filtering can be quite complex, especially for complex queries involving joins, subqueries, aggregations, etc. Clearly, the techniques based on this approach are impractical as they require continuous online presence of the data owner to enforce access control.

Order-Preserving Encryption Scheme

Encryption is a widely used technique to ensure privacy of outsourced data. However, once encrypted, data can no longer be easily queried from exact matches. Order-preserving encryption scheme (OPES, Agrawal et al., 2004) is a scheme for numeric data which allows any comparison operation to be directly applied to the encrypted data. Also, the integration of existing encryption techniques with database systems causes undesirable performance degradation because current encryption techniques do not preserve order, and therefore database indices such as B-tree can no longer be used. This makes query execution over the encrypted databases unacceptably slow. OPES allows comparison operations to be directly applied on encrypted data, without decrypting the operands. The following sections illustrate the approach followed by OPES technique to arrive at the solution.

Assumptions

(1) The storage system used by the database software is vulnerable to compromise; (2) the database software is trusted; (3) all disk-resident data are encrypted; and (4) the database consists of a single table containing a single column.

Proposed OPES

The basic idea of OPES is to take a user-provided target distribution as input and transform the plaintext values in such a way that the transformation preserves the order while the transformed values follow the target distribution. Experimental results show that running OPES in different input distributions produces the same target distribution.

Overview of OPES

When encrypting a given database P, OPES makes use of all the plaintext values currently present in P, and also uses a database of sampled values from the target distribution. Only the encrypted database C is stored on the disk. Simultaneously, OPES also creates some auxiliary information K, which the database system uses to decrypt encoded values or encrypt new values. Thus, K encrypted using conventional encryption techniques serves the function of the encryption key.

Three Stages of OPES

The stages of OPES include (1) modeling the input and target distributions, (2) flattening the plaintext database into a flat database, and (3) transforming the flat database into the cipher database. These stages are explained in the following paragraphs. The input and target distributions are modeled as piecewise linear splines. In the flattening stage, the plaintext database P is transformed into a "flat" database F such that the values in F are uniformly distributed. In the transform stage, the flat database F is transformed into the cipher database C such that the values in C are distributed according to the target distribution.

Advantages. Query results produced are sound (no false hits) and complete (no false drops). This feature of OPES differentiates it from the previously discussed scheme that produces a superset of answers, which requires filtering of extraneous tuples. The scheme handles updates efficiently, and new values can be added without requiring changes in the encryption of other values. Standard database indexes can be built over encrypted tables and can easily be integrated with existing database systems. The encryption is also robust against the estimation of true values of the encrypted data, in environments in which the intruder can get access to the encrypted database, but does not have any idea about the domain such as distribution of values and thus, cannot encrypt or decrypt values of his choice.

Limitations. The proposed scheme is focussed only on developing order-preserving encryption techniques for numeric values.

User-Side Policy Enforcement Solutions

Approaches based on shifting access policy enforcement to the client side have been proposed to overcome the limitations of owner-side policy enforcement. Query-processing approaches focusing on access-policy enforcement to the client side require an offline processing stage where data sets accessed by each user, or more accurately, a user profile are identified. An appropriate key is then used to encrypt each data set before its outsourcing. Finally, a set of keys are provided to each user for accessing the data sets according to his/her privileges. In this case, the data owner's task is to identify the set of keys necessary for each user or group to access the data they are authorized to view. The main advantage of this idea is that the data owner is not directly involved when the user queries the encrypted database.

Enforcing access control at the user level faces two main problems: (1) the determination of the key set and (2) their efficient generation. Each user may be accessing several data sets, which themselves may be intersecting with those of other users who require the data sets to be subdivided into nonintersecting data sets, thereby increasing complexity. Also, the key generation algorithm must be as simple as possible to be scalable in practice, and flexible to adapt to frequent changes in a user's access rights. The following solutions proposed for user-side policy enforcement attempt to address these issues.

TUH (Tree Based on User Hierarchy)-Based Solution

This method uses a greedy algorithm that tries to solve the nondeterministic polynomial-time (NP)-hard problem of minimizing the number of keys directly communicated to users. The authors consider a matrix representation of user's rights on sets of data, representing the users as rows and the access configuration (AC) associated with data tuples as columns (Damiani et al., 2006). A sample access matrix is represented by Figure 5.1. A key derivation tree (TUH), as shown in Figure 5.2, is built from the access matrix A enforcing the access control policy. With an empty vertex as root, the proposed algorithm starts including in the TUH all the vertices representing AC by scanning the matrix A. Again, the vertices should satisfy a partial order where each vertex is pointed by an edge from a parent vertex p whose access configuration AC_p is included in AC_v. That is why each vertex will be

	t_1	t_2	t_3	t_4	t_5	t_6
A	0	1	1	0	1	1
B	1	0	1	1	1	1
C	0	0	1	1	0	1
D	0	1	0	1	1	1

FIGURE 5.1
A sample access matrix.

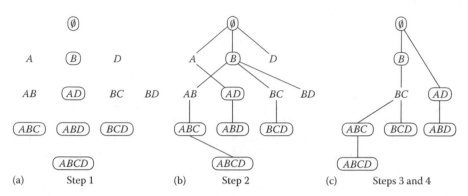

FIGURE 5.2
An example of building a TUH. (a) The set of vertices wherein material vertices are circled and nonmaterial vertices are not circled. (b) Vertices are chosen and connected for tree user hierarchy (TUH) construction. (c) TUH tree is pruned and key is assigned.

associated with a secret key k such that a key for vertex v will be given to a user u if and only if u belongs to a v_s AC and u does not belong to parent (v) AC. All the other keys for which user u has rights are obtained by using a derivation mechanism using one-way hash functions from the assigned keys. The building of the key derivation tree is done using a transformation algorithm that includes steps such as (1) select vertices, (2) TUH construction, (3) prune tree, and (4) key assignment.

Advantages. The solution has the advantage of outsourcing access control enforcement and it does not require the constant online presence of the data owner for encrypting the queries and decrypting the results. In addition to that, this mechanism reduces the number of private keys that each client has to keep with him/her.

Limitations. The flexibility of the structure to adapt to frequent modifications in access control policies, especially for the insertion and deletion of users, is extremely limited. After several updates, the edges between vertices are modified such that TUH is unable to maintain key derivation efficiency. Consequently, TUH must be completely rebuilt, AC rekeyed, and data deciphered and enciphered.

Diffie–Hellman-Based Key Generation Scheme

This work represents an alternate key management scheme based on a partial access order configuration-directed hierarchy (Zych and Petkovic, 2007). The levels of such a structure are the ACs for single users and the edges are constructed bottom-up such that each intermediate vertex is pointed by exactly two edges, and its AC is the union of the fits of two children. The resulting hierarchy is then run by the Diffie–Hellman (DH) key generation algorithm, which assigns secret keys to vertices in a bottom-up fashion such that a user with a single key could derive all the secret keys assigned to its data sets. The authors arrive at the key management solution by the following two steps: (1) creating a V-graph whose elements are ACs and (2) generating the keys using DH key generation scheme for the V-graph. These steps are explained in the following paragraph.

Key Management Solution

The authors first propose an algorithm for constructing the partial order on ACs. The partial order is represented by a reduced directed graph, where nodes are ACs and arcs (directed edges) connect comparable elements: from greater to smaller. The constructed graph, called V-graph, has two additional properties called V-conditions: (1) the number of arcs coming into a node is either 2 or 0 and (2) for any two nodes, there is at most one node, from which arcs go to both of them. The design of the proposed key management system is presented in Figures 5.3 and 5.4.

Figure 5.3 shows the protection phase, performed by the Central Authority (CA) at the data server side. First, an access control table containing access rights of the users (1) is processed by an ordering module, which constructs the order relation on ACs (2), represented by a directed reduced V-graph. The key assignment module, which takes as an input the access control V-graph

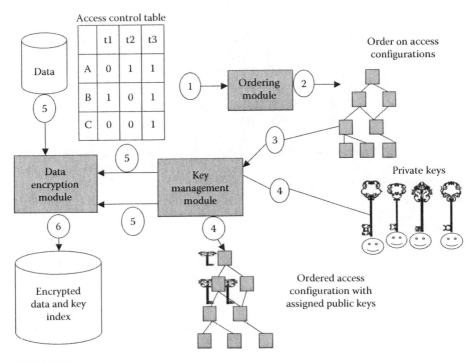

FIGURE 5.3
Data protection phase.

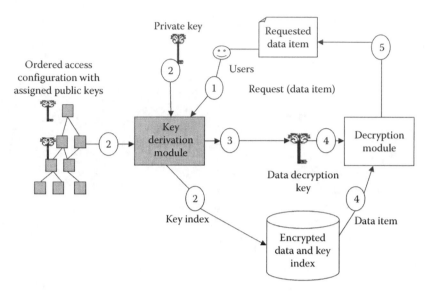

FIGURE 5.4
Query phase.

(3) and returns the V-graph of public keys and secret keys distributed to the users of the system (4) the data encryption module, encrypts the data, using the access control table and the constructed graph of keys (5). The final result of this phase (6) is the encrypted data provided with the indexes of the corresponding decryption keys.

Figure 5.4 shows the query phase, performed by the users at the client site. First the request (1) is processed by the key derivation module. In order to derive a decryption key for the requested data, the index of the desired key (2) is retrieved from the database, and then the key with this index is generated (3) based on the DH scheme, using the private key of the user and the hierarchy of public keys (2). The obtained decryption key (4) is used by the decryption module to decrypt the requested data (4) received from the database.

Advantages. The work is first of its kind to propose a key generation solution designed especially for access control. The proposed solution eliminates the need for multiple copies of data keys and reduces the storage required per user to a single key. The improvement is achieved by applying the proposed DH key generation scheme. The solution also reduces the required size of public information.

Limitations. Construction of algorithm and generation schemes is very complex and costly. Again, access rights updates cause a rebuilding of the hierarchy and mandatory key generation leading, in most cases, to data rekeying.

Trie-Based Approach

In this binary trie-based key generation and management approach (El-Khoury et al., 2009), each category is considered as a binary string representing the path from the root of the trie to a leaf. The management of structure does not suggest a partial order of vertices, even though the idea of key derivation to minimize the number of distributed keys is preserved. This approach reduces the key management complexity using the intrinsic properties of the binary trie. This structure has many properties that make it suitable for access control. As the keys are not tightly coupled to ACs, most of the data access policy updates do not require significant changes to the structure which reduces key generation, and thus data rekeying.

The Binary Trie-Building Algorithm

In real databases, the data are represented by a set of views V that overlap. In the binary trie-based approach, the authors assume that there is a process used to partition these views into separate categories, so that $C_i \cap C_j = \emptyset$ if $i \neq j$. Further, the users having the same privileges on categories are organized in a single group. Like all other works related to access control in outsourced databases, given a system with a set G of groups and a set C of categories, this approach also assumes that access control policies are represented in a matrix A having $|G|$ rows and $|C|$ columns.

A trie (Fredkin, 1960) is an ordered tree data structure that has been used for various applications such as the construction of database systems, natural language dictionaries, networking (Medhi and Ramasamy, 2007), etc. The following is the illustration of the trie construction algorithm that takes the access matrix A as an input and returns the binary trie as output.

Given the access matrix A, $\forall$ I, $j \in N$ for $j: = 1: |c|$ the algorithm scans $a[g, j]$, and for $i|g|$ each category is scanned bit per bit. Then for each entry $A[i, j]$, the function Insertvalue (value, position) to insert the bit value at the current position in the binary trie is applied. Here, two cases exist: (1) for the bit value 1, if the right child node exists, the filed AccessNb will be incremented by 1. Otherwise, a new right node is created having AccessNb value equal to 1, and its content will be the concatenation of the content value of its parent and bit 1; and (2) similarly, for the bit value 0, if the left child node exists, the field AccessNb will be incremented by 1. Otherwise, a new left node is created having AccessNb equal to 1, and its content will be the concatenation of the content value of its parent and bit 0 (Figure 5.5).

Advantages. This structure is flexible toward access rights granted dynamically to the users and the growth of the accessed data set. It efficiently handles all these changes without need for a symmetric reconstruction. The approach follows a key-derivation algorithm, so that the

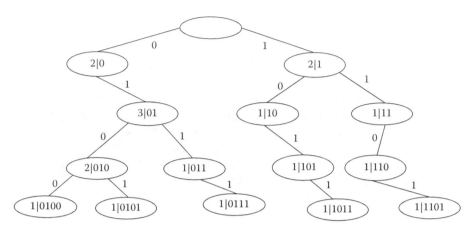

FIGURE 5.5
Example of a binary trie.

binary trie is scanned from left to right in-depth—first manner. The approach also offers a good flexibility in the case of access rights update. The only case where categories should be rekeyed is in the case of user key revocation.

The mechanism allows an offline key generation and exploits hierarchical key derivation methods using the one-way hash function. The approach has the advantage of efficiently generating the group keys ring from the binary trie structure according to their privileges and of reducing the cost of the key management in the case of access control policies modifications.

Limitations. In this approach, the increase in the depth of the binary trie with increase of group numbers will have an impact on the key generation time. But the approach does not reduce the number of keys held by a group, and it is observed that the groups at the low level of the structure hold a large number of keys. Again, the binary trie-based approach does not completely resolve the well-known data rekeying problem due to user rights revocation.

Client- and Server-Side Policy Enforcement Solutions

In the case of user-side policy enforcement solutions, key revocation poses a problem when the user's rights are revoked. To overcome the limitations user-side policy enforcement solutions, shared key management solutions were proposed in which the access policy is shared among the actors. The basic idea is that the user should not possess the entire key which might still help access data even after access rights are denied. The following approaches attempt to reduce the complexity of key delivery.

Overencryption

The problem of outsourcing resource management to service providers who are honest in providing their services but may be curious in knowing the content has received much attention in the recent past and many solutions have been proposed (Di Vimercati et al., 2007). Most proposals assume that data are encrypted with a single key only (Ceselli et al., 2005; Hacigumus et al., 2002). In such a situation, either the authorized users are assumed to have the complete view of the data; or if different views need to be provided to different users, the data owner needs to participate in the query execution to possibly filter the result of the service provider. The owner's participation can be minimized by selective encryption where different keys are used to encrypt different portions of the data. In outsourcing scenario, all these proposals in the case of updates of the authorization policy would require the reencryption of resources.

The work proposes a solution that addresses the issues in outsourcing, thereby making data outsourcing a successful paradigm. The proposed solution focuses on the problem of defining and assigning keys to users by exploiting hierarchical key assignment schemes (Akl and Taylor, 1983) and of efficiently supporting policy changes. The authors propose the solution by following three steps: (1) a formal base model for the correct application of selective encryption is proposed. The model allows the definition of an encryption policy equivalent to the authorization policy to be enforced on resources; (2) building on the base model, a two-layer approach to enforce selective encryption without requesting the owner to reencrypt the resources every time there is a change in the authorization policy. The first layer of encryption is done by the data owner at initialization time, while releasing the resource for outsourcing and the second layer of encryption is done by the service provider itself to take care of dynamic policy changes. The two-layer encryption is assumed to allow the owner to outsource, besides the resource storage and dissemination, the authorization policy management, while not releasing data to service provider; and (3) the authors provide a characterization of the different views of the resources by different users and characterize potential risks of information exposure due to dynamic policy changes. The proposed solution uses the key derivation method proposed by Atallah et al. (2005) based on the definition and computation of public tokens that allow the derivation of keys from other keys.

Advantages. The proposed solution does not substitute the current proposals, rather it complements them, enabling them to support encryption in a selective form and easily enforce dynamic policy changes.

Limitations. The proposed method does not evaluate whether the approach is vulnerable to attacks from users who access and store all information offered by the server, or from collusion attacks, where different users combine their knowledge to access resources that they would not otherwise be able to access.

Cloud-Based Key Management

The work proposes the usage of cloud for decoupling the management of local user-specific encryption keys from one of the role-specific protection keys (Nadia et al., 2010). This approach enables simple key management and revocation schemes. In the cloud-based model of DaaS, a data owner calls in a cloud service provider to host his or her database and provide clients with seamless access to data. But outsourcing databases to an honest but curious third party can lead to several issues related to confidentiality, privacy, authentication, and data integrity. Encrypting the database has been proposed as a solution even when the service provider cannot be trusted. The encryption-related solutions lead to key management concerns. This work suggests a solution to reduce the complexity of key management wherein the source database is replicated n times, where n is the number of different roles having access to the database. Each database replica is a view entirely encrypted using the key k created for the corresponding role. Each time a new role is created, the corresponding view is created with a new key expressly generated for the new role. This ensures that no changes are required on the database in case of dynamic role update. Role revocation leads only to the creation of a new materialized view which will involve the rekeying of corresponding database. Data reencryption in the much more frequent case of user revocation is avoided by not delivering the key used to encrypt the role-associated database to the users associated with that role. Rather, each user receives a token that allows her to address a cipher demand to a set of key servers Ks on the cloud. The proposed approach for key management is explained in the following paragraphs.

Joint Encryption Scheme

In this scheme, a subset of parties is enabled to encrypt a given message using their own individual keys, and the cooperation of all the current parties is requested to decrypt the same message. A joint encryption scheme for a group of n parties is a collection of encryption functions $\{E_s: s \subseteq [1 \ldots n]\}$ and a collection of decryption functions $\{D_i: i \in [1 \ldots n]\}$ such that a given message M, $\forall_i: D_i \, (E_s(M)) = E_{s-\{i\}} \, (M)$ and M can be directly computed from $E_\emptyset(M)$. In this approach, the encryption algorithm needs to be homomorphic.

Cloud-Based Key Management Protocol

In the proposed protocol, the data owner creates a Vr, the materialized view assigned to the role r. Then she generates a virtual key α (a set of shadows s_1, $s_2, \ldots s_n$) and uses α to encrypt the view Vr before outsourcing the individual shadows s_i to a server ks_i belonging to the set ks on the cloud. Also the data owner shares with the servers in ks its policy, by sending to each server ks_i the association between α, represented by the shadow s_i each server holds and a set of users, for example, in the form of identification tokens. Each user u_i assigned to a role r receives from the data owner a different key βu_i, while the shadows $(s_1, s_2, \ldots s_n)$ of the role key α are placed in the key servers chosen on the cloud. A user u_i having role r_i to get the result of a query q, first encrypts q with her local key βu_i and sends it to the first server of ks, the one detaining the s_1 shadow. The query is then superencrypted in a collaborative manner by the servers holding the set of shadows $s_1, s_2, \ldots, s_n$. The last server ks_n sends back the encrypted query to the client, who decrypts it with βu_i and sends the result $E_\alpha(Q)$ to the untrusted DBMS and returns the result to the user u_i. The user encrypts it with βu_i and sends it to the servers of ks in order to decrypt it with respect to α, making it readable for the final user. Revocation of user u_i can be performed by the data owner by broadcasting a message kill—token(ui) message to all servers in ks. Then, servers in ks will forget about decrypting queries encrypted with βu_i. Ignoring the instructions of the data owner would require agreeing on the part of all servers on ks.

Advantages. The proposed protocol uses static key distribution that reduces the number of keys held by users. In addition, the approach simplifies revocation by decoupling local keys that are held by users and protection keys corresponding to roles.

Limitations. One data view per role solution can lead to excessive redundancy even in the presence of less storage costs.

Summary

This section includes a summary of the key management solutions for DaaS (Table 5.1), discussed in the sections "Owner-Side Policy Enforcement Solutions," "User-Side Policy Enforcement Solutions," and "Client- and Server-Side Policy Enforcement Solutions."

Conclusion

In this chapter, a survey of key management solutions for outsourced databases, particularly with regard to key generation, distribution, and revocation are examined. As part of the survey, several research works were reviewed and the solutions were categorized into three main classes: (1) owner-side policy enforcement solutions where the access

TABLE 5.1

Comparison of Key Management Solutions for DaaS

Type of Solution	Scheme/ Feature	Advantage	Limitation
Owner-side policy enforcement solutions	NetDB2	Addresses performance issues	Needs filtering of false hits
	OPES	1. No false hits, requiring no filtering of extraneous tuples 2. Handles updates of new values, without changing the encryption of other values 3. Standard database indexes can be integrated with existing database systems. 4. Robust against estimation of true values of encrypted data	Considers only order-preserving encryption of numeric values
User-side policy enforcement solutions	TUH-based approach	Reduces the number of private keys that each client has to keep with him/her	Flexibility of the structure to adapt to frequent modifications in access control policies is extremely limited.
	DH-based key generation	1. Eliminates the need for multiple copies of data keys 2. Reduces the storage required per user to a single key 3. Reduces the required size of public information	1. Algorithm construction and generation schemes are very complex. 2. Access rights updates cause a major rebuilding of the hierarchy and mandatory key generation.
	Trie-based approach	1. Flexible toward access rights granted dynamically to the users 2. Offers flexibility in the case of access rights update 3. Allows offline key generation 4. Reduces cost of key management	1. Increase in the number of groups impacts the key generation time. 2. Does not reduce the number of keys held by a group 3. Does not completely resolve the well-known data rekeying problem

(Continued)

TABLE 5.1

Continued

Type of Solution	Scheme/ Feature	Advantage	Limitation
Shared policy enforcement solutions	Overencryption	Complements the existing key management solutions	Does not evaluate the vulnerability to collusion attacks and attacks from users who store all information on the server
	Cloud-based key management	1. Uses static key distribution that reduces the number of keys 2. Simplifies key revocation	One data view per role solution can lead to excessive redundancy

control policy is enforced by the owner itself; (2) user-side policy enforcement solutions where the access policy is enforced by the user itself; and (3) shared-policy enforcement solutions where all the actors, namely, owner, user, and server are involved in enforcing access control policies. Every solution has its particularities, presenting different features, requirements, and goals.

The analysis of various solutions made it clear that there is no unique solution that can achieve all requirements. While owner-side policy enforcement solutions are easy to implement, it requires the online presence of data owner to enforce access control. User-side policy enforcement solutions have the advantage that the data owner is not directly involved when the user queries the encrypted database. But these solution need to address the issues related to key distribution and key revocation. Additionally, the best solution for a particular application may not be best for another, hence it is important to understand fully the requirements of the application before selecting a security solution.

References

Agrawal, R. et al. "Order preserving encryption for numeric data." *Proceedings of the 2004 ACM SIGMOD International Conference on Management of Data*, June 13–18, ACM, Paris, pp. 563–574, 2004.

Akl, S. and Taylor, P. "Cryptographic solution to a problem of access control in a hierarchy." *ACM Transactions on Computer Systems*, 1(3): 239–248, 1983.

Atallah, M. et al. "Dynamic and efficient key management for access hierarchies." *Proceedings of the 12th ACM CCS*, November, Alexandria, VA, 2005.

Ceselli, A. et al. "Modeling and assessing inference exposure in encrypted databases." *ACM Transactions on Information and System Security*, 8(1): 119–152, 2005.

Damiani, E. et al. "Selective data encryption in outsourced dynamic environments." *Electronic Notes in Theoretical Computer Science*, 168: 127–142, 2007.

DES. Data Encryption Standard. *Federal Information Processing Standards Publication No. 46*, January 15, National Bureau of Standards, 1977.

Di Vimercati, S. D. C. et al. "Over-encryption: Management of access control evolution on outsourced data." *VLDB'07: Proceedings of the 33rd International Conference on Very Large Databases*, VLDB Endowment, September 23–27, ACM, University of Vienna, Austria, pp. 123–134, 2007.

El-Khoury, V. et al. "Distributed key management in dynamic outsourced databases: A trie-based approach." *DBKDA'09: Proceedings of the 2009 First International Conference on Advances in Databases, Knowledge and Data Applications*, March 1–6, IEEE, Cancun, Mexico, pp. 56–61, 2009.

Fredkin, E. "Trie memory." *Communications of the ACM*, 3: 490–499, 1960.

Gupta, D. et al. "A formal framework for on-line software version change." *IEEE Transactions on Software Engineering*, 22(2): 120–131, 1996.

Hacigumus, H. et al. "Executing SQL over encrypted data in the database-service provider model." *Proceedings of the ACM SIGMOD*, June, Madison, WI, 2002a.

Hacigumus, H. et al. "Providing database as a service." *Proceedings of the 18th International Conference on Data Engineering (ICDE)*, February, IEEE, San Jose, CA, pp. 29–38, 2002b.

Hamilton, G. and Cattell, R. "JDBC: A Java SQL API." O'Reilly, 1996, http://citeseer.uark. edu:8080/citeseerx/showciting;jsessionid=55C9192A07D170ED4B6536139463B4FD?cid=1622505.

Karlton, P. et al. "The SSL protocol v3.0." Internet draft. November 1997, http://tools. ietf.org/html/draft-ietf-tls-ssl-version3-00

Kyawt, K. K. and Khin, M. M. "Secured key distribution scheme for cryptographic key management system." *ARES'10: International Conference on Availability, Reliability, and Security*, IEEE, Kraków, Poland, February 15–18, 2010, pp. 481–486, 2010.

Medhi, D. and Ramasamy, K. *Network Routing: Algorithms, Protocols, and Architectures.* Morgan Kaufmann Publishers, Burlington, MA, April 12, 2007.

Nadia, B. et al. "Toward cloud based key management for outsourced databases." *Proceedings of the 34th Annual IEEE Computer Software and Applications Conference Workshops*, IEEE, Seoul, Republic of Korea, July 19–23, 2010.

Rivest, R. L. et al. "A method for obtaining digital signatures and public key cryptosystems." *Communications of the ACM*, 21(2): 120–126, 1978.

Sandro, R. and David, H. "A survey of key management for secure group communication." *ACM computing Surveys*, 35(3): 309–329, 2003.

Schneier, B. *Applied Cryptography: Protocols, Algorithms and Source Code in C*, 2nd edition. New York: Wiley, 1996.

Zych, A. and Petkovic, M. "Key management method for cryptographically enforced access control." *5th International Workshop on Security in Information Systems (WoSIS)*, INSTICC Press, Madeira, Portugal, June, pp. 9–22, 2007.

6

Trustworthy Coordination in Ad Hoc Networks

Aakanksha and Punam Bedi

CONTENTS

Introduction

In the past few years, there has been a rapid expansion in the field of mobile computing due to the proliferation of inexpensive, widely available wireless devices leading to the trend toward mobile *ad hoc* networking. A mobile *ad hoc* network is an autonomous collection (or group) of mobile devices (laptops, smart phones, sensors, etc.) that communicate with each other over wireless links and cooperate in a distributed manner in order to provide the necessary network functionality in the absence of a fixed infrastructure. Application scenarios include, but are not limited to, emergency and rescue operations, conference or campus settings, car networks, personal networking, etc.

Apart from the benefits, *ad hoc* networks also have some inherent challenges. These challenges are the outcome of the dynamic nature of these networks and include a vast complexity of networking problems such as group coordination, load-balancing, routing and congestion control, service and resource discovery, etc. Out of these challenges, group coordination is the most difficult as it incorporates in it challenges of load-balancing, routing, and congestion control, etc. The dynamicity in the topology of an *ad hoc* network makes it difficult to coordinate the nodes (devices) participating in the network efficiently and reliably. The devices may leave or join the local groups while on the move. Various coordination models have been proposed in the literature such as tuple spaces, EgoSpaces, and publish/subscribe (pub/sub). This chapter discusses the various coordination models in detail and introduces Mobile Process Groups (MPGs), a coordination model to coordinate nodes in an *ad hoc* network efficiently and reliably by forming trustworthy groups.

MPG is a set of mobile processes (agents) on mobile hosts, within a local radio transmission range, which reliably coordinate and communicate with each other in a self-organized manner using group views. For ensuring the credibility of the participating nodes and to curb selfish and unreliable behavior, trust on the nodes (participating in group formation) is computed. In this chapter, *trust* is defined as a quantifiable belief (either subjective or objective) established between two entities formed over a period of interaction.

Thus, this chapter introduces "trustworthy group coordination in *ad hoc* networks" and discusses the following topics in detail:

- Issues in *ad hoc* networks
- Coordination models for *ad hoc* networks
- Mobile Process Groups
- Distributed trust in MPGs

Issues in *Ad Hoc* Networks

The technological challenges that designers of *ad hoc* networks are faced with include a vast complexity of networking problems such as group coordination, load-balancing, routing and congestion control, service and resource discovery, etc. Out of these challenges, group coordination is the most difficult one as it incorporates in it challenges of load-balancing, routing, and congestion control, etc. The dynamicity in the topology of an *ad hoc* network makes it difficult to coordinate the nodes (devices) participating in the network efficiently and reliably. The devices may leave or join the local groups while on the move. Various coordination models have been proposed in the literature such as tuple spaces, EgoSpaces, and pub/sub. The next section discusses the various coordination models in detail and introduces MPGs, a coordination model to coordinate nodes in an *ad hoc* network efficiently and reliably by forming trustworthy groups.

Coordination Models for *Ad Hoc* Networks

The performance of any *ad hoc* network depends on the coordination mechanisms that are used to organize and coordinate the nodes in the network. Various coordination middlewares and frameworks have been proposed in the literature that claim to use decentralized and distributed approaches for coordination but in some way or the other they used centralized mechanisms. The following subsections describe some of the prominent coordination middlewares for *ad hoc* networks.

Tuple Spaces

A "space" is a shared persistent memory in which clients may read, write, and take objects. Clients selectively choose which objects they read and take by using template matching. A "tuple" can be thought of as a set of attributes to be used as a template for matching. Tightly coupled communication, as opposed to space-based communication, is restricted in that a sender must know the receiver's location and both sender and receiver must be available at the same time. In contrast, tuple spaces are based on the communication model: "anyone, anywhere, and anytime," adopted by agents.

Using tuple spaces as a middleware enables applications to manipulate a shared collection of data objects, called tuples, to communicate. Allowed operations on tuple space are putting, reading, and taking tuples or shared objects. Tuple spaces offer a coordination mechanism for communication between autonomous entities by providing a logically shared memory along with data persistence, transactional security as well as temporal and spatial decoupling [1]. These properties make it a desirable technique in distributed systems for e-commerce and pervasive computing applications. Tuple space implementations, by employing type-value matching of ordered tuples, object-based polymorphic matching, or XML-style pattern matching for tuple retrieval.

Tuples on the Air (TOTA) [2] is a novel middleware for supporting adaptive context-aware application in dynamic network scenarios, relies on spatially distributed tuples for both representing contextual information and supporting uncoupled and adaptive interactions between application components. Tuples are propagated via TOTA middleware across a network on the basis of application-specific patterns. As the network topology changes, it adaptively reshapes the resulting distributed structures. These distributed structures that are created dynamically are locally sensed by the application components and then exploited to acquire contextual information, so that complex coordination activities are carried out in an adaptive way.

Linda in a Mobile Environment (LIME) [3] extends the concept of tuple spaces with a notion of location. It supports the development of applications that exhibit physical mobility of hosts, logical mobility of agents, or both [4]. LIME adopts a coordination perspective inspired by work on the Linda model [5]. Linda model provides both spatial and temporal decoupling and adopts the concept of a globally accessible, persistent tuple space, which is refined in LIME to transient sharing of identically named tuple spaces carried by individual mobile units. Programs are given the ability to react to specified states. This model provides a minimalist set of abstractions that promise to facilitate rapid and dependable development of mobile applications.

Publish/Subscribe

Publish/subscribe (or pub/sub) is an asynchronous messaging paradigm. System gets information about events from publisher and event requirements from subscriber, and sends events to subscribers whose requirements match the events. The senders (also called publishers) do not send their messages to specific receivers (called subscribers). Rather, published messages are characterized into classes, without information about the subscribers if there are any. This paradigm provides the subscribers the ability to express their interest in an event or a pattern of events, in order to be notified of any event, generated by a publisher, which matches their registered interest [6]. The event is asynchronously propagated to all subscribers that have registered their interest in that event. Subscribers express interest in one or more classes, and only receive messages that are of interest, without knowledge about the publishers if there are any. The basic publish/subscribe model is given in Figure 6.1. The heart of this interaction model is an event notification service that acts as a mediator between publishers and subscribers. It facilitates storage and management of subscriptions and efficient delivery of events. The subscribers call *subscribe()* to register their interest in an event, without knowing the actual source or publisher of the event. *Unsubscribe()* operation terminates the subscription. Publisher calls *publish()* to publish or generate an event. Every subscriber is notified of every event that is of interest to the subscriber using *notify()* operation.

The advantage of the pub/sub paradigm is decoupling of publishers and subscribers which facilitates greater scalability and a more dynamic network topology. The decoupling between publishers and subscribers in this event-based interaction service can be done in time, space, and synchronization.

Event service

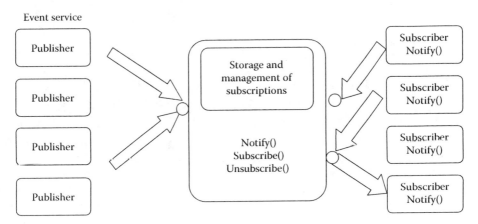

FIGURE 6.1
A simple publish/subscribe system.

The parties involved in interaction need not know each other. Publishers do not know how many and which are all the subscribers involved in interaction and vice versa; this also holds true for subscribers. This is called space decoupling (Figure 6.2).

Time decoupling enables the publishers while the subscriber is disconnected, and conversely the subscriber may get notified of an event while publisher is disconnected (Figure 6.3).

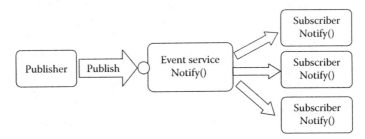

FIGURE 6.2
Space decoupling.

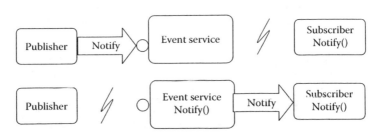

FIGURE 6.3
Time decoupling.

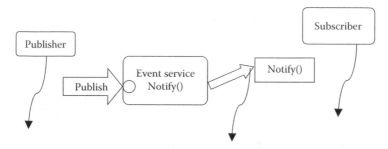

FIGURE 6.4
Synchronization decoupling.

Synchronization decoupling allows subscribers to be notified of some event while performing some concurrent activity asynchronously and publishers are also not blocked while producing events (Figure 6.4). Hence, production and consumption of events do not happen in synchronous manner.

Some applications of pub/sub paradigm are stock information delivery, auction system, air traffic control, etc.

Advantages of Publish/Subscribe

1. *Loosely coupled.* Publishers are loosely coupled to subscribers and thus may not even know about their existence. With the topic being the focus, publishers and subscribers remain ignorant of system topology. Each can continue to operate normally regardless of the other. In the traditional tightly coupled client–server paradigm, the client cannot post messages to the server while the server process is not running, nor can the server receive messages unless the client is running. Many pub/sub systems decouple not only the locations of the publishers and subscribers, but also decouple them temporally.

2. *Scalable.* For relatively small installations, pub/sub gives better scalability than traditional client–server paradigm. This is done through parallel operation, message caching, tree-based, or network-based routing, etc. However, as systems scale up with thousands of servers sharing the pub/sub infrastructure, this benefit is not achieved as expected.

Disadvantages of Publish/Subscribe

The major disadvantage of pub/sub systems is a side effect of their main advantage: the decoupling of publisher from subscriber. The problem is that it can be hard to specify stronger properties that the application might need on an end-to-end basis.

As a first example, many pub/sub systems will try to deliver messages for a little while, but then give up. If an application actually needs a stronger message delivery guarantee, such that either the messages will always be delivered or, if delivery cannot be confirmed, the publisher will be informed, the pub/sub system does not have a way to implement such properties.

Another example arises when a publisher "assumes" that a subscriber is listening. Suppose that a pub/sub system is used to log problems in a factory: any application that senses an error publishes an appropriate message, and the messages are displayed on a console by the logger daemon, which subscribes to the errors "topic." If the logger happens to crash, publishers would not have any way to see this, and all the error messages will vanish.

Hence, it is observed that while pub/sub scales very well with small installations, however, the technology often does not scale easily for large systems. As a result, problems such as instabilities in throughput (load surges followed by long silence periods), slowdowns as more and more applications use the system (even if they are communicating on disjoint topics), and so-called IP broadcast storms (which can shut down a local area network by saturating it with overhead messages that choke out all normal traffic, even traffic unrelated to pub/sub) emerge as a bottleneck for these systems.

Extensive research has been done in this area and several extensions to pub/sub model have been discussed in the context of mobile networks and for dealing with unreliable connections. JMS [7] permits messages to be stored within persistent memory until all registered subscribers have connected and received them or until a given timeout is reached and they are discarded. It deals with disconnections through durable subscriptions, which are stored and can be activated by resubscribing.

Scopes [8] limit the visibility of notifications published by a producer and confines these notifications only to those consumers that are in the defined scope or in the same scope as defined by the publisher. The same concept of scope is also applied to sensor networks [9], as a generic abstraction for a group of nodes.

Middleware Based on Object Places and Views for Context Awareness

The middleware coordination using ObjectPlaces is a hybrid of the above two approaches: pub/sub and tuple spaces. An application component in mobile *ad hoc* networks (MANETs) requires information from other nodes in order to accomplish a distributed task. But due to the mobility of all the nodes and the limited transmission or communication range of these *ad hoc* networks, a number of nodes may leave or enter a node's transmission range, thereby changing its context. This is a common scenario in *ad hoc* networks and thus the context of every node changes frequently. For the purpose of coordinating these nodes, a coordination middleware using ObjectPlaces

[10] was proposed that takes care of the context information of the nodes. The context of a node or an application component refers to the aggregate information that is available for all currently reachable nodes in the MANET. Thus, application-context changes because information on a reachable node is changed or the set of reachable nodes changes. Views [11,12] are an abstraction in the context of the application component. Therefore, a node can gather context by defining a view. A view describes "how far" it reaches over the network and what information the node is interested in. It is a subscription to events on ObjectPlaces in the vicinity of a client. A view is an actively maintained structure and when the context of a node changes the view also changes.

Each node can maintain "viewable data" in a local collection of objects called ObjectPlace. Objects that are of interest to other nodes and can be shared are placed in ObjectPlace.

An ObjectPlace is a tuple space variant. It is a set of objects that can be manipulated by operations such as put, read, and take, and is an asynchronous interface. The operations return control to the client immediately and the results are returned as they are available via a callback. Views are abstractions that provide the means for clients to observe the contents of remote ObjectPlaces. Clients can create ObjectPlaces at will, but for simplicity only one ObjectPlace is created on each node. Templates are used to read or take objects from an ObjectPlace, as they are used in tuple spaces to read shared data. Templates are used to indicate the objects of interest of a particular client or node. A template is therefore defined as a function from the set of objects to a Boolean value. For every matching object the template returns true. The following operations are defined on ObjectPlace:

- *Put* (Set, Callback) puts the given set of objects in the ObjectPlace and returns true to the callback if all objects were successfully added.
- *Take* (ObjectTmplt, Callback) removes the objects matching the template from the ObjectPlace and returns them to the callback.
- *Watch* (ObjectTmplt, EventTmplt, Lease, Callback) observes the content of the ObjectPlace. Returns copies of objects matching the object template to the callback (the ObjectPlace is not changed). The event template is used to indicate which events the client is interested in on the matching object.

A watch operation's event template can match with three possible events: isPresent, isPut, or isTaken.

A view is a local collection of objects, reflecting the contents of multiple ObjectPlaces on reachable nodes in the network based on a declarative specification. The middleware continuously changes this collection of objects (i.e., view), both with respect to changing the contents of the ObjectPlaces currently in the view, as with respect to changes in the network topology. Hence, a view represents the context of the viewing node.

To build a view, a client or node specifies the following:

A distance metric and a bound. This determines how far the client's view
   will reach over the network. For example, hop count metric with a
   bound of four: the view will span ObjectPlaces reachable in a maximum of four hop counts from the viewing node (where the view is
   built).

An object template. It defines and constrains what objects will be included
   in the view.

Using these properties and parameters, the middleware searches the network for nodes satisfying the constraints given by the application, gathers
copies of relevant data objects, and returns them to the application in the
form of a local collection of objects. The view formation and changes are
depicted in Figure 6.5.

To enable the middleware to function properly, a distributed view protocol
[10] is developed that constructs the views in a mobile *ad hoc* network as the
topology and the context of a node changes. This view protocol primarily
builds a minimal spanning tree over the part of the network which is limited
by the distance metric defined in the view. This tree is then used to route the
gathered data to the builder of the view (i.e., viewing node which is the route
of the spanning tree).

Protocol-Based Coordination Using Roles

A different middleware approach is based on the concept of roles. This
approach offers roles as first-order abstractions of the behavior of one side in
an interaction protocol. It handles the distributed instantiation of roles in an
interaction session, and maintains the session as nodes in the mobile network

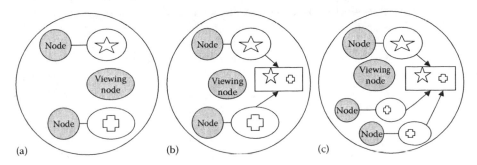

FIGURE 6.5
Circles denote nodes, ovals denote tuple spaces, and rectangles denote a view. The other shapes
represent data objects. (a) A viewing node declares a new view. The circle denotes the bound
on the distance metric from which the middleware will gather data for this particular view.
(b) The middleware has gathered the appropriate data, and the view is built. (c) A new node
enters the zone, so the view is updated with the new data object that the node carries.

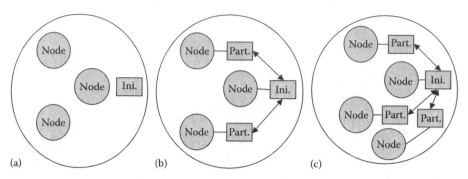

FIGURE 6.6
(a) A role called initiator is activated. The circle denotes the zone in which participants for this particular interaction should be activated. (b) The middleware activates the necessary participants in the protocol. The protocol can begin execution. (c) A new node enters the zone, so a new participant role is instantiated on the node. The initiator role is notified of this new participant and the new participant starts partaking in the protocol as well.

move (Figure 6.6) [13]. This approach is based on the view-based middleware, but the issue of coordination here is concerned with maintaining behavioral relationships between application components as opposed to gathering and/or spreading information in existing middleware approaches. The behavioral relationship between the applications components is maintained by the exchange of multiple, related messages, that is, a protocol. In this approach, though spreading and gathering of information is necessary, but it is not a necessary condition.

Existing middleware approaches (pub/sub and tuple-based) tender anonymous communication (messages are delivered based on their content and the location of the sender or receiver), and offer 1 to n communication (one sender can reach any number of receivers, and vice versa).

Collectively, these properties ensure that the application components remain loosely coupled, and rely on the middleware to deal with the low-level issues involved in delivering a message to the intended receivers. But due to these properties, only current middleware techniques provide little support when dealing with protocols.

The reason why these approaches provide little support for protocols is as follows: First, protocols rely on identification of their interaction partners, whereas in pub/sub systems; simply an event is raised based on the receiver's interest without revealing the identity of the publisher. Second, communication is not always 1 to n as opposed to the communication provided by pub/sub which is always 1 to n. Lastly, protocols are structured as multiple, related messages, and current middle wares do not provide support for maintaining these structures.

For protocol-based coordination roles are used for interactions. A *role* specifies the behavior of one class of interaction partners in a particular interaction. While a role describes the behavior of an interaction partner in general terms, a *role instance* is a runtime entity that represents a specific component in a particular interaction session [13] and a *role type* describes the view one object holds of another object. An object which conforms to a given role type plays a role specified by that role type and acts according to the type specification. At any given time, an object may act according to several different role types. Thus, different clients may have different distinct views on an object. Also, different objects may provide behavior specified by a common role type.

For every interaction, there is one role that is responsible for the lifetime of the interaction session. This role is called *initiator role*. An initiator role is instantiated by the application when a new interaction session is needed. After the interaction session is completed or has reached its goal, the initiator role instance can stop the session.

Another dual type of role, called the *participant role*, is reactive. The middleware instantiates a participant role on a node when an initiator instance indicates, that it wants to start a session with a participant role on the node. The participant role instance stays instantiated until the initiator breaks up the session or the participant goes out of the interaction's zone.

An initiator role can specify on which nodes participants must be instantiated by declaring the type of participant role, identified by a name, which the initiator expects to interact with, and a constraint on the properties of the nodes on which a participant role should be instantiated, called a zone.

A *zone* can be viewed as a distributed communication group consisting of initiator and participant instances. Thus, *zone* is a declarative construct on top of a multicast group communication primitive. This mechanism relieves the application programmer from having to deal with explicitly joining and leaving groups, and enables the programmer to declare the groups based on meaningful application-level semantics.

While the session is in progress, the middleware monitors the network and maintains appropriate instantiation of the participants as the properties of the nodes change. If a node enters the zone (i.e., its properties change such that it complies with the constraint of the zone), a participant instance is created on the node (if the node deploys the participant role). Also, if a node leaves the zone, the participant instance is notified that it is outside the zone, and subsequently removed. In both the situations, the initiator instance is notified as well. If the middleware detects that a participant instance is not able to communicate with its initiator instance, it is removed and notified that it is outside the zone. These operations can well be performed by traditional group communication protocols.

EgoSpaces

It focuses on the needs of application development in *ad hoc* environments by proposing an agent-centered notion of context, called a view, whose scope extends beyond the local host to data and resources associated with hosts and agents within a subnet surrounding the agent of interest [14].

An agent may operate over multiple views whose definitions may change over time. An agent uses declarative specifications to constrain the contents of each view by employing a rich set of constraints that take into consideration properties of the individual data items, the agents that own them, the hosts on which the agents reside, and the physical and logical topology of the *ad hoc* network.

EgoSpaces allow agents to assign behaviors to their personalized views. Behaviors consist of actions that are automatically performed in response to specified changes in a view. Behaviors include reactive programming, transparent data migration, automatic data duplication, and event capture.

Need for Trust

The coordination models presented in the previous section use different ways to coordinate mobile devices or nodes in an *ad hoc* network. These models prove to be beneficial in one situation or the other but the issue of trust remains unresolved in all of them. The opportunistic and ubiquitous interaction of the devices makes them vulnerable to various attacks and subsequent degradation of performance. The notion of trust is introduced to make these networks more reliable and for decreasing the chances of attacks. The concept of trust is not a new one. Various authors have used trust in different ways for solving different problems. A secure trusted environment is needed in order to ascertain the credibility of the participating nodes working together to achieve a goal. Though trust negotiation and establishment is challenging and provides a promising approach for any node joining or leaving a cluster dynamically, few research proposals exist to date.

Establishing trust in a distributed computing environment is one of the most challenging and important aspects of cluster computing. Achieving optimal performance in clusters forces the migration of processes to other nodes. This is where trust comes into play. An approach to build a trusted distributed environment for load balancing by incorporating two more server processes—registration and node authentication module in the process migration server (PMS) has been proposed in Ref. [15]. The proposed approach makes an attempt to design a dynamic trust management system that detects malevolent nodes at the run time environment.

A coordination model for MANETs, which reasons about trust groups to decide who to interact with, has been proposed in Ref. [16]. It characterizes the trust groups, and formalizes the dynamics of trust group creation, evolution, and dissolution, based on an agent's history of interactions and on the ontology used to encode the context of trust.

The following section presents an approach that uses trust groups to coordinate the mobile nodes of an *ad hoc* network. The approaches tries to resolve the issue of coordination in a fully decentralized and self-organized manner using distributed trust computation within the framework of MPGs.

Trustworthy Coordination in *Ad Hoc* Networks

As discussed in the previous section, there is need for trust in the coordination models for *ad hoc* networks because establishing trust in a distributed computing environment is one of the most challenging and important aspects of cluster computing. This section discusses the formation of a trusted distributed environment using MPGs. The following subsections discuss the view formation and update of MPGs and distributed trust computation in MPG.

Mobile Process Groups

MPGs are defined as the collection of mobile nodes within a local transmission range [17] forming group views using consensus. Formally, MPG is a collection of processes or mobile nodes used for reliable coordination and communication between mobile processes [18–21].

Let P be the set of all possible mobile nodes or processes. A mobile group is denoted by the set of nodes $g = \{p_1, p_2, \ldots, p_n\}$, $g \subseteq P$. Join, leave, move, and send are four operations on members of mobile group defined as follows:

- Join (g, p): issued by process p, when it wants to join group g
- Leave (g, p): issued by process p, when it wants to leave group g
- Move (g, p, l): issued when a mobile process p in group g, wants to move from its current location to location l
- Send (g, p, m): issued by process p, when it wants to multicast a message to the members of group g

A group view is also installed by each member of the group. A group view is a mapping between processes of group g and their respective IP addresses. The group view at any instance of time is represented by the set of group members which are mutually operational at a particular radio frequency within a defined distance (local area). For example, let us assume that nodes $p_1, p_2, \ldots, p_n$ are present in the radio range of frequency f at time t, then all of these nodes will form a group $g(t) = \{(p_1, ip_1), (p_2, ip_2), \ldots, (p_n, ip_n)\}$. Correspondingly the group view will be stored at all the nodes as follows:

$$v_1^1(g) = v_2^1(g) = \ldots = v_n^1(g) = \{(p_1, ip_1), (p_2, ip_2), \ldots, (p_n, ip_n)\}.$$

The group view changes dynamically on the occurrence of one of the events: join, leave, and move. Whenever a change occurs in the view, a new view is installed by all the operational group members at that moment. A number is associated with each group view. This number increases monotonically with group view installations. A view $v_i^j(g)$ represents view number j installed by a process p_i in a group $g = \{p_1, p_2, \ldots, p_n\}$.

The example below shows group views at three different time instances for the group $g = \{p_1, p_2, p_3\}$. At time $t = 1$, the group contains processes p_1, p_2, and p_3. So, the group view installed by each member of group g is the same and is represented as follows:

$$v_1^1(g) = v_2^1(g) = v_3^1(g) = \{(p_1, l_1), (p_2, l_2), (p_3, l_3)\}.$$

At time $t = 2$, the group view changes and a new group view is installed as follows:

$$v_1^2(g) = v_2^2(g) = v_3^2(g) = \{(p_1, l_1), (p_2, l_2), (p_3, l_4)\}.$$

At time $t = 3$, the group view changes and new group view is installed as follows:

$$v_2^3(g) = v_3^3(g) = \{(p_2, l_2), (p_3, l_4)\}.$$

View 1 at time $t = 1$ contains three processes p_1, p_2, and p_3 at locations l_1, l_2, and l_3, respectively. View 2 contains the same three processes, but now the third process has moved to location l_4, therefore, the group view is changed and installed by all the three processes accordingly. Similarly, view 3 also has a different set of processes mapped to different locations. The group in view 3 contains only two processes p_2 and p_3 at locations l_2 and l_4, respectively, and the process that was running at location l_1 has now crashed.

Consider an *ad hoc* network consisting of 20 nodes lying in four radio frequency ranges and forming overlapping MPGs {G1, G2, G3, G4} at time t (Figure 6.7). In the example, node ids {A, B,..., T} represent IP addresses. Some nodes may lie in two or more radio ranges thus forming overlapping

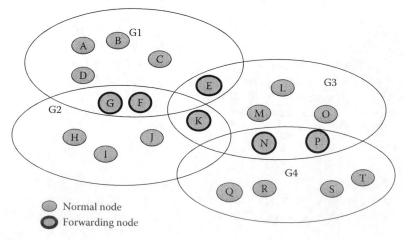

FIGURE 6.7
Ad-hoc network with four group views at time t.

groups. The nodes that are part of two or more overlapping groups (nodes E, F, G, K, N, and P) act as routers or forwarding nodes.

Using this framework, a group of agents can ensure message delivery guarantees and coordinate with each other using some sort of virtual synchrony while moving.

Distributed Trust in MPGs

This section introduces MPG-based trustworthy group coordination for *ad hoc* networks [22]. This section presents the inclusion of trust in MPG.

Each member of the MPG maintains three tables: *view table, forwarding table,* and *overlapping groups table.* The formats of the three tables are shown in Tables 6.1 through 6.3.

The *view* table stores the opinion of a node about all the forwarding nodes in its group at time *T*. This opinion value is called local trust value (LTV) and is computed locally within a local MPG. Initially, the value is set to 0.01 and is updated periodically (for each forwarding node i). The periodic update is done with the ratio of the number of packets forwarded to the number of packets received by that forwarding node.

$$\text{LTV}_i = \text{LTV}_i + N_f^i / N_r^i. \tag{6.1}$$

Forwarding table (Table 6.2) stores the group reputation (GrTV) of a forwarding node f. Group reputation is computed by aggregating (summing up) the opinions (LTVs) of all the members of a particular MPG as shown below:

$$\text{GrTV}_f(g) = \sum_{\forall i \in g} (\text{LTV}_i). \tag{6.2}$$

The confidence parameter, γ/η, for a forwarding node represents the physical characteristics of the mobile node (device) and is calculated as the amount

TABLE 6.1

View Table

View No.	IP Address	Opinion (LTV)	Timer

TABLE 6.2

Forwarding Table

IP Address	Group ID	Group Reputation (GrTV)

TABLE 6.3

Overlapping Groups Table

Group ID	Forwarding Node ID	Confidence

of remaining battery power of a node. This value is a representation of the confidence on a node whether it can be trusted to serve all the overlapping groups (of which the node is a part) with a particular amount of remaining battery power.

These three tables are used to compute and store three trust values: Opinion (LTV), Group Reputation (GrTV), and Confidence (GTV). The three trust values represent route reliability, trustworthiness, and confidence parameters used to evaluate the overall trust on a forwarding node. Finally, the overall trust is computed and stored in $T_f(g)$ representing trust on a forwarding node f in group g. Overall trust $T_f(g)$ is computed using the following equation:

$$T_f(g) = \alpha \cdot \delta + \beta \cdot \tau + \frac{\gamma}{\eta}, \qquad (6.3)$$

where:
   α represents the route reliability of a forwarding node and is represented by the LTV on a node
   δ represents the mobility rate which is inversely proportional to the distance traveled per unit of time by node i
   β is the trustworthiness factor and is measured in terms of the ratio of the number of data packets forwarded correctly to the number of data packets received by a forwarding node, as shown below:

$$\beta = N_f / N_r. \qquad (6.4)$$

τ defines the time interval for which a forwarding node has been a part of a local group and η defines the number of overlapping groups of that node.

Thus, the three trust factors *route reliability* (α), trustworthiness (β), and *confidence* (γ) contribute toward computing overall trust on a forwarding node as discussed earlier.

Using Equation 6.4 for computing trust on a forwarding node is intuitively justified as the node can be trusted only if it has forwarded a significant number of data packets in the past and is comparatively static and less loaded. Otherwise, there are chances of packet drops due to movement of the node or its selfish and/or malicious behavior. Hence, to reduce the chances of packet drops and ensure reliability of packet forwarding and routing, MPG-based trustworthy coordination helps in reliably coordinating nodes based on their behavioral (forwarding behavior) and physical characteristic (such as battery power and number of overlapping groups) and is therefore better than those that do not apply trust.

Summary

In this chapter, various coordination models for *ad hoc* networks have been presented and discussed at length such as tuple spaces, EgoSpaces, Pub/Sub, and MPGs. One of the fundamental requirements for coordination in *ad hoc* networks is trust. This chapter highlights the need for trust in *ad hoc* network coordination. A trustworthy coordination ensures performance increase and decreases chances of security breach. The existent trustworthy coordination scheme based on trusted MPGs is presented. The scheme uses formation of trustworthy groups based on the interaction of the members of the group.

References

1. Deepali, K., Tim, F., and Anupam, J. "Semantic tuple spaces: A coordination infrastructure in mobile environments." *Proceedings of the 2nd International Semantic Web Conference*, IEEE, pp. 268–277, 2003.
2. Marco, M., Franco, Z., and Letizia, L. "Tuples on the air: A middleware for context-aware computing in dynamic networks." *Proceedings of the 23rd International Conference on Distributed Computing Systems (ICDCSW)*. IEEE Computer Society, pp. 342–347, 2003.
3. Amy, L. M., Gian Pietro, P., and Roman, G. C. "LIME: A coordination model and middleware supporting mobility of hosts and agents." *ACM Transactions on Software Engineering Methodology*, 15(3): 279–328, 2006.
4. Radu, H., Jamie, P., Christine, J., and Roman, G.-C. "Coordination middleware supporting rapid deployment of ad hoc mobile system." *Proceedings of the 23rd International Conference on Distributed Computing Systems Workshops (ICDCS)*, IEEE, pp. 362–368, 2003.
5. Gian Pietro, P., Amy, L. M., and Roman G. C. "LIME: Linda meets mobility." *Proceedings of the 21st International Conference on Software Engineering*, ACM, pp. 368–377, 1999.
6. Patrick, E. T., Felber Pascal, A., Rachid, G., and Anne-Marie, K. "The many faces of publish/subscribe." *ACM Computing Surveys*, 35(2): 114–131, 2003.
7. Hapner, M., Burridge, R., Sharma, R., Fialli, J., and Stout, K. Java message service. Sun Microsystems Inc., Santa Clara, CA, 2002.
8. Steffan, J., Fiege, L., Cilia, M., and Buchmann, A. "Scoping in wireless sensor networks." *Proceedings of the Workshop on Middleware for Pervasive and Ad-Hoc Computing*, ACM, pp. 167–171, 2004.
9. Fiege, L., Mezini, M., Muhl, G., and Buchmann, A. P. "Engineering event-based systems with scopes." *Proceedings of the European Conference on Object-Oriented Programming (ECOOP)*, Springer-Verlag, Heidelberg, vol. 2, pp. 309–333, 2002.
10. Kurt, S., Tom, H., and Yolande, B. "Views: Customizable abstractions for contextaware applications in MANETs." *ACM SIGSOFT Software Engineering Notes*, 30(4): 1–8, 2005.

11. Kurt, S. and Tom, H. "Co-ordination middleware for decentralized applications in dynamic networks." *2nd International Doctoral Symposium on Middleware*, France, 2005.
12. Kurt, S., Tom, H., and Yolande, B., "Views: Middleware abstractions for context-aware applications in MANETs." *Software Engineering for Multi-Agent Systems IV*, Springer, Berlin/Heidelberg, pp. 17–34, 2005.
13. Kurt, S., Danny, W., and Tom, H. "Middleware for protocol-based coordination in dynamic networks." *Proceedings of the 3rd International Workshop on Middleware for Pervasive and Ad-Hoc Computing (MPAC)*, pp. 1–8, 2005.
14. Julien, C. and Roman, G. C. "EgoSpaces: Facilitating rapid development of context-aware mobile applications." *IEEE Transactions on Software Engineering*, 32(5): 281–298, 2006.
15. Mishra, S., Kushwaha, D. S., and Misra, A. K. "A cooperative trust management framework for load balancing in cluster based distributed systems." *International Conference on Recent Trends in Information, Telecommunication and Computing (ITC)*, pp. 121–125, 2010.
16. Capra, L. "Reasoning about trust groups to coordinate mobile ad-hoc systems." *International Conference on Security and Privacy for Emerging Areas in Communication Networks*, pp. 142–152, 2005.
17. Aakanksha, V. and Bedi, P. "A self-organizing self-healing on-demand loop-free path routing protocol using mobile process groups for mobile ad-hoc networks." *International Conference on Advances in Recent Technologies in Communication and Computing (ARTCom)*, pp. 396–400, 2009.
18. Assis Silva, F. M. and Macedo, R. J. A. "Reliability requirements in mobile agent systems." *Proceedings of the 2nd Workshop on Tests and Fault Tolerance II (WTF)*, Curitiba, Brazil, 2000.
19. Macedo, R. J. A. and Assis Silva, F. M. "Mobile groups." *19th Brazilian Symposium on Computer Networks (SBRC)*, 2001.
20. Macedo, R. J. A. and Assis Silva, F. M. "Coordination of mobile processes with mobile groups." *Proceedings of the IEEE/IFIP International Conference on Dependable Systems and Networks (DSN)*, pp. 177–186, 2002.
21. Macedo, R. J. A. and Assis Silva, F. M. "The mobile groups approach for the coordination of mobile agents." *Journal of Parellel and Distributed Computing*, 65: 275–288, 2005.
22. Aakanksha, V. and Bedi, P. "MPG-TAR: Mobile process groups based trust aware routing protocol for MANETs." *International Conference on Advances in Recent Technologies in Communication and Computing (ARTCom)*, pp. 131–135, 2010.

7

Toward a Computational Trust Model for Secure Peer-to-Peer Communication

Shyam P. Joy and Priya Chandran

CONTENTS

Introduction

Group communication is invariably used for services that demand high availability, such as distributed databases, replicated servers and clouds, and collaborative applications, such as multimedia conferencing, shared white-boards, and multiparty games [1]. Peer-to-peer (P2P) architecture is preferred over hierarchical network architectures in such distributed applications for achieving availability. As data are stored at multiple peers in such architectures, assuring data integrity and confidentiality is important [2].

Group applications on the P2P architecture, which need to maintain confidentiality of communication, seldom use computationally expensive and thereby slow, public key cryptography for routine communication. They use symmetric session keys, which are generated in a distributed manner using key generation or agreement protocols. Group key agreement (GKA) [3] protocols allow members of a group to contribute and/or agree on a key value without explicitly transmitting it over insecure channels in any form. Several GKA protocols have been proposed in the literature [4].

In applications that involve secure transactions, the verifiability [5] of *all* messages, actions, and transactions by group members is difficult due to exponential time computations. P2P communications are no exception to this problem. Moreover, P2P applications are inherently dynamic and require being secure even when members join and leave the group while the application is running, apart from remaining secure for the usual computational and communication activities. As verifiability would not be practical [5], group members require some information based on which they would be able to take decisions such as joining a new group or allowing a new member into the group. This calls for a notion of *trust* for groups and group members, based on which decisions such as joining a group, passing information, allowing new members into the group, leaving groups, allowing members to leave the group, or generating new keys for information passing, could be taken by the group members. Trust models for group communication are a very important issue, which the research community needs to address so as to realize the full potential of the group communication paradigm, and its effective adaptation in various domains.

This chapter outlines a proposal for a probability-based trust model for group communication, and describes a set of computational trust metrics for groups, which are based on the concept of forward secrecy of protocols for key establishment in a group. Assurance of a very high probability of maintaining secrecy of information, even after joining the selected group for long or short periods of time, would motivate users to participate in group communication and applications based on them. The trust metrics proposed would enable users to make such decisions, as also motivate them to achieve high trust scores by resorting to secure protocols, and adapting secure practices at their workplaces. The probability of maintaining secrecy of information depends on the behavior of the users, as well as the key used for communication, and the protocols used for establishing the key. Therefore, the basis for the trust metrics proposed in this chapter is the computation of trust from the trust value of the principal components in the protocol as well as the mode of key computation.

As a first step toward building a trust model, a framework for assessing the long-term secrecy of keys used in group communication is proposed. The degree of partial forward secrecy (DPFS) is introduced as a metric to assess the long-term secrecy of established keys. The description of a framework for assessing DPFS, and the analysis of several protocols on their DPFS, is presented.

The next contribution in this chapter is a metric for trust, the group trust score (GTS), based on the DPFS, as well as the trustworthiness of individual members—a concept that is explained later in this chapter. From the GTSs of protocols the concept of *trust strength* evolved. Trust strength is a measure of the stability or the long-term standing of GTS. An analysis of the suitability of different protocols for different P2P applications, based on trust strength and computational difficulty of keys in the protocol, is subsequently presented.

The analyzed protocols use different cryptographic primitives, illustrating that the technique is general and not specific to any cryptographic category.

The following section provides the background knowledge required for understanding the rest of this chapter.

Background

GKA protocols are designed to provide several properties to the session key generated for use by the participants, such as key privacy, key confirmation, key contributiveness, immunity from impersonation attacks, and perfect forward secrecy [6]. Formal security analysis establishes the first four properties for protocols, and such a proof is essential for the acceptance of protocols for use by groups. However, perfect forward secrecy is a computationally cumbersome property to achieve in a protocol, and hence several protocols proposed in the literature do not satisfy it or, rather, satisfy similar properties to a weaker extent.

The authors of this chapter propose a security metric, the DPFS, to analyze and compare the forward secrecies offered by different group key authentication protocols. As the metric is based on probabilities of the loss of secrecy of different components in the protocol, computation of this metric could be used as a trust metric, on which prospective and existing members could base their decisions on joining and participating in the group. Because an understanding of forward secrecy is crucial to follow the rest of this chapter, the concepts are described next.

Perfect Forward Secrecy

Participants of a GKA protocol (i.e., members of a group) possess long-term as well as short-term secrets. Short-term secrets are valid only for a session of the protocol, whereas long-term secrets are valid across several sessions, or even the existence period of the participant on the Internet. These long-term secrets (e.g., private keys of the participants, hash functions, signatures, or any other key valid across multiple sessions with different participants), as well as short-term secrets, form part of the contribution of participants in computing the session key. Because the long-term secrets are valid across sessions, they have a higher probability of getting compromised over a long interval of time. Moreover, a group member can migrate to a different group, resulting in compromise (i.e., knowledge by a person outside the group) of long-term keys.

An issue of concern to group members would be whether the compromise of long-term keys can possibly lead to the disclosure of old short-term secrets or keys, because the compromise of such secrets along with old stored message replays would reveal all the old messages in their unencrypted form. This question is addressed by the property termed as the *forward secrecy* of GKA protocols, which is defined in the following way in Ref. [7]: "A protocol is said to have perfect forward secrecy if compromise of long-term keys does not compromise past session keys."

The definition of perfect forward secrecy in Ref. [7] is with respect to compromise of long-term keys. But short-term keys can also be retrieved from hardware at any later point of time as long as they are not explicitly erased. Therefore, the possibility exists of the penetrator or intruder, or outsider getting these ephemeral secrets. The attacks that could reveal the ephemeral secrets in addition to the long-term keys of the participants are called strong *corruptions* in security analysis terminology.

GKA protocols that do not compromise the session key, even when the ephemeral secrets are compromised, satisfy *strong perfect forward secrecy* [8]. In this chapter, for the analysis of GKA protocols, intruders are considered possible of being highly corrupted, and hence there is no differentiation between short-term and long-term secrets and so both are referred to as secrets.

Perfect forward secrecy requires the secrecy of the previously established session key even when all long-term secrets have been compromised. Not all protocols have been able to achieve perfect forward secrecy. When protocols are designed to optimize other parameters such as computation time and number of rounds, perfect forward secrecy has to be compromised. To cite an example, Boyd and Nieto proposed a conference key protocol [9], which was optimized for a number of rounds, but does not satisfy perfect forward secrecy. A weaker form of perfect forward secrecy is called *partial forward secrecy* and is explained as follows [10]: "A protocol provides partial forward secrecy if the compromise of long-term keys of one or more specific principals does not compromise the session keys established in previous protocol runs involving those principals."

Perfect forward secrecy considers the secrecy of the previously established session key when all long-term secrets have been compromised, whereas partial forward secrecy considers the secrecy of the established session key when some long-term secrets are compromised. Hence, perfect forward secrecy can be viewed as a special case of partial forward secrecy.

Intuitively, among protocols that satisfy partial forward secrecy, those that allow larger number of secrets to be compromised, without compromising the session key, would be stronger than those that allow lesser number of secrets to be compromised. A formal metric for comparison of protocols based on their partial forward secrecy, that is, their level of tolerance to compromise of long-term keys, would help in protocol selection and in evolving trust values for groups that use the protocol.

Later in this chapter, a formal framework to grade different GKA protocols satisfying partial forward secrecy is proposed. The framework includes a metric for comparison of partial forward secrecy of protocols and an algorithm for computing the metric.

The proposal is further illustrated in two collections of protocols. The first collection consists of GKA protocols that use the Diffie–Hellman (DH) cryptographic primitive. Perfect forward secrecy is considered as an attribute of DH-based cryptographic protocols, as the ephemeral keys are discarded once the session key is computed. But under strong corruptions, protocols that otherwise satisfy perfect forward secrecy satisfy only partial forward secrecy, as the assumption is that the ephemeral short-term keys could have been stored and/or leaked. Protocols based on DH have varying strengths with respect to partial forward secrecy, under strong corruptions, and depending on their design. Our analysis helps to grade these protocols and could be used by protocol designers to gain an insight into whether protocol designs would meet the required levels of partial forward secrecy. The second collection in the test set consists of GKA protocols that use different cryptographic primitives, illustrating that the proposed technique for analysis is independent of the cryptographic primitives used by protocols.

At this point, let us examine an example of a group-based application to understand the requirement for forward secrecy. The members of a group may want to store confidential data in a remote location, such as a cloud, encrypting it with a key, known only to them. This key would be generated by a GKA protocol. Selecting a GKA protocol with perfect forward secrecy or strongest partial forward secrecy would minimize the risk of compromising the session key, and the resulting loss of confidentiality of stored data, in the event that some of the secrets are compromised.

Several models for analyzing the perfect forward secrecy of protocols have been proposed [11–13] in the literature, but models for assessing and measuring partial forward secrecy have not been framed yet. In this chapter, the *strand space* model for protocols is extended to formulate a framework for assessing partial forward secrecy. As strand spaces form the backbone of the proposed framework, they are described in what follows next.

Strand Spaces

Strand spaces [14–16] provide a general framework of modeling protocols and proving the correctness of protocols. Failure to prove correctness gives insights into possible flaws in protocols. The model uses a graphical representation to show the causal precedence relationships between the terms in the protocol. Tools such as Athena [17], ASPECT [18], and others, used for security analysis of protocols are based on strand spaces. Such tools analyze protocols represented in strand space notation and compute whether the protocols satisfy the security conditions or not. If not, the computations give an insight into the types of attacks that could be possible in these protocols. The availability of such tools and the simplicity of representation are the primary reasons for our choice of strand spaces for the framework proposed in this chapter.

Group members in a protocol would be referred to as *principals* in the strand space representation of the protocol. Let us assume set A to represent all possible terms that can be sent and received in a protocol. The trace of all activities performed by an instance of a principal, or a penetrator, is represented as a *strand*. Each strand consists of a sequence of nodes connected by the symbol "⇒." Each node represents *action* taken by a principal. The action can be that of *send* or *receive*. If a principal sends a term t from node n_1, and the same is received by node n_2, then $term\ (n_1) = +t$ and $term\ (n_2) = -t$ and node n_1 is connected to node n_2 by the symbol "→."

Strand spaces model the capability of the adversary (i.e., the penetrator or the intruder) using penetrator strands. The penetrator, as defined in Ref. [16], is rendered capable of *guessing a text message* (M), *guessing a nonce*, known only to him at the time (R), *concatenating* two of the received terms (C), *separating* the received terms into components (S), *guessing a key* (K), *encrypting* a term with a known key (E), *decrypting* a term with a key known to the penetrator (D), *guessing a DH term* (F), *signing* a term with a known signature key (σ), *extracting* plain text from signed terms (X), and *hashing* any term (H). Hence, the traces of the penetrator strands would contain nodes such as the following:

1. M, text message: $<+t>$, where $t \in T$
2. R, fresh nonce: $<+r>$, where $r \in R_P$ denotes the set of nonces known to the penetrator
3. C, concatenation: $<-g, -h, +gh>$
4. S, separation into components: $<-gh, +g, +h>$
5. K, key: $<+K>$, where $K \in K_P$ denotes the set of keys known to the penetrator
6. E, encryption: $<-K, -h, +\{h\}_K>$
7. D, —decryption: $<-K^{-1}, -\{h\}_K, +h>$
8. F, fresh DH value: $<+g^p>$
9. σ, signing: $<-K, -h, +[h]_K>$, where $K \in K_{Sig}$. K_{Sig} being the set of keys used by participants to sign messages
10. X, extraction of plain text from signatures: $<-[h]_K, +h>$
11. H, hashing: $<-g, +hash(g)>$.

An *attack* on the protocol is modeled by interleaving the strands of the regular participants with that of the penetrator. The set of all nodes along with the set of all edges, $\rightarrow \cup \Rightarrow$, form a directed graph. A finite, acyclic subgraph of the above-directed graph is called a *bundle* if the following two conditions are satisfied:

1. For all receiver nodes present, the corresponding sender nodes are also present in the bundle.
2. For all nodes present, their immediate causal predecessors are also present in the bundle.

If a set S is a proper subset of the set $\rightarrow \cup \Rightarrow$, then the symbol $\leq_S$ stands for the reflexive, transitive closure of S. Suppose C is a bundle. Then $\leq_C$ is a partial order relation and every nonempty subset of nodes in C has $\leq_C$ minimal members. A term is considered to *originate* on a node if any of the terms from which the term can be deduced originates on the node [15]. Therefore, using strand spaces, a term is proved to be secret by also proving that it does not originate in the bundle.

Strand space representation of protocols can be used in conjunction with tools such as Athena [17] to analyze the correctness of protocols and to find possible attacks in a protocol. The basic strand space model has not been used for the assessment of forward secrecy of protocols. This chapter presents extensions to the strand space model to create a framework for analyzing the forward secrecy of protocols. The extensions are stated in the following section as a part of the description of the framework for the analysis of forward secrecy of protocols.

Framework for Analyzing Partial Forward Secrecy

A GKA protocol that satisfies partial forward secrecy guarantees the secrecy of the session key, even if some of the secrets are compromised. Consequently, the number of secrets that can be compromised without compromising the session key seems an obvious choice of a metric to evaluate the GKA protocols with respect to their partial forward secrecy. But a deeper look reveals that sometimes it may be easier for the penetrator to get more secrets than to get some single secret. For example, in the case of asymmetric group Diffie–Hellman 2 (A-GDH.2) protocol [19], it is easier for the penetrator to obtain a pair of secrets rather than a single secret. Therefore, a finer metric, namely, DPFS is introduced. The DPFS is defined as follows: "The DPFS of a GKA protocol is the highest probability of obtaining the session key, expressed in terms of the probability of loss of secrets."

A protocol with lowest value of DPFS has the minimum risk of compromising the session key, when the secrets used in its computation are compromised. As strand spaces do not have the capability to model the probability of events, the strand space model is extended to provide it. The proposed model is described in the following section.

Extensions to the Strand Space Model

Let set A be the set of terms used in the strand space representation of the protocol. In the extended model, each element t of A has an associated value, the probability with which the penetrator knows the term t. If $t \in A$ and if t may be known to the penetrator with a probability p, then the term is represented as $t(p)$. All terms that are transmitted and received will be known to the penetrator with certainty, and hence in such cases the probability is not explicitly represented (the probability is 1). The penetrator model for strand spaces is defined, based on the above proposal, as follows:

1. M, generate atomic message $<+t(p)>$
2. C, concatenation $<-g(p), -h(q), +gh(pq)>$
3. S, separation $<-gh(p), +g(p), +h(p)>$
4. K, generate keys $<+K(p)>$
5. E, encryption $<-K(p), -h(q), +\{h\}_K(pq)>$

6. D, decryption $<\!\!-K^{-1}(p), -\{h\}_K(q), +h(pq)\!\!>$

7. F, generate DH values $<\!\!+\alpha^x(p)\!\!>$

8. σ, signing $<\!\!-K(p), -h(q), +[h]_K(pq)\!\!>$

9. X, extraction of plain text from signature $<\!\!-[h]\}_K(p), +h(p)\!\!>$

10. H, hashing $<\!\!-g(p), +hash(g)(p)\!\!>$

11. EXP, DH exponentiation $<\!\!-x(p), -\alpha^y(q), +\alpha^{xy}(pq)\!\!>$

An attack on the protocol, which results in the loss of the session key, can be represented by interposing penetrator strands with strands of the regular participants. Hence, the probability of compromising the session key can be computed from the strand space representation of the attack. For example, the interpretation of the EXP strand is as follows: if the penetrator knows a secret x with a probability p and α^y with a probability q, then the penetrator knows α^{xy} with a probability pq. As the penetrator certainly knows α^y, $q = 1$, so the penetrator knows α^{xy} with a probability p.

A GKA protocol can be represented in the extended strand space model and analyzed for the secrecy of specified keys using practical tools for protocol analysis. The idea is represented as an algorithm for computing DPFS.

Algorithm for DPFS Computation

Let P denote the strand space representation of the protocol, Q denote the set of secrets held by the participants, I represent the initial information available with the penetrator, and P_Q represent the probability with which the penetrator knows the elements of Q.

The algorithm (named DPFS-I) takes the above parameters as input and returns DPFS of the protocol.

The set R, the power set of secrets less the null set, with the elements arranged in the increasing order of cardinality, is computed first. The variable E successively denotes each element of R. The algorithm loops through R, checking for attacks on the protocol, when the secrets in set E are combined with the penetrator's initial information. CHECK_SECRECY performs the above function and returns either an attack or NIL. CHECK_SECRECY function can be realized using any practical tools, such as Athena [17].

DPFS-I (P, Q, P_Q, I)

```
1.  m ← 0
2.  R ← 2^Q\{φ}
3.  while R ≠ φ
4.      do E ← head(R)
5.          if (t ← CHECK_SECRECY(P, I ∪ E)) ≠ NIL)
6.              then p ← PROB_OF_FAILURE(P, Q, P_Q, t, I ∪ E)
```

```
7.                    m ← MAX (p, m)
8.        R ← R\{E}
9. return (m)
```

The function CHECK_SECRECY returns NIL if the leakage of elements in *E* does not compromise session keys. It returns an attack, a non-NIL value, if the leakage of elements of *E* compromises session keys. If an attack is returned, the probability for compromise of session key is computed using the function PROB_OF_FAILURE. The function takes the protocol representation, the set of secrets, the probability of compromise of each of the secrets, the attack representation, and the penetrator's initial information, and it returns the probability of compromise of the session key. The extended strand space model, as explained in the section "Extensions to the Strand Space Model," provides details on the computation of probability of session key compromise.

The model is based on the assumption that the penetrator would choose the easiest option (with highest probability) to attack the protocol, if there was more than one option available. Hence, the above actions are repeated for all elements in *R* and the highest probability of compromise of the session key is computed as the DPFS.

The complexity of DPFS-I is exponential in the number of secrets, $|Q|$, because the algorithm loops through power set of secrets, *R*. However, this is not a serious issue, as secrets can be classified into types, in most of the protocols, and then leaking of a particular type of secret alone needs to be considered, instead of leaking of all secrets of the type. Then the complexity would be exponential in terms of the number of types, which is a constant in most protocols.

The roles played by different participants in a protocol are proposed to be used for identifying the types of secrets. An instance of a protocol can be considered as a collection of roles and instances of the roles. For example, the A-GDH.2 protocol [19], which is analyzed in the section "Analysis of GKA Protocols on Partial Forward Secrecy," works with *n* participants. It can be modeled as a single instance of initiator's role, $n - 2$ instances of intermediary roles and a single instance of controller's role. Each role along with the nature of key, whether it is short term or long term, is defined as a type. Therefore, A-GDH.2 protocol has five types of keys: (1) initiator's short-term key, (2) initiator's long-term key, (3) intermediary's short-term key, (4) intermediary's long-term key, and (5) controller's short-term key. Thus, the number of types would be constant. DPFS-II is a revised algorithm, applying the notion of types.

The algorithm DPFS-II takes the protocol, *P*, as input represented using strand spaces, the set of types of secrets, Q_T, and the initial information with intruder, *I*. The set of secrets Q_T contains elements that are tuples corresponding to each secret type. The format of a tuple is *secret type, number of instances of type*, and *probability with which an instance of the secret of this type may be known to the intruder*. Q_I represents the set of all instances of all secrets. The algorithm first checks if there are attacks when the initial information of the penetrator is enhanced with set Q_I, that is, the protocol does not satisfy strong perfect forward secrecy. If it does not satisfy perfect forward secrecy then the algorithm proceeds to find the DPFS.

The set R_T represents the power set of secret types other than null set, arranged in increasing order of cardinality, with E_T being the first element. The algorithm iterates through the set (R_T), checking for an attack, and calculating the probability of failure whenever an attack happens. For each set of secret types that results in an attack, its supersets are removed from R_T. The above steps are repeated for all elements of set R_T and the highest probability of compromising the session key is returned as the DPFS. DPFS-II has exponential time complexity in the number of types. As the number of types is practically constant for a protocol, the exponential time complexity is not a serious issue.

DPFS_II (P, Q_T, I)

```
1.  Q_I ← set of all secrets
2.  if Q_I ≠ φ
3.       then
4.       if (t ← Check Secrecy (P, I ∪ Q_I)) ≠ NIL
5.              then
6.                       R_T ← 2^Q_T\{φ}
7.       m ← φ
8.       while R_T ≠ φ
9.              do
10.                 if (t = Check_Secrecy (P, I ∪ E_T)) ≠ NIL
11.                         then p ← Prob_of_Failure(P, Q_T, t, I ∪ E_T)
12.                              m ← Max (p, m)
13.                         while R_T ≠ NULL
14.                                 do if F ⊃ E_T | F ∈ R_T
15.                                        then R_T ← R_T\{F}
16.                       U ← U\{E_T}
17.                 return (m)
```

The classification of secrets into types would have to be protocol specific, and generalization would fail to capture the nuances of all protocols. A *type-based* classification of secrets in GKA protocols calls for some interesting and challenging research in itself. For instance, could it be possible in some protocol that more than one instance of the same role type could reveal the session key, whereas instances of different types could not? Checking for such cases would require reverting to DPFS-I.

Failure to meet perfect forward secrecy, followed by a run of DPFS-II returning 0 as the probability of compromise of the session key, implies that there is a need to execute DPFS-I.

Computed in this manner, DPFS can be used to compare various GKA protocols, and thereby allow the users to know what the chances of getting their session keys compromised are. Several GKA protocols published in the literature are analyzed as given in the following section.

Analysis of GKA Protocols on Partial Forward Secrecy

A set of nine GKA protocols have been chosen for analysis and comparison. They are the Burmester–Desmedt (BD), Just–Vaudenay (JV), Becker–Willie (BW), Steiner–Tsudik–Weidner (STW), Kim–Perrig–Tsudik (KPT), Kim Lee–Lee (KLL), Boyd–Nieto (BN), A-GDH.2, and ID-based bilinear pairing protocols. Six of these protocols are based on the DH cryptographic primitive, and three of them follow different cryptographic primitives. The protocols in the set are analyzed on the basis of the DPFS. The proposed metric and framework are applicable to protocols irrespective of their cryptographic primitives.

An accurate analysis of DPFS requires that the probability of compromise of the various terms or secrets be known. As the probability would also depend on the actual application scenario, the members who participate, and the underlying network [20], the analysis for an actual application should be done as explained in the section "Framework for Analyzing Partial Forward Secrecy." For the purpose of the analysis presented here, it is assumed that each term has an equal probability, p, of getting compromised. Moreover, it is assumed that there are no attacks on these protocols other than those published in the literature. The analysis of each of the protocols in the set, under the assumptions stated above, is described in this section.

Burmester–Desmedt Protocol

Burmester et al. proposed a conference key exchange protocol [21]. The BD protocol works in the following manner. The n users, denoted as $u_1 \ldots u_n$, respectively, are arranged in a logical ring. Let G be a cyclic subgroup of Z_p^* of order q and α be a generator of G.

In the first round, every user u_i, randomly selects $r(i) \in Z_q$ and broadcasts $\alpha^{r(i)}$ denoted as z^i. In the second round, every user computes and broadcasts $X_i = (z^{i+1}/z^{i-1})^{r(i)}$. Each user computes the common conference key as $z_{i-1}^{nr(i)} X_i^{n-1} X_{i+1}^{n-2} \ldots X_{i-2}$. The computed session key is of the form $\alpha^{r(1)r(2)+r(2)r(3)+ \ldots r(n)r(1)}$.

It can be proved that the penetrator can compute the session key if the short-term key, that is, $r(i)$ of at least one user is known. If the probability that the intruder knows a user's short-term key is p and that there are n users, then the intruder knows the session key with probability $1-(1-p)^n$.

Therefore, DPFS is $1 - (1 - p)^n$. When individual probabilities are not equal, DPFS algorithms could be used for the computation of DPFS.

Even if one user's short-term key gets leaked, the session key is compromised.

Just–Vaudenay Protocol

Just et al. proposed an authenticated key agreement protocol (JV) based on decisional DH assumption [22].

Let G be a cyclic subgroup of Z_p^* on order q and α be a generator of G. The group members, $u_1 \ldots u_n$, are assumed to be arranged in a logical ring. Each user u_i has a long-term private key $r(i)$, such that $r(i) \in G$, and a public key $\alpha^{r(i)}$.

The protocol consists of two stages. In the first stage, two adjacent members engage in a key exchange protocol and establish a shared key in the following manner: User A selects $x(A) \in Z_q$ at random and sends $[\alpha^{x(A)}, I(A)]$ to B, where $I(A)$ is the identity of A. B selects $x(B) \in Z_q$ and computes the key, K_i as $[\alpha^{x(A)}]^{[x(B)+r(B)]} \cdot [\alpha^{r(A)}]^{x(B)} = \alpha^{x(A)x(B)+x(B)r(A)+x(A)r(B)}$. B sends to A, $(\alpha^{x(B)}, I(B), h[\alpha^{x(B)}, \alpha^{x(A)}, I(A), I(B)])$, where h is a public one-way hash function. A computes the key $K_i = [\alpha^{x(B)}]^{x(A)+r(A)} \cdot [\alpha^{r(B)}]^{x(A)} = \alpha^{x(B)x(A)+x(B)r(A)+r(B)x(A)}$.

In the second stage, each member broadcasts $W_i = K_i/K_{i-1}$ where K_i is the key shared between the ith and the $i + 1$th user. Each user computes the group key as $K_{i-1}{}^n W_i{}^{n-1} W_{i+1}{}^{n-2} \ldots W_{i-2} = K_1 K_2 \ldots K_n$.

In this protocol, a penetrator can compute the key exchanged between a pair of users if he knows both the long-term and short-term keys of at least one of the n users. The intruder can compute the session key if at least one key between a pair of users is known. If the short-term key of the ith user and the short-term keys of either the $i - 1$th or the $i + 1$th user be known with a probability p, then the key exchanged between a pair of users can be known to the intruder with a probability $1-(1 - 2p^2)^n$. Hence, the probability of knowing the session key is $1-(1 - 2p^2)^n$ and so DPFS is $1-(1 - 2p^2)^n$. Thus, it can be seen that secrets belonging to at least two principals must be leaked here for the session key to be compromised. One compromised user cannot spoil the show.

Becker–Willie Protocol

Becker and Willie proposed the Octopus protocol (BW) for computation of group keys [23]. Let G be a cyclic group on order q and α be a generator of G. The protocol assumes the existence of a bijection $\Phi: G \to Z_q$.

Participants are divided into four groups: $I(A)$, $I(B)$, $I(C)$, and $I(D)$ with A, B, C, and D as group leaders. A performs a DH key exchange with each group member $P(i)$ resulting in a key $k(i)$, where $P(i) \in I(A)$. A computes a key $K(A) = \Pi_i \varphi[k(i)]$. B, C, and D perform the same sequence of steps as A.

A performs a DH key exchange with B resulting in the key $\alpha^{K[I(A)\cup I(B)]}$. Similarly, C performs a DH key exchange with D. The group key is computed as follows:

$$K = \alpha^{\Phi\left(\alpha^{K[I(A)\cup I(B)]}\right) \cdot \Phi\left(\alpha^{K[I(C)\cup I(D)]}\right)}$$

A sends $\alpha^{K[I(A)\cup I(B)]/[j]}$ and $\alpha^{K[I(A)\cup I(B)]}$ to $P(j) \in I(A)$ so that $P(j)$ can calculate K.

It is assumed that each group leader uses different short-term keys for DH exchange with each group member to compute the key $k(j)$. In such a case, if the intruder knows the short-term keys of a member of the group, say I_A, with a probability p then assuming there are $(n - 4)/4$ members, the probability of knowing the keys of all members is $p^{(n-4)/4}$. It suffices to know the key of either $I(A)$ or $I(B)$ or $I(C)$ or $I(D)$ for knowing the session key. Therefore, the probability that the intruder knows the session key and DPFS is $1-(1 - p^{(n-4)/4})^4$.

It can be inferred that in this protocol, the keys of all members of a subgroup have been known to compromise the session key. Malicious group-based attacks are possible in such a protocol.

Steiner–Tsudik–Waidner Protocol

Steiner et al. proposed the STW protocol for P2P key exchange protocols, namely, IKA.1 [4]. The protocol, with n participants, works as follows: let G be a cyclic group on order q and α be a generator of G. The participants of the protocol are arranged in the form of a logical ring. The nth member of the group has a dedicated role of a *group controller*. The protocol consists of $n - 1$ rounds of unicast messages, collecting contributions from group members, followed by one round of broadcast messages. The protocol is presented below using the style followed in Ref. [11].

- Round i ($1 \leq i < n$)
  $U_i \rightarrow U_{[i+1]}$: $\{\alpha^{(r(1)\cdots r(j))/(r(j))}| \; j \in [1,i], \alpha^{r(1)\cdots r(i)}\}$
- Round n
  $U_n \rightarrow$ All U_i: $\{\alpha^{(r(1)\cdots r(j))/(r(j))}| \; i \in [1,n)\}$
- Upon receipt of the above, every U_i computes: $K_n = \alpha^{((r(1)\cdots r(j))/(r(j)))/r(i)} = \alpha^{(r(1)\cdots r(n))}$

The group key is computed as $\alpha^{r(1)\cdots r(n)}$, where each $r(i)$ is the short-term contribution from each group member. α is the generator of a cyclic group of prime order q and $r(i) \in Z_q$.

The penetrator can compute the session key if he is in possession of at least one of the short-term secrets. Therefore, if the intruder knows the short-term secret of a user with a probability p, and if there are n users then the probability with which he can know the session key is $1-(1-p)^n$. Therefore DPFS is $1-(1-p)^n$. It can be observed that only one of the secret keys of a participant needs to be compromised for losing the secrecy of the session key.

Kim–Perrig–Tsudik Protocol

Kim et al. proposed a tree-based group DH protocol [24], referred to as KPT. The key computation is assumed to proceed along the branches of a logical binary tree from leaves to the root with the leaves representing the users. Each user selects a random integer $r(i)$ and broadcasts $\alpha^{r(i)}$. The user $u(i)$ or its sibling $u(j)$ can calculate the key corresponding to the parent, as $\alpha^{r(i)r(j)}$ and broadcast $\alpha^{\alpha^{r(i)r(j)}}$. The process is repeated till the key corresponding to the root is computed, which will form the group key.

If the penetrator knows at least one short-term key, then the group key can be computed by the penetrator. Hence, DPFS can be calculated as $1-(1-p)^n$. The compromise of a single user's secret is sufficient to compromise the session key.

Kim Lee–Lee Protocol

Kim et al. proposed a GKA protocol [25] similar to BD in terms of message exchanges. In this protocol (referred to as KLL), the structure of the key is different from that used in BD. The participants, ID = $\{U_1, U_2, \dots U_n\}$, of the protocol are arranged in a logical ring. Each user U_i has a signature key pair $[sk(i), pk(i)]$.

In the first round, each U_i chooses a nonce $N(i)$ and $r(i)$ and computes $z(i) = g^{r(i)}$. U_n computes $H(N(n)|0)$, where H represents a hash function. U_i and U_n broadcasts the following messages:

- $U_i \rightarrow *$: $[z|ID|0]_{sk(i)}, z(i)$
- $U_n \rightarrow *$: $[H(N(n)|0)|\ z(n)\ |\ ID|0]_{sk(n)}, H(N(n)|0)$

In the second round, each U_i computes $t_i^L = H(z(i-1)^{r(i)}\ |D|\ 0)$, $t_i^R = H(z(i+1)^{r(i)}\ |D|\ 0)$, and $T_i = t_i^L \oplus t_i^R$.
U_n computes $T' = N(n) \oplus t_1^R$ and broadcasts as follows:

$U_i \rightarrow *$: $[N(i)\ |T_i|ID|\ 0]_{sk(i)}, z(i)$
$U_n \rightarrow *$: $[T'|T_n|ID|\ 0]_{sk(n)}, T'|\ T(n)$

Each U_i computes the group key as $H[N(1)\dots N(n)]$ by using equations $t_{i+1}^R = T_{i+1} \oplus t_i^R$.

It can be observed that the secrecy of the computed group key depends only on the secrecy of $N(n)$, which in turn depends only on the secrecy of $r(i)$. Hence, the DPFS for the protocol is $1-(1-p)^n$. The secrecy of the session key depends on the secrecy of all the participants' short-term keys, and one leaky member can compromise the session key.

Boyd–Nieto Protocol

The Boyd–Nieto (BN) protocol [9] is a GKA protocol, which runs in a single round, and is hence computationally efficient. The protocol is quoted from [9] as follows: Let U denote a set of n users U_1, U_2, ..., U_n and N_1, N_2, ..., N_n denote their respective nonces and $K(1)$, $K(2)$, ..., $K(n)$ denote their respective public keys. U_1 has a designated role of group controller. $S_{d(i)}(X)$ represents the signature over X using key $d(i)$. All messages are broadcast.

- $U_1 \rightarrow {}^*{:}U, S_{d(1)}[U, \{N_1\}_{K(2)}, \{N_1\}_{K(3)}, \ldots, \{N_1\}_{K(n)}]$
- $U_1 \rightarrow {}^*{:}\{N_1\}_{K(i)}$ for $2 \leq i \leq n$
- $U_i \rightarrow {}^*{:}U_i, N_i$

The group key is $K_U = h(N_1 || N_2 || N_3 || .. || N_n)$, where h is a public one-way hash function. The secrecy of the session key depends only on the secrecy of N_1. Compromise of private key of any user other than U_1 would reveal the session key as N_1 is encrypted by the public keys of participants except U_1. Hence, the protocol does not satisfy partial forward secrecy.

If the secret N_1 has a probability p_n of getting compromised, then the DPFS would be p_n. If p_n is very low, because of external factors such as policy matters, which prevent a group controller from becoming a leaky participant, then the private keys of other participants would become the vulnerable points to be considered.

Assume that the penetrator knows the private key of any participant other than group controller, $K(i)^{-1}$, with a probability p. Then the penetrator can deduce the nonce N_1 with a probability p. Because there are $n - 1$ such cases, and knowledge of at least one K_i^{-1} would reveal N_1, the probability of knowing the session key is $1-(1 - p)^{n-1}$. The DPFS of BN protocol is $\min[1-(1 - p)^{n-1}, p_n]$.

A-GDH.2 Protocol

A-GDH.2 protocol [19] is a GKA protocol that provides perfect forward secrecy, provided the short-term keys are explicitly erased. However, under the assumption of strong corruption, the short-term secrets are available to the penetrator. The protocol described below is quoted from Ref. [11].

Let $M = \{M_1 \ldots M_n\}$ be a set of n users wishing to share a key S_n. The group of users agrees on a cyclic group G and generator α. Each user M_i, $i = 1 \ldots (n - 1)$ shares a key, K_{in} with user M_n.

Let p be a prime and q a prime divisor of $p - 1$. Let G be the unique cyclic subgroup of Z_p^* on order q, and let α be a generator of G, where G and α are public. Each group member M_i is assumed to select a new secret random value $r(i) \in Z_q^*$ during each session of the protocol.

Round i ($1 \leq i < n$):

- M_i selects $r(i) \in Z_q^*$
- $M_i \rightarrow M_{i+1}$: $\{\alpha^{[r(1) \cdots r(i)]/r(j)} \mid j \in [1, i]\}$, $\alpha^{r(1) \cdots r(i)}$

Round n

- $M_n \rightarrow$ All M_i: $\{\alpha^{[r(1) \cdots r(n)]/r(i)K_{in}} \mid i \in n \, [1, n)\}$
- Upon receipt of the above, every M_i computes $S_n = \alpha^{[r(1) \cdots r(n)/}$ $^{r(i)]K_{in}K_{in-1}r(i)} = \alpha^{[r(1) \cdots r(n)]}$

The penetrator can compute the session key if the short-term key of M_n is compromised. Let the probability of compromise of the short-term key of M_n be p_n. Then the probability of loss of the session key would be p_n.

Another possibility is the compromise of a short-term key of one user and the corresponding long-term key of some other user. In that case, the probability of compromise of the session key is $1-(1-p^2)^{n-1}$. DPFS would be min $[p_n, 1-(1-p^2)^{n-1}]$.

ID-Based One-Round Authenticated GKA with Bilinear Pairing Protocol

ID-based one-round authenticated GKA with bilinear pairing (IDBP) protocol [26] is a GKA protocol based on bilinear pairing and the discrete logarithm problem. Each participant has one long-term secret and one short-term secret. The protocol description below is quoted from Ref. [26].

Let G_1 be a cyclic additive group and G_2 be a cyclic multiplicative group, both on order q. Let P be the generator of G_1. Let e: $G_1 \times G_1 \rightarrow G_2$ be a bilinear map satisfying the following properties:

- Bilinear: $\forall [(P,Q \in G_1) \wedge (a,b \in Z_q^*)]$, $e(aP,bQ) = e(P,Q)^{ab}$
- Nondegenerate: $\exists P \in G_1$, such that $e\,(P,Q) \neq 1$
- Computable: $e(P,Q)$ is computable in polynomial time.

The discrete logarithm problem over elliptic curve is assumed to be hard in G_1 and G_2.

A key generation center (KGC) chooses $s(1)$, $s(2) \in Z_p^*$ as KGC's private keys and computes $s(1)P$ and $s(2)P$ as the corresponding public keys. Let H: $\{0,1\}^* \rightarrow Z_p^*$ be a cryptographic hash function.

Let $U_1, U_2, .., U_n$ be the participants of the protocol with unique IDs, $ID(i)$. Let $I(i) = H[ID(i)]$. KGC generates users public key, $Q(i) = [I(i)s(1) + s(2)]P$ and the corresponding private key $S(i) = [I(i)s(1) + s(2)]^{-1}P$.

Each user U_i selects $a(i) \in Z_p^*$ as his short-term secret, computes $T(i, j) = a(i)\,Q(j)$, and sends to U_j. U_i computes the group key $K(i) = e(T(1, i) + T(2, i) + \ldots + T(i-1, i)$ $+ a(i)Q(i) + T(i+1, i) + \ldots + T(n, i)$ and $S(i) = e(P, P)\,[a(1) + a(2) + \ldots + a(n)]$.

There are two ways in which secrecy of session key can be compromised—penetrator's knowledge of any two long-term keys or a long-term key and the corresponding short-term key. In the first case, because there are n number of long-term keys and any two would reveal the session key, the probability of compromise of a session key would be $1-(1-p)^n - np(1-p)^{n-1}$. In the second case, because there are n long-term and corresponding short-term key pairs the probability of compromise of a session key would be $1-(1-p^2)^n$. However, compromise of one long-term key does not result in the compromise of the session key, and hence the protocol satisfies partial forward secrecy. Because $1-(1-p)^n - np(1-p)^{n-1} > 1-(1-p^2)^n$, DPFS is $1-(1-p)^n - np(1-p)^{n-1}$.

Summary of the Analysis of GKA Protocols

In the section "Analysis of GKA Protocols on Partial Forward Secrecy," the analysis of the test set of protocols, on the assumption of uniform probability of compromise of all keys, and strong corruptions, was presented.

Consider BD, JV, BW, STW, KPT, and KLL, the protocols in the test set which are based on the DH cryptographic primitive. DPFS for these protocols is summarized in Table 7.1. A plot of the probability of compromise of the session key against the number of participants (8–100) for different protocols and a given probability is shown in Figure 7.1. The maximum range of n is selected based on the typical number of participants in a P2P application, which is 100, as stated in Ref. [27]. The plots are indicative of DPFS, that is, the secrecy of the session keys for larger number of participants. The minimum has been assumed as eight, so that there is at least one member in the group under each of the four leaders for BW protocol. The probability of compromise of secrets (p) is assumed to be 0.01 for all protocols.

It can be observed that protocols BD, STW, KPT, and KLL show similar strength with respect to partial forward secrecy under the uniform probability assumption. The DPFS value for these protocols is higher compared to that of

TABLE 7.1

DPFS of DH-Based GKA Protocol

Protocol	DPFS
BD	$1-(1-p)^{n-1}$
JV	$1-(1-2p^2)^n$
BW	$1-(1-p^{(n-4)/4})^4$
STW	$1-(1-p)^n$
KPT	$1-(1-p)^n$
KLL	$1-(1-p)^n$

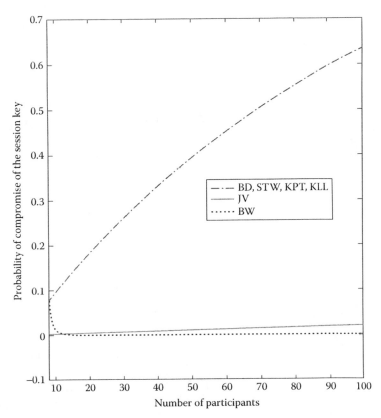

FIGURE 7.1
Probability of compromise of session key versus number of participants in DH-based protocols.

TABLE 7.2

DPFS of GKA Protocols with Different Cryptographic Primitives

Protocol	DPFS
BN	$1-(1-p)^{n-1}$
IDBP	$1-(1-p)^{n} - np(1-p)^{n-1}$
A-GDH.2	$1-(1-p^2)^{n-1}$

JV and BW, indicating that they are weaker than JV and BW. For larger values of n, BW has the least value for DPFS indicating that BW is stronger than JV.

Consider BN, IDBP, and A-GDH.2 protocols, each protocol in this set uses a different cryptographic primitive. DPFS for these protocols is summarized in Table 7.2. A plot of DPFS versus the number of users shown in Figure 7.2 reveals that A-GDH.2 protocol has the lowest value of DPFS and is hence the strongest, followed by IDBP and BN protocols, under the assumption of uniform probability of compromise of terms. The analysis demonstrates that the framework allows comparison of protocols that use different cryptographic primitives.

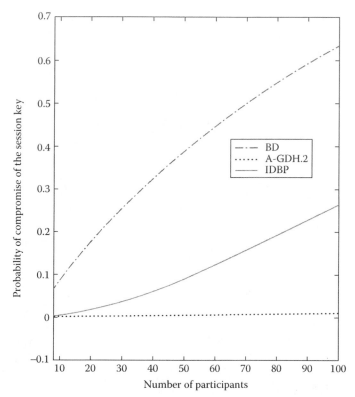

FIGURE 7.2
Probability of compromise of session key versus number of participants in different GKA protocols.

   The proposed analysis and framework pave the way for formalizing the concept of trust in group applications. Trust concerns the belief of a member on the safety of his information or transactions in the hands of other group members [28]. Therefore, the probability of loss of secrets is a logical computational paradigm to assume or assign trust. In the following section, the proposal for a trust model for P2P communications is presented and explained with three sample applications. The method in which the trust model could be used by the users of those applications to base their decisions is also illustrated.

Trust Model

The core of the trust model is the concept of *trust values for a group* as an entity, and *trust values for members* of the group individually. An *individual*, in this context, is a user, a program, an institution, or a community with a single logical membership in a group application. An individual constitutes the role of a single member, or potential member of the group. Individuals would participate in a GKA protocol, and would have a trust value dependent on the probability of leaks or compromises of keys from the individual.

The trust value of an individual would depend on several things, such as the behavioral trust [28] and the member, the nature of security mechanism at the location in which the member usually participates in and accesses information, and the network used at the user's end [28–30]. This can be assessed using underlay awareness in groups [31] and through behavioral history of the individuals [29].

In the proposed framework, the individual trust score (ITS) is a value between 0 and 100, assigned to a member, by a central trust computing application (CTCA), based on the underlying network features and the behavioral history of the individual in the past. The CTCA would be an underlay-aware [20] group application designed securely for the purpose of assessing and certifying the trustworthiness of individuals. The value is initialized when a new individual joins the set of potential members in different group applications, and is periodically revised by the CTCA. The score is in the form of a certificate signed by the CTCA, whose public key is known to all individuals. It represents the probability of not compromising keys known to that individual through the individual, that is, $(1 - p) \times 100\%$, where p is the probability of a compromise or leak.

The GTS is computed by the members of the group and published by the group members, and is verifiable by all individuals who know the group key exchange protocol used by the group. The trust value is the probability that the session keys used by the group members would not be compromised, and is expressed as a percentage. It can be computed by the group members using the DPFS algorithms, outlined in the previous section, that is, the framework for computing the DPFS. Thus $GTS = (1 - DPFS) \times 100$.

GTS could be utilized by individuals to take decisions such as joining or leaving a group and renewing subscriptions or choosing not to renew them, among others. For example, a new individual may choose not to join a group if the GTS of a group is below a desired trust threshold. The individual may also decide to leave a group whose GTS falls below the threshold. Similarly, the individual may take a decision to renew a subscription to a group publication only if the group trust value meets the expected threshold levels.

The group members may also use ITS to take decisions regarding an individual. For example, an individual who is already a member of a group, or wants to join the group, and whose trust certificate shows a lesser trust score than an expected threshold, may not be allowed to join the group, or allowed to continue as a member, or allowed to do tasks such as renewing subscriptions. A user may be asked to leave, that is, kept out of session key computations, if the ITS falls below expected thresholds, or the existing group threshold. In this manner, in a dynamic P2P application, users would be able to use trust values to take their decisions judiciously. Such decisions may have repercussions that might entail computational cost. For example, after a member is asked to leave, or when a new member joins, the key may have to be recomputed, depending on the application. To illustrate the above points, and analyze the working of the trust-based scheme, consider three different yet simple P2P-based applications, which would have different dynamic behaviors, and therefore pose different requirements or trust expectations from users and groups. These applications are described in the following section.

Applications Based on P2P Communications

This section presents three simple examples of group application requiring confidentiality and secrecy with varying dynamic behavior. The first application in the set is a cloud-based article collection, that is, a *subscribed journal* application. In this application, individuals from a group of members individually contribute articles (signed by them) to a common collection hosted on a cloud. The journal may be published once in a given time period, for example, once in a month. All the members in the group would have access to the journal, that is, the current month's set of articles, using a common session key.

The second example is an *interactive whiteboard*, an application in which all the members involved in the preparation of a common document update and edit the document, with a session key generated for each document that a particular group generates.

The third application is a *boardroom meeting*, in which the members of a committee or board discuss the items on their agenda and arrive at decisions. Their discussion may entail voting, anonymous or open, or interactions with invited individuals for advice, comments, interrogation, or any other relevant purpose. The statements made by the members have to remain confidential, whereas the minutes of the meeting or proceedings passed by the committee would be released as an open document. These applications requirements are discussed in detail next.

Cloud-Based Subscribed Journal

In a cloud-based subscribed group journal, a unit of stored information, such as the set of articles contributed in a month, is to be allowed access to by the members of the group. In the beginning of the time period, the group members (subscribers and contributors) may agree on a common key. The contributors would encrypt their article with the session key, sign it, and store it on the cloud. The subscribers would access these from the cloud and decrypt using the same session key. As long as there are no changes in membership, the group could continue to use the same session key for every new edition.

But the joining of a new member or leaving of an old member would call for change of keys in the following manner.

Joining the group

Assuming that the new member's access has to be restricted to publications after the member's joining, the group would have to work out a new key for the month, by running the GKA protocol in use. All the new articles contributed would have to be encrypted with the new session key. This may give rise to the necessity of storing session keys by the individual members, as the monthly subscriptions they view may each have a different session key. The assumption of loss of session keys, that is, strong corruptions would be justified in this application, due to the possible storage of session keys.

Leaving the group

Assuming that members may decide to leave the groups due to various reasons, a new key would have to be generated for the succeeding month if a member decides to leave. This would ensure that subsequent subscriptions would not be available to the member who left, but the old subscriptions would still be available to that individual.

To further illustrate the role of trust here, let us consider a domain for this application, for example, assume there is a set of researchers in a particular field working on related areas, who informally share their research ideas and results in stages much earlier than their formal publication. New members would choose to join the group only if the group has a convincing group trust value, otherwise the individual would be concerned about losing the secrecy of his/her work through leakage from other group members. The decision taken by the new member would be based on behavioral trust [28] of the other members which he/she may assess based on aspects such as professional reputation and affiliation. Apart from the behavioral trust, he may be concerned with issues such as whether a single member's compromise, or negligence, or insecure computing practices may lead to loss of secrecy of his/her work, and whether there are members who are likely to do that. Hence, the ITS of the members and the GTS would be useful for the prospective new member to take his/her decision.

On the other hand, the new individual's trust score would help the group to decide whether to let the individual join as a member or not. Once the new member joins, the GKA protocol would be executed again to get new session keys, and the old session keys would not be revealed to the old member. The old publications, hence, would also not be available to the new member by default.

In the same manner, if the GTS falls below a particular point, an individual could choose to leave the group. Alternatively, if an individual's trust score falls too low, perhaps because he is in the habit of accessing from insecure, remote locations, giving rise to a low ITS for the individual, the group may decide to exclude a member.

Interactive Whiteboard

The set of individuals who collaborate on a document is likely to be known before the starting of the document in the *interactive whiteboard* application. The preparation of each document would require a new session key to be generated and used by all the members.

Joining the Group

New members joining the group would be mainly by invitation, that is, when the original group feels that some individual's expertise is necessary for preparation of the document, they may choose to invite that member into the group. The new member would require access to the whole document and its history, and not just new access. This could be done in two ways, either a new session key be generated, and the whole document and its history encrypted with the new session key. This would be a secure method as it does not involve transmission of keys. Alternatively, the session key could be informed to the new individual, and the document be kept as it was. This would bring down the trust value of the group to $(1 - p_n) \times 100$, where p_n is the probability of loss of private key of the new member. Hence, the first method is more likely to be the accepted one, unless the new member's private key has a very low probability of leakage (corresponds to a highly trusted individual). However, it may be recalled that the new member was an invited one, so he/she/it is likely to have a very high trust score. Under such circumstances, the computational complexity of the GKA agreement may be a decisive factor in the choice of adopting either of the methods illustrated earlier.

Leaving the Group

Members need not leave the group, as they would be members of the group till the task is over. However, if the ITS of an individual falls below a threshold, the group may decide to exclude the member, calling for the generation of a new key. The trust score of the individual may fall for various reasons, for example, he/she was in the habit of accessing from insecure, remote locations.

Here, individual and GTSs could be used by the group members to decide which protocol to use for key computation. The decision to join, on the part of the invited members, would be aided by GTSs and ITSs. Moreover, ITSs could be used by the group to exclude probable compromisers and leakers.

Boardroom Meetings

In the *boardroom meeting* application, each meeting can be considered as a session. The members would be known beforehand, and they would plan to meet for a particular duration of time, discussing the items on the agenda, referring documents, if necessary, as a part of their discussions. The discussion and the documents could be text or multimedia. The session key would be agreed upon, using a GKA protocol before the meeting. All the comments and other elements of the discussion would be broadcast using the session keys for encryption, and received using the same session key for decryption. In board room meetings, the discussions are to be confidential, and opinions expressed by the members have to be kept secret. They could have additional features such as polling, either anonymous or named, sub-group discussions or dissents. Only the final minutes, agreed upon by all the members, would be passed on to public domain.

Joining and Leaving the Group

The number of members are fixed in such an application. However, the members may call for the inclusion of other individuals for limited interaction such as advice, interrogation, or report presentation, and these members are required to be present in the group for varying amounts of time. For these subsets of the total time interval, different session keys may be required, and after the leaving by the new member, the original key could be resumed. Moreover, purposes such as subgroup discussion and polling may call for newer keys for limited periods of time.

As the members become regular, known members of the boardroom, their trust scores are likely to be very high. When new members are called, they are invited for a purpose. They have to be allowed to join, irrespective of their trust scores. However, depending on the application, members may be allowed to take the option of not joining the boardroom. If they are allowed to avail the option, they could base their decisions on the trust value of the group. The group could use trust values to decide whether a member should be included into the meeting or not, and whether a new key needs to be computed or not for interaction with that member.

The section "Trust-Based Analysis of Protocol Suitability to Applications" provides an analysis of the suitability of the protocols, analyzed in the section "Analysis of GKA Protocols on Partial Forward Secrecy," for use in group applications described in this section, based on the DPFS, and the computability requirements in the protocol.

Trust-Based Analysis of Protocol Suitability to Applications

The nine protocols discussed in this chapter, and analyzed in the section "Analysis of GKA Protocols on Partial Forward Secrecy," namely, BD, JV, BW, STW, KPT, KLL, BN, IDBP, and A-GDH.2 protocols, are assessed here on their suitability to the three group-based applications discussed in the section "Summary of the Analysis of GKA Protocols," which are the *subscribed journal, interactive whiteboard*, and *boardroom meetings*. The parameters for assessment are as follows:

1. The trust strength τ, the ratio of the maximum GTS to the lowest ITS
2. The numbers of rounds, ρ, required for computing one session key

Trust strength, τ, would determine whether high GTSs, as desired by the application, could be achieved even in the presence of individuals with low trust scores. For a protocol in which the session key could be compromised due to the presence of one leaky, or low trust member, this ratio would be low, whereas it would be high for a protocol in which the trust value for a group, that is, the probability of not compromising the session key remains high despite the presence of a leaky member. *Trust strength* is an indicator of the strength or the robustness of the trust in the event of leaky members joining the group.

The number of rounds, ρ, would determine the implications of the use of the protocol in a dynamic scenario, where the members leave and join during sessions, and possibly require new keys to be generated.

The analysis of the trust strength and the number of rounds for each protocol in the test set is presented next, with resulting comments on the suitability of the protocol three P2P applications in the set of applications.

Burmester–Desmedt Protocol

In BD protocol [21], each user computes the common session key which has the form $\alpha^{r(1)r(2)+r(2)r(3)+ \ldots r(n)r(1)}$, where $r(i)$ is the short-term key of user i. The penetrator can compute the session key if the short-term key, that is, $r(i)$ of at least one user is known. The maximum trust score would be obtained if all the members except one had ITSs of 100%, making the trust strength τ equal 1, as $t/t = 1$, where t is the trust score of the least-trusted member. The number of rounds, ρ, is 2.

For *subscribed journal*, n, the number of members is possibly large. In addition, as users could be very different and with different backgrounds, a large number of low trust individuals are likely to be there, making τ low, less than 1.

Here, the number of rounds is just two, but recomputations need not be frequent, as there would be an application-specific time period (a month in our example) to allow joining and leaving. Hence, the computational advantage of less number of rounds for key agreement is not very significant for this application.

To sum up, BD can give rise to low group trust values, and hence is not recommended for the *subscribed journal* application.

For *interactive whiteboard*, the number of users is likely to be less, and the members would be cooperating closely to create the document. It can be expected that individuals in this group application are likely to be more trusted ones. Therefore, the actual GTS is more likely to reach high values closer to the maximum. Moreover recomputations could be frequent, as new members may join or leave frequently. Thus, two rounds of computation would make key generation easier and prove more suitable to this dynamic application.

Considering the two aspects discussed, BD would be a recommended protocol for *interactive whiteboard* application. In the *boardroom meeting* application, the initial members are likely to have high trust values, resulting in high τ, but the members who may be invited for advice, comments, interrogation, or similar reasons may have varying trust scores. Moreover, their trust values may be too low, but they have to be invited due to the essentiality of their presence. If low trust members are present, τ may fall, but we can observe that such members are likely to be just one or two at a time, and much less in comparison to the highly trusted regular members. Hence, τ tends to retain higher values, keeping the GTS high as well. But the high trust requirement of the application and possibly low DPFS are not recommended qualities in this application.

On the other hand, low ρ ($= 2$) makes frequent key computations possible, which is also an important requirement in the *boardroom meeting* application.

Considering both aspects, BD would be a weakly recommended protocol for *boardroom meetings*.

Just–Vaudenay Protocol

In JV protocol [22], based on DH primitive, in the first stage, two adjacent members engage in a key exchange protocol and establish a shared key. In the second stage, each member broadcasts $W_i = K_i/K_{i-1}$, where K_i is the key shared between ith and $i + 1$th user. Each user computes the group key as $K_{i-1}{}^n W_i{}^{n-1} W_{i+1}{}^{n-2} \ldots W_{i-2} = K_1 K_2 \ldots K_n$.

In this protocol, a penetrator can compute the key exchanged between a pair of users if he knows both the long- and short-term keys of at least one of the n users. Hence, the probability of knowing the session key is $1-(1 - 2p^2)^n$ and so DPFS is $1-(1 - 2p^2)^n$. The number of rounds is two.

If there is one weak (low trust) participant, then both his keys can get compromised. Hence, τ would be 1.

For *subscribed journal*, the possibility of compromise of keys due to one compromised individual is not a desired property. Hence, even though the number of rounds is two, it is not recommended.

For *interactive whiteboard*, $\rho = 2$ is a desirable property as frequent key recomputations would be rendered possible. Group trust values may be typically smaller than BD, as lower trust values for any participant can decrease the trust value immensely below 1. But, as discussed earlier, invited members to this application are likely to have a high trust score. Considering both aspects, JV would be recommended for interactive whiteboard.

In the *boardroom meeting* application, where the regular members have high trust values, JV would provide a high GTS, as there would be only one user with a lower trust score, and the chances of the *long-term* key of even a leaky participant getting compromised are probably small.

Recomputation of keys is easy as the number of rounds is two. Considering the two aspects, JV is a weakly recommended protocol for *boardroom meeting*.

Becker–Willie Protocol

The BW (Octopus) protocol [23] is a DH-based protocol. Participants are divided into four groups $I(A)$, $I(B)$, $I(C)$, and $I(D)$ with A, B, C, and D as group leaders. In the first round A, B, C, and D perform a DH key exchange with each of their respective group members. Group member $P(i)$ resulting in a key $k(i)$. In the second round, A performs a DH key exchange with B resulting in the key $\alpha^{K[I(A) \cup I(B)]}$. Similarly C performs a DH key exchange with D. Then the group key is computed. The probability that the intruder knows the session key and DPFS was computed earlier as $1-(1 - p^{(n-4)/4})^4$.

We had inferred that, in this protocol, the keys of all members of a subgroup have to be known for compromising the session key. Malicious groupbased attacks are possible in such a protocol.

Here, the number of rounds is large, because though there are only three logical rounds, the first round consists of $n/4$ sequential DH exchanges for the group leaders. Hence, facilitating the computation of keys on a frequent basis is not possible with the adoption of this protocol. The number of rounds can be counted as $n/4 + 2$.

However, trust strength is very high. If all participants except one had 100% trust score, the probability of loss of session key would be 0, making GTS = 100. So, τ would typically be greater than 1 in the presence of a leaky participant.

Due to high ρ, *boardroom meeting* and *interactive whiteboard* applications would not find it suitable to use this protocol. However, for *subscribed journal*, BW is recommended, owing to two reasons. The computation of keys need not be a frequent affair, so the number of rounds for a key computation can be tolerated. In addition, trust strength is high, indicating that the protocol is strong in the face of low trust members too. The underlying reason for the high trust strength is that at least $n/4$ members have to be compromised in order to compromise the session key. Considering the two aspects, BW is a recommended protocol for *subscribed journal*.

Steiner–Tsudik–Waidner Protocol

STW or IKA.1 protocol [4] with n participants assigns the nth member of the group the dedicated role of *group controller*. The protocol consists of $n - 1$ rounds of unicast messages, collecting contributions from group members, followed by one round of broadcast message.

The penetrator can compute the session key if he/she/it is in possession of at least one of the short-term secrets. Therefore, if the intruder knows the short-term secret of a user with a probability p and if there are n users then the probability with which the intruder can know the session key is $1 - (1 - p)^n$. Thus, DPFS is $1 - (1 - p)^n$. It can be observed that only one of the secret keys of a participant needs to be compromised for losing the secrecy of the session key in this case.

Due to the large number of rounds, ρ, *interactive whiteboard* and *boardroom meeting* cannot be recommended. The dependency of the trust value of a group on a single member's short-term key makes GTS small and τ small as a consequence. Hence, the protocol is not recommended for *subscribed journal*, where n could be large, and all the members need not have comparable high trust values.

Kim–Perrig–Tsudik Protocol

KPT is a tree-based group DH protocol [24] in which the key computation proceeds along the branches of a logical binary tree from leaves to the root with the leaves representing the users. Each user selects a random integer $r(i)$ and broadcasts $\alpha^{r(i)}$. The user $u(i)$ or its sibling $u(j)$ can calculate the key corresponding to the parent, as $\alpha^{r(i)r(j)}$ and broadcast $\alpha^{\alpha^{r(i)r(j)}}$. The process is repeated till the key corresponding to the root is computed, which will form the group key.

If the penetrator knows at least one short-term key, then the group key can be computed by the penetrator. Hence, DPFS can be calculated as $1 - (1 - p)^n$. The compromise of a single user's secret is sufficient to compromise the session key. The protocol is not recommended for *subscribed journal*, as the trust strength is not high and single user can compromise the session key. Due to the same reason, it is not recommended in *boardroom meeting* where the trust levels of new members may be low. *Interactive whiteboard* may find the n rounds (as each node's computation is broadcast) unsuitable as frequent key computations would be rendered impossible.

Kim Lee–Lee Protocol

KLL protocol [25], similar to BD protocol in terms of message exchanges, differs in terms of key structure. There are two rounds in which every user participates.

It was observed that the secrecy of the computed group key depends only on the secrecy of a short-term key of a user. Hence, the DPFS for the protocol is $1 - (1 - p)^n$. The secrecy of the session key depends on the secrecy of all the participants' short-term keys and one leaky member can compromise the session key. The analysis for its applicability is similar to that of BD.

Boyd–Nieto Protocol

BN protocol [9] is a GKA protocol that runs in a single round and is hence computationally efficient. It has been seen earlier that the protocol did not satisfy partial forward secrecy.

If the secret N_1 has a probability p_n of getting compromised, then the DPFS would be p_n. If p_n is very low, because of external factors such as policy matters, which prevent a group controller from becoming a leaky participant, then the private keys of other participants would become the vulnerable points to be considered.

Assume that the penetrator knows the private key of any participant other than group controller, $K(i) - 1$, with a probability p. Then it would be able to deduce the nonce N_1 with a probability p. Because there are $n - 1$ such cases, and the knowledge of at least one K_{i-1} would reveal N_1, the probability of not revealing the session key is $1-(1 - p)^{n-1}$. The degree of partial forward secrecy of the BN protocol is thus computed as $(1-(1 - p)^{n-1}, p_n)$.

As the group controller would be a highly trusted member, the probability of losing N_1 would be small. However, dependency on losing the secret of just one member makes it unsuitable for *subscribed journal* type of an application. It is recommended for *interactive whiteboard*, where the probability of leaky member is less and computationally efficiency due to low ρ is a useful property owing to frequent key computations. For *boardroom meeting*, the protocol is weakly recommended due to its low ρ, despite low trust strength.

A-GDH.2 Protocol

A-GDH.2 protocol [19] is a GKA protocol that provides perfect forward secrecy, provided the short-term keys are explicitly erased. It requires n rounds, and is hence not recommended for *interactive whiteboard* and *boardroom meeting* applications, where frequent key computations are required. However, for *subscribed journal*, it is recommended as the probability of losing the key (i.e., a low ITS) of an important member are low, otherwise the key is compromised only if the keys of two members are compromised (two members have low trust scores).

ID-Based One-Round Authenticated GKA with Bilinear Pairing Protocol

IDBP protocol [26] is a GKA protocol based on bilinear pairing and the discrete logarithm problem. Each participant has one long-term secret and one short-term secret and there are only two rounds of computation.

There are two ways in which secrecy of the session key can be compromised—penetrator's knowledge of any two long-term keys or a long-term key and the corresponding short-term key. In the first case, because there are n long-term keys and any two would reveal the session key, the probability of compromise of session key would be $1 - (1 - p)^n - np(1 - p)^{n-1}$. In the second case, because there are n long-term and corresponding short-term key pairs, the probability of compromise of session key would be $1 - (1 - p^2)^n$. However, compromise of one long-term key does not result in the compromise of the session key and hence the protocol satisfies partial forward secrecy. In this protocol, ρ is low as there are only two rounds of computation. The trust strength is low because even if one member has a low ITS, the GTS would be less than 1.

For *subscribed journal*, the dependence on the keys of a single member is not a desirable situation. Moreover, the loss of two long-term keys has a higher probability if there are two or more less trusted members. Therefore, the protocol is not recommended for *subscribed journal*.

For *interactive whiteboard*, it is a recommended protocol, as trust values of individuals is probably higher, and loss of long-term keys is a very less probable event and low ρ is desirable in the context of recomputations of the key.

For *boardroom meetings*, it is a weakly recommended protocol as ρ is small but a single member's low trust score can bring down the GTS.

Table 7.3 provides a summary of our discussion on the applicability of the protocols in the protocol set to the applications. The abbreviations R, NR, and WR stand for recommended, not recommended, and weakly recommended, respectively.

The above analysis provides some insight into the application requirements and protocol characteristics. An important observation is that the protocol suitability for an application is not dependent on the cryptographic primitives used, when the computational time is counted by the granularity of the number of rounds. Protocols with the same cryptographic base behave differently with respect to group trust and trust strength for different applications and application scenarios.

Applications with a restricted dynamism, but high sensitivity to low trust participants, as well as a high probability of a low trust participant, such as *boardroom meeting*, are the hardest to satisfy. An actual application of this nature would probably have to use a costly, computation-intensive protocol such as the STW, to satisfy its secrecy requirements and maintain its GTS.

Applications with a possibly large number of participants, such as *subscribed journal*, many of which could probably get a dip in their trust scores, is also difficult to satisfy. A large number of participants decrease trust strength. Protocols in which one low trust member can substantially bring down the GTS would not be suitable for such applications. However, protocols with higher GTSs can be used, and the complexity of key computations is not a decisive factor.

Applications with low probability of low trust members, such as interactive *whiteboard* are easier to satisfy, despite their requirement for key recomputations. Protocols with low trust strengths could also be used, as the probability of getting low trust members is low.

This analysis brings to the fore the point that when an application is designed, the criterion for choosing GKA protocols for use in the application should not be the cryptographic technique, or the computational complexity of keys or the number of rounds alone. The choice should follow a systemic assessment of the application requirements and scenario, and the computation of the trust strength. The ITSs of the prospective users should be used, in conjunction with various GKA protocols available in the literature,

TABLE 7.3

Assessment of Suitability of GKA Protocols for the Applications in the Sample Set

	Protocol	Number of Rounds (ρ)	Minimum Number of Compromised Members Tolerated	Ratio of Highest GTS to Lowest ITS (τ)	Suitability for Subscribed Journal	Suitability for Interactive Whiteboard	Suitability for Boardroom Meeting
1	BD	2	1	≤ 1	NR	R	WR
2	JV	2	1	≤ 1	NR	R	WR
3	STW	$n/4 + 2$	$n/4$	≥ 1	R	NR	NR
4	KPT	$n - 1$	1	≤ 1	NR	NR	NR
5	BW	N	1	≤ 1	NR	NR	NR
6	KLL	2	1	≤ 1	NR	R	WR
7	BN	1	1	≤ 1	NR	R	WR
8	A-GDH.2	N	1 or 2	≥ 1	R	NR	NR
9	IDBP	2	1	≤ 1	NR	R	WR

Note: NR, not recommended; R, recommended; WR, weakly recommended.

to compute the DPFS and the GTSs on using each protocol. The algorithm proposed in this framework, in conjunction with tools such as Athena [17], could be used for this purpose. The trust strength computed would indicate the appropriate protocol for the scenario. The protocol chosen should meet the computational requirements and resource constraints, as posed by the application domain and resource constraints. A procedure for computation of ITSs and a statistical approach to approximating ITSs at application design time are topics for further research.

Related Work

A model for analyzing the secrecy of the session key of A-GDH.2 protocol was proposed in Ref. [11]. The model establishes the secrecy of a term by checking the consistency of a collection of linear equations. The model can also be used to analyze forward secrecy by assessing the secrecy of the session key that has included the participants' secrets in intruder information.

Another model for analyzing the property of forward secrecy of protocols was proposed in Ref. [12]. The protocol is modeled as a set of messages, a set of events operating on messages, and set of traces, where a trace is a sequence of events. A message can be generated in a protocol from an initial set of messages, if a trace from the initial set of messages to the message exists. Therefore, a message remains a secret if there is no trace of the penetrator to the message from the initial information. The protocol satisfies forward secrecy if there are no traces for the current session of protocol such that a secret message can be generated from the initial set of messages.

The model uses rank functions to assign positive ranks to all messages that are not secret and zero otherwise. Honest participants and events of a sound protocol are expected to generate only positive messages from positive set of messages. To verify forward secrecy, the rank of zero is assigned to the past session key in addition to the secrets of the current session. Therefore, a group protocol satisfies forward secrecy if all the messages that can result from all possible traces of the current session using the messages known to the penetrator have positive ranks.

A communicating sequential processes (CSPs)-based model for analyzing forward secrecy was proposed in Ref. [13]. The role of each user is defined in the CSP model. The processes are described in terms of events. Trace of a process is a finite sequence of events in which the process has engaged up to some moment in time. A specification is a predicate on the traces of a process. To model forward secrecy, an event *leak* corresponding to revealing the long-term key is defined. The time instant at which the leak event occurs is defined as the number of events in the trace before leak plus 1.

According to Herzog [13], the practice of assigning positive and non-positive ranks for messages that are public and private, respectively, is not applicable while analyzing forward secrecy. This is because long-term keys must remain a secret until they are leaked and thereafter become public. Therefore, the rank of long-term keys must remain nonpositive till the leak event and thereafter become positive.

To overcome this difficulty, Herzog [13] proposes temporal ranks. The messages that must remain secret are assigned rank ∞. The initial information of the intruder is assigned a rank 0. All the information that the intruder gains before the leak event is assigned a rank $0 < t < n$. The terms that are leaked will have a rank n. The rank of a message corresponds to the time at which the message can be generated by the intruder.

All of the above models are concerned with verifying whether a given protocol satisfies perfect forward secrecy, whereas the proposed formal framework is for comparing GKA protocols with respect to partial forward secrecy. The framework uses strand space extensions for modeling and analyzing the protocol.

Conclusions and Future Work

This chapter illustrates a computational trust model for P2P applications, on which decisions regarding the group membership and participants of group-based applications are based. In P2P applications that require security of information, transmission and storage of information is done using symmetric session keys, rather than computation-intensive public key cryptography. The leakage or compromise of these keys would result in the loss of secrecy of valuable data. Hence the security of these keys has to be ensured. The forward secrecy of protocols used for GKA, for establishing the session keys, is an indicator of the security of these keys. A formal framework for comparing GKA protocols with respect to partial forward secrecy, under strong corruption, is proposed in this chapter, and used to develop the trust model.

The proposed framework for partial forward secrecy consists of a metric, the DPFS, and algorithms to compute it. The framework is illustrated through the analysis of a set of nine protocols. Six of these protocols—BD, JV, BW, STW, KPT, and KLL use the DH cryptographic primitive. It is observed that BD, STW, KPT, and KLL protocols all have similar strength with respect to partial forward secrecy. While JV protocol is better, BW gives the best results, for large n, that is, the number of participants. The analysis gives insight into designs of GKA protocols with improved partial forward secrecy. The analyzed set also contains BN, IDBP, and A-GDH.2 protocols. It is inferred that, with respect to partial forward secrecy, A-GDH.2 protocol is the strongest, followed by IDBP and BN protocols.

The members in a P2P application have to take many dynamic decisions involving joining and leaving groups. As verifiability of the consequences of each action is not computationally feasible, a concept of trust or trustworthiness is needed for basing these decisions. This chapter proposes a trust model for P2P applications based on the behavioral and computational trust of the members, and that of the group as an entity. The trust value for individuals depends on the probability of compromise of secrets from the individual, and is assessed by a central trust certifying application, which bases its assessment on the underlying network behavior and history of the individual. The DPFS can then be computed using the framework described in this chapter. The GTS, based on the DPFS, is the probability that the session key would remain a secret. Individuals and group members could take decisions such as whether to join a group, allow a new joining, leave a group, or exclude an individual, based on the GTS and ITSs, respectively. The GTS of an application depends on the ITSs and the GKA protocol used. The stability of the GTS depends on the maximum value it can maintain in the presence of individuals with low trust values. The trust strength metric is introduced to denote this quality in an application. The trust strength solely depends on the GKA protocol used.

Nine protocols in the test set are analyzed with respect to the trust strength. The trust strength and computational requirements of a protocol would determine its suitability for a P2P application. The suitability of these protocols for three different P2P applications, having different requirements, was analyzed. The analysis results in the conclusion that the choice of a protocol has to be done using a detailed analysis of trust strength requirements and computational resources of the application. The suitability of GKA protocols to applications is independent of their cryptographic primitives.

This chapter opens several avenues for future research. The development of an application which can function as a central trust certifier to assess ITSs, in a universe of groups is an interesting research topic. Another interesting research requirement is to develop a statistical approach to compute the GTSs for applications at design time, when the actual set of users is not known. A *type-based* classification of secrets in GKA protocols is also a topic that is worth exploring.

References

1. Chockler, G. V., Keidar, I., and Vitenberg, R. "Group communication specifications: A comprehensive study." *ACM Computing Surveys*, ACM, New York, December, pp. 427–469, 2001.
2. Leibnitz, K., Hoßfeld, T., Wakamiya, N., and Murata, M. "Peer-to-peer vs. client/server: Reliability and efficiency of a content distribution service." *Proceedings of the 20th International Teletraffic Conference on Managing Traffic Performance in Converged Networks*, Springer-Verlag, Berlin/Heidelberg, June 17, pp. 1161–1172, 2007.

3. Amir, Y., Kim, Y., Nita-Rotaru, C., and Tsudik, G. "On the performance of group key agreement protocols." *Proceedings of the 22nd International Conference on Distributed Computing Systems*, IEEE Computer Society Press, July, pp. 463–464, 2002.

4. Steiner, M., Tsudik, G., and Waidner, M. "Key agreement in dynamic peer groups." *IEEE Transactions on Parallel and Distributed Systems*, 769–780, 2000.

5. Manulis, M. "Survey on security requirements and models for group key exchange." *Technical Report TR-HGI-2006-002*, Horst Gortz Institute for IT Security, Bochum, Germany, 2008.

6. Menezes, A. J., van Oorschot, P. C., and Vanstone, S. A. *Handbook of Applied Cryptography*. CRC Press, 1996.

7. Boyd, C. and Mathuria, A. *Protocols for Authentication and Key Establishment.* Maurer, U. and Rivest, R. L., eds., Springer-Verlag, Berlin, 2003.

8. Bresson, E. and Manulis, M. "Securing group key exchange against strong corruptions and key registration attacks." *Proceedings of ACM Symposium on Information, Computer and Security (ASIACCS)*, ACM, New York, March, pp. 249–260, 2008.

9. Boyd, C. and Nieto, J. M. G. "Round-optimal contributory conference key agreement." *Lecture Notes in Computer Science*, 2567: 161–174, 2003.

10. Pereira, O. "Modelling and security analysis of authenticated group key agreement protocols." PhD Thesis, Universite Catholique de Louvain, Belgium, 2003.

11. Gawanmeh, A. and Tahar, S. "Rank theorems for forward secrecy in group key management protocols." *Proceedings of the 21st International Conference on Advanced Information Networking and Applications Workshop*, IEEE Computer Society Press, Los Alamitos, CA, May, pp. 18–23, 2007.

12. Delicata, R. and Schneider, S. "Temporal rank functions for forward secrecy." *Proceedings of the 18th IEEE Computer Security Foundations Workshop (CSFW)*, IEEE Computer Society, Washington, DC, June, pp. 126–139, 2005.

13. Herzog, J. C. "The Diffie-Hellman key-agreement scheme in the strand-space model." *Proceedings of the 16th IEEE Computer Security Foundations Workshop*, IEEE Computer Society Press, Los Alamitos, CA, June, 2003.

14. Fabrega, F. J. T., Herzog, J. C., and Guttman, J. D. "Honest ideals on strand spaces." *Proceedings of the 11th IEEE Computer Security Foundations Workshop*, IEEE Computer Society Press, Rockport, MA, June, pp. 66–78, 1998.

15. Fabrega, F. J. T., Herzog, J. C., and Guttman, J. D. "Strand spaces: Proving security protocols correct." *Journal of Computer Security*, 7(2/3): 191–230, 1999.

16. Song, D. "Athena: A new efficient model checker for security protocol analysis." *Proceedings of the 12th IEEE Computer Security Foundations Workshop*, IEEE Computer Society, Washington, DC, June, pp. 192–202, 1999.

17. Joy, S. P. and Chandran, P. "A formal framework for comparing group key agreement protocols with partial forward secrecy." *Proceedings of the 4th International Conference on Internet Multimedia Services Architecture and Application (IMSAA)*, IEEE, December, pp. 1–6, 2010.

18. Ateniese, G., Steiner, M., and Tsudik, G. "New multiparty authentication services and key agreement protocols." *IEEE Journal on Selected Areas in Communications*, 18(4): 628–639, 2000.

19. Burmester, M. and Desmedt, Y. "A secure and efficient conference key distribution system." *Lecture Notes in Computer Science*, Vol. 1189, pp. 119–129, Springer, Berlin/Heidelberg, 1994.

20. Just, M. and Vaudenay, S. "Authenticated multi-party key agreement." *Proceedings of Advances in Cryptology (ASIACRYPT)*, Springer, Berlin/Heidelberg, November, pp. 36–49, 1996.
21. Becker, K. and Wille, U. "Communication complexity of group key distribution." *Proceedings of the 5th ACM Conference on Computer and Communications Security*, ACM, New York, November, pp. 1–6, 1998.
22. Kim, Y., Perrig, A., and Tsudik, G. "Tree-based group key agreement." *ACM Transactions on Information and System Security*, 7: 60–96, 2004.
23. Kim, H. J., Lee, S. M., and Lee, D. H. "Constant-round authenticated group key exchange for dynamic groups." *Proceedings of Advances in Cryptology (ASIACRYPT)*, Springer, Berlin/Heidelberg, November, pp. 245–259, 2004.
24. Shi, Y., Chen, G., and Li, J. "ID-based one round authenticated group key agreement protocol with bilinear pairings." *Proceedings of the International Conference on Information Technology: Coding and Computing (ITCC)*, IEEE Computer Society, Washington, DC, April, pp. 757–761, 2005.
25. Bresson, E., Chevassut, O., and Pointcheval, D. "Dynamic group Diffie-Hellman key exchange under standard assumptions." *Proceedings of Advances in Cryptology EUROCRYPT, International Conference on the Theory and Applications of Cryptographic Techniques*, Springer-Verlag, Amsterdam, The Netherlands, April 28–May 2, pp. 321–336, 2002.
26. Gligor, V. and Wing, J. M. "Towards a theory of trust in networks of humans and computers." *Technical Report CMU-CyLab-11-016*, CyLab, Carnegie Mellon University, 2011.
27. Monahan, B. "Introducing ASPECT—A tool for checking protocol security." Hewlett-Packard Company Technical Report from Trusted E-Services Laboratory, HP Labs, Bristol, UK, 2002.
28. Marsh, S.P. "Formalizing trust as a computational concept." PhD Thesis, University of Stirling, Stirling, 1994.
29. Dong, P., Wang, H., and Zhang, H. "Probability-based trust management model for distributed e-commerce." *Proceedings of the International Conference on Network Infrastructure and Digital Content (IC-NIDC) IEEE*, Beijing, China, November, 2009.
30. Sun, Y. L., Yu, W., Han, Z., and Liu, J. R. "Information theoretic framework of trust modeling and evaluation for ad hoc networks." *IEEE Journal on Selected Areas in Communications*, 24(2): 305–317, 2006.
31. Abboud, O. et al. "Underlay awareness in P2P systems: Techniques and challenges." *Proceedings of the 23rd IEEE Symposium on Parallel and Distributed Computing (IPDPS)*, IEEE Computer Society Press, Washington, DC, May 2009.

8

Trust in Autonomic and Self-Organized Networks

Karuppanan Komathy

CONTENTS

Introduction

Autonomic networks are adaptive and self-organizing, and as a consequence, securing such networks is crucial. Most security schemes suggested for autonomic networks, for example, MANETs tend to build upon some fundamental assumptions regarding the trustworthiness of the participating hosts and the underlying networking system. If autonomic network is to achieve the same level of acceptance as traditional wired and wireless network infrastructures, then a framework for trust management must become an integral part of its infrastructure. The main goal of this chapter is to highlight critical issues that impinge on trust management and to detail the approaches used for establishing trust that dynamically assesses the trustworthiness of the participating nodes in the autonomic networks.

With increasing dependence on the Internet and the integration of the various computing and networked services, the disruption of Internet connectivity and the availability of networked services may have a profound impact on the lives of many individuals as well as the economic viability of businesses and organizations. Similarly, the security of nations is directly linked to the availability, survivability, and dependability of the Internet and the many Internet-based data networks. Therefore, prolonged or unpredictable unavailability of networked services is unacceptable. One of the great challenges in the twentieth century was to devise approaches to ensure that networked services can survive unexpected security and availability challenges, such as attacks, large-scale natural disasters, and faults, in a timely manner. Autonomic networks and systems, especially self-managed wireless *ad hoc* networks, powered with situational-aware proactive computing, will play an important role in meeting this challenge. Autonomic network is a network that can automatically extend, change, configure, and optimize its topology, coverage, capacity, cell size, and channel allocation, based on changes in location, traffic pattern, interference, and the situation or environment. Wireless *ad hoc* networks is a special class of autonomic networks, where capabilities or existence of links and capabilities or availabilities of nodes or network services are considered as a random function of time.

Definition of terminologies used:

- *Autonomic system.* It is a system with autonomic functions to provide self-managing capabilities such as self-configuring, self-protecting, self-healing, and self-optimizing.
- *Autonomic node.* A node that employs autonomic functions. It may operate on any layer of the networking stack. For example, routers, switches, personal computers, call managers, etc.
- *Autonomic network.* It is a collection of autonomic nodes that undertake self-management and self-optimizing policies and functions dictated by the network.

An autonomic node does not depend on external input to operate; it needs to understand its current situation and surrounding, and operate according to its current state. Autonomic nodes communicate with each other through an autonomic control framework, which provides a robust and secure communications overlay. The autonomic control framework is self-organizing and autonomic itself.

The key features of an autonomic network are the following:

- *User intervention.* It is an autonomic network, which minimizes the need for user intervention.
- *Decentralization and distribution.* The goal of autonomic networking is to minimize dependence on central elements.

- *Modularity.* It is to provide independence of function and layer.
- *Autonomy.* Autonomic functionality is independent of the function of a node.
- *Consistency.* Autonomic network seeks to improve the ability of network and services to cope with unpredicted change including changes in topology, load, task, physical and logical characteristics of the networks that can be accessed, and so forth.

Broad-ranging autonomic solutions require designers to account for a range of end-to-end issues affecting programming models, network and contextual modeling and reasoning, decentralized algorithms, trust acquisition, and maintenance. These issues, and probably their solutions, may draw many approaches and results from a surprisingly broad range of disciplines. Autonomic communications imply a stronger degree of self-management and self-optimization than is found in conventional networks that are dissociated from human intervention. To provide self-management and optimization capabilities, it is necessary to investigate the context-aware approach to improve networking properties. Software entity, network components, and software agents are used to collect context information related to the presence, location, identity, and profile of users and services. A typical context use involves locating services and users, calling-up services according to user behavior, providing information for service composition, facilitating *ad hoc* communication mechanisms between users, and adaptation of the qualities of service to changes in the environment as a result of user and service mobility [1]. Context is raw information that, when correctly interpreted, identifies the characteristics of an entity. An entity can be a person, place, device, or any object that is relevant to the interaction between a user and the services. A data model and dissemination protocol represent, store, and manage context information. A context-level agreement protocol can provide automatic context matching with the user's profile, terminal capabilities, and service requirements and offering. The primary aim of such a protocol is the adaptive distribution of context information among multiple mobile and fixed sources and destinations (e.g., devices, service components) using (negotiated) specific dissemination attributes such as power saving and cost. Context dissemination can be achieved in both *pull* and *push* modes.

Self-Organizing Wireless *Ad Hoc* Networks

A wireless *ad hoc* network is a collection of communications devices or nodes that wants to communicate without demanding on fixed infrastructure and predetermined topology of links. Individual nodes are responsible for dynamically discovering other nodes that they can directly communicate with. A key assumption is that not all nodes can directly communicate with each other, so nodes are required to relay packets of other nodes to deliver data across the network, which results in a multihop *ad hoc* network. A significant feature of *ad hoc* networks is rapid changes in connectivity and link characteristics that are introduced due to node mobility and power control practices. *Ad hoc* networks can be built around any wireless technology including infrared and radio frequency. The simplest *ad hoc* network can be seen as a wireless radio network between a collection of vehicles, ships, aircraft, or even people on foot, operating in a geographical area with no networking infrastructure. Some examples of such scenarios are cars and trucks on country highways or freeways and persons on field outings where no infrastructures are available. A robust *ad hoc* networking scheme frees the individual from the geographical constraints of the fixed network.

Technological Challenges of Wireless *Ad Hoc* Routing

The topology will vary because some nodes will move in and out of coverage of wireless network. Therefore, the number of hops between source and destination will vary as well. Network throughput will also fluctuate because the larger the number of hops, the greater the routing delay will be. When the distance between wireless nodes increases the signal-to-noise ratio (SNR) decreases and the achievable throughput is reduced. Routing problems fall into two categories: route discovery and route maintenance. Routing can also be divided into two different models: proactive routing, where the routing tables in all nodes are continuously updated; reactive routing, where the routing tables are only updated on demand. Proactive routing should be used when the node mobility is low and the utility traffic has real-time demands. Reactive routing is used when the demand for real-time transmission is low. Route discovery is only carried out when a new route is needed or when an old route is no longer in working order. Another technical challenge is scalability. In Ref. [2], Gupta and Kumar have shown a very interesting result based on a simple interference model. If there are N nodes in a bounded region, the total throughput capacity of an *ad hoc* wireless network grows at $\sqrt{N}$ which implies that the throughput per node decreases at $1/\sqrt{N}$. Thus, with a large number of nodes, the performance per node approaches zero. The number of route updates will also increase with the number of nodes. This is obviously a scaling problem as specified in Ref. [3].

The third challenge is the power supply of the mobile nodes using battery. Forwarding packets for routing updates or packets of other nodes may tempt users to turn off their equipment, and only turn it on when they want to communicate themselves. Solutions to make the batteries last longer are by making better batteries, making the transmitter power adjust to the length of the wireless path, and letting the equipment go into sleep mode when there is no traffic.

The most serious challenge is security. Some of the works found in the literature [3–10] have attempted security for wireless *ad hoc* networks as their primary theme. All of them describe the principles and methods for adding security to existing *ad hoc* routing protocols, where security was not an issue from the beginning. A very limiting factor today is the capacity of the terminals. When N terminals, each having a capacity C, work closely together to form an *ad hoc* network, the useful capacity CU of each terminal (because of interference) is $CU = C/\sqrt{N}$. With 100 terminals in the network, the useful capacity of each terminal will be only 10% of its original capacity. The routing protocol may use 70%–80% of this capacity if the terminals are highly mobile. To make the *ad hoc* network secure, some form of authentication is necessary, which easily can use the rest of the capacity, especially if it is based on public key infrastructure and threshold cryptography.

Security in self-organizing networks is characterized by availability, integrity, confidentiality, authenticity, and accountability. The basic challenge of maintaining security and reliability of self-organizing networks is to handle trust and have efficient and working security and networking mechanisms under ever-changing conditions in *ad hoc* networks, where nodes roam freely; communicate with one another via multihop, error-prone wireless communication; and may join, leave, or fail dynamically. Figure 8.1 illustrates the challenges that an autonomic architecture faces.

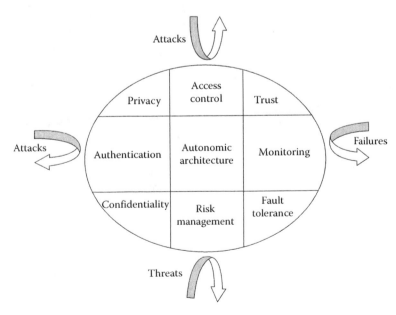

FIGURE 8.1
Attacks and threats to autonomic architecture.

Graphical Algebraic Interpretation of Trust Establishment in Autonomic Networks

Trust estimation in resource-constraint networks such as mobile *ad hoc* networks (MANETs), sensor networks, vehicular *ad hoc* networks (VANETs), and ubiquitous computing systems is significantly more complex than in traditional networks such as the Internet and wireless LANs with base stations and access points. Autonomic networks, which have neither preestablished infrastructures, nor centralized control servers, depend on their peers with which they form the network. The salient features of distributed trust management in MANETs against traditional and centralized networks are listed below:

- *Uncertainty and incompleteness.* Trust value ranges between −1 and 1, where −1 stands for complete distrust and 1 for complete trust. Trust evidence is provided by peers, which can be incomplete and even incorrect.

- *Locality.* Trust information is focused locally. Global exchange of trust values requires high communication costs as fast network changes lead to frequent updating of trust values.

- *Distributed computation.* Each node performs trust evaluation independently.

However, distributed trust computation offers several advantages. Locality property saves network resources such as power, bandwidth, computations, and storage. It avoids single point of failure as well. Moreover, autonomic networks are dynamic with frequent topology and membership changes and distributed trust computation involves contact with only a few and easy-to-reach nodes. In other words, trust evaluation is confined to information provided by directly connected nodes, that is, it is based on local interactions. Indirect relations between two entities that have not previously interacted is done by using direct trust relations that intermediate nodes have with each other. Several research works on trust computation are based on relations with one-hop neighbors. In Ref. [11], direct observations are exchanged between neighboring nodes. Assume that node A receives its neighbors' opinions about X node in the network. Node A merges its neighbors' opinions if they are close to the opinion of A on that node. This paper provides a method to link nodes' trustworthiness with the quality of the data they provide. Jiang and Baras [12] study the inference of trust value rather than generation of direct trust using Eigen Trust values introduced by Kamvar et al. [13]. In Eigen Trust, neighbors' opinions are weighted by the trust place on them:

$$t_{ik} = \sum_j c_{jk}\, c_{jk}, \qquad (8.1)$$

where:

t_{ik} means i's local trust value computed for node k from neighbors j

Trust values are normalized to satisfy $\Sigma_j c_{ij} = 1$. To address the adversary collusion problem, they assume that there are peers in the network that can be pretrusted. It has been proposed in Refs. [14,15] that similar algorithms evaluate trust by combining opinions from a selected group of users. One possible selection for the selected users is the one-hop neighbors. In addition to the above works, an extensive amount of research has focused on designing decentralized trust protocols, such as given in Refs. [16,17]. Formally, all these protocols categorize trust information into direct trust and recommendations and are evaluated by aggregating trust opinions along a particular path.

Jiang and Baras [12] framed the local interaction rule using algebraic graph theory and provide a theoretical justification for network management that facilitates trust propagation. The model employs direct trust connections among nodes as a directed graph (*digraph*) $G(V,E)$, called the *trust graph*. The nodes of the graph are the users/entities in the network.

Suppose that the number of nodes in the network is N, that is, $|V| = N$ and nodes are labeled with indices $\{0, 1,\ldots, N-1\}$. A directed arc from node i to node j, denoted as (i, j), corresponds to the *trust relation* that entity I, also referred to as *trustor*, has on entity j, also referred to as *trustee*. Each arc also comes with a weight called *confidence value*. The weight function c_{ij}: $V \times V \to W$, where $W = [-1, 1]$ represents the degree of belief i has on j. In a digraph, arcs joining two nodes in the same direction are called *parallel arcs*, and a *loop* is an arc that joins a node to itself. Assume the trust graph is a simple graph without parallel arcs or loops because a trust relation is considered to be between two distinct entities and it is unique within a given context. The set of neighbors for the node i is given by the following equation:

$$N_i = \{j | (i, j) \in\} \subseteq \{0,\ldots, N-1\}/\{i\}. \tag{8.2}$$

Figure 8.2 illustrates an example of a trust graph. For a homogenous distributed network, all nodes are treated equal. If node i wants to estimate the trustworthiness of node j, it is required to aggregate all its neighbors' opinions on j, which is the trust relations the neighbors have on j. It could be interpreted as the following general rule:

$$t_{ij} = f(c_{ij}, \{(c_{ik}, t_{kj}) | \forall k \in N_i, k \neq j, \text{and } c_{ik} > 0\}). \tag{8.3}$$

The mandatory rule is that we do not consider the opinion from the target node itself. The function $f(.)$ should satisfy the following properties:

- Nodes trust themselves, that is, $t_{ii} = 1$.
- $-1 \leq f(.) \leq 1$, because our trust value is in the range of -1 to 1.
- Opinions from nodes with high confidence values are more realistic, so they should carry larger weights.

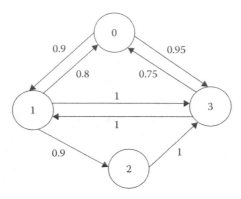

FIGURE 8.2
An example of a trust graph.

Trust and Distrust

Distrust is usually ignored in most work on trust management. However, the real world considers distrust to be at least as important as trust. In the case of networks with adversaries, systems should pay even more attention on distrust information than on trust information, because the damages made by adversaries are usually so severe that the network may be wholly shut down. In the absence of distrust, it is not clear whether a trust value of 0 means distrust or "no opinion." Jiang and Baras [12] explicitly set trust values between −1 and 1 having distrust marked as −1 and "no opinion" as 0. Modeling distrust as negative trust raises several challenges, for instance, if it is required to combine distrusts in a trust chain. Guha et al. in Ref. [18] gave three models of propagation of trust and distrust: *trust only*, where distrust is completely ignored; *one-step distrust*, where when i distrust j, i distrust all opinions made by j, thus, distrust propagates only one single step; and *propagated distrust*, where trust and distrust propagate together. Experimental results have shown that one-step distrust propagation performs better when compared to others. Therefore, this graph model also incorporates distrust using the one-step distrust model. Thus, if $c_{ij} < 0$, i will not ask any opinion from j, whereas the negative trust i has on j influences others' trust evaluation on j. Then, the general evaluation rule given in Equation 8.3 has been modified to incorporate distrust as follows:

$$t_{ij} = f(c_{ij}, \{(c_{ik}, t_{kj}) \mid \forall k \in N_i, k \neq j, \text{ and } c_{ik} > 0\}). \tag{8.4}$$

It is understood that if t_{kj} falls as negative, it represents distrust opinion.

Weighted Voting

In this section, a simple evaluation rule called *simple voting* rule for evaluating confidence value is presented. Let us assume that direct confidence values are fixed, that is, for any time $\tau \geq 0$, $c_{ik}(\tau) = c_{ik}$, and it is not true always because nodes are always willing to adjust their opinions based on new information. This section is concerned with the convergence of evaluation rule. By varying the voting values, the convergence time would vary, but eventually trust value converges to the same steady state, given that opinions will be fixed finally. $t_{ij}(\tau)$ is used as the trust value of i on j at time τ. Then

$$t_{ij}(\tau) = \begin{cases} 1 & \text{if } i = j \\ \sum_{k \in Ni, k \neq j} c_{ik}^+ t_{kj}(\tau - 1)/z_i & \text{if } i \neq j \end{cases}, \quad (8.5)$$

where:

$$c_{ij}^+ = \begin{cases} c_{ij} & \text{if } c_{ij} > 0 \\ 0 & \text{otherwise} \end{cases}$$

because only opinions from neighbors with positive confidence are considered, and

$$z_i = \sum_{k \in N_i} c_{ik}^+.$$

In the voting rule, node k, a neighbor of i, votes for target j with the current trust value it has t_{kj} $(\tau - 1)$. Node i combines all those votes using weighted average, where weights are equal to i's direct confidence values on its neighbors. Observe that the computation of t_{ij} is independent of the trust values on any other nodes. The only values that matter are trust values on j and direct confidence values. Let us evaluate trust on a particular node, say node 0. Let us define an $(N - 1) \times (N - 1)$ matrix, $c^{+0} = \{c_{ij}^+\}$, $i, j = 1, \ldots, N - 1$, and the trust vector on 0 as $T^{(0)} = [t_{10}, t_{20}, \ldots, t_{(N-1)0}]$. Then Equation 8.5 is rewritten in matrix format as follows:

$$T^{(0)}(\tau) = (Z^{(0)})^{-1} C^{+(0)} \cdot T^{(0)}(\tau - 1),$$

where:

$$Z(0) = \text{diag}[z_1, \ldots, z_{N-1}].$$

The above equation works for evaluation on other nodes as well. The equation can be used by omitting the index 0 as below:

$$T(\tau) = Z^{-1} C^+ \cdot T(\tau - 1). \quad (8.6)$$

As τ becomes large, the trust vector will converge to the steady state which decides trust values on the target. Let $t_{i0} = \lim_{\tau \to \infty} t_{i0}(\tau)$, denote the trust value of i on 0 at the steady state. Trust values at the steady state are called *ultimate* trust values and those at any time before convergence are called *provisional* trust values. A threshold rule is applied on the ultimate trust values. The threshold rule is defined through parameters η^- and η^+ such that a trust value is neutral between η^- and η^+, whereas trusted condition prevails when 1> trust value >η^+ and distrusted condition occurs when η^-< trust value <–1.

Diffusion of Trust within a Network

In order to evaluate a particular node trust opinions start from nodes that have direct trust relations with the target. Those opinions spread throughout the network and finally reach all the nodes in the network. Let us now consider the convergence of the system and also investigate the diffusion of trust as the system reaches steady state. Consider a virtuous network without adversaries, where all nodes behave rationally, that is, all nodes follow the predesigned evaluation rule and this case results in $C^+ = C$. The analysis given below shows that trust cannot be established under the simple voting rule.

Define a matrix $M = Z^{-1}C$. Then the state update Equation 8.6 can be written as follows:

$$T(\tau) = MT(\tau-1) = M^T T(0). \tag{8.7}$$

M can be viewed as a weighted adjacency matrix of the trust graph. It is essential to verify that M is a *stochastic* matrix. A matrix is called a stochastic matrix, if the sum of the elements of each row is 1. Because M is a stochastic matrix, the largest eigenvalue of M is 1. A trust graph is said to be strongly connected if every node can be reached from every other node following the direction of the edges. It can be shown that a graph, whose (weighted) adjacency matrix is M, is strongly connected if and only if the matrix M is irreducible. Let row vector π be the left eigenvector (a row vector) corresponding to eigenvalue 1 with $|\pi| = 1$, then $\pi M = \pi$. It could be proven by ergodicity [19] that says

$$\lim_{\tau \to \infty} M^{\tau} = \begin{pmatrix} \pi \\ \pi \\ \vdots \\ \pi \end{pmatrix}.$$

Thus $\forall i, \quad 1 \le i \le N - 1$

$$t_{i0} = \lim_{\tau \to \infty} t_{i0}(\tau) = \pi \times T(0) = \sum_{j=1}^{N-1} \pi_j t_{j0}(0). \tag{8.8}$$

Because t_{i0} is independent of i, every node reaches the same trust value at the steady state. It is seen that trust values purely depend on the initial trust vector $T(0)$. If a node is trusted by a large portion of nodes in the network at the beginning, its trust values are high at steady state. However, in distributed networks, initially only a small number of nodes have direct trust relations with the target, that is, $T(0)$ is sparse with a few nonzero entries. In addition, entries of π are usually small when N is large. For instance, consider a k-regular graph in which all arcs are with confidence value 1. Then, the left eigenvector $\pi = [1/(N - 1),...,1/(N - 1)]'$.

Because the trust graph is sparse, $k << N$, we have

$$t_{i0} = \frac{k}{N - 1} << 1.$$

The result indicates that even in a network with all neighbors' confidence values equal to 1, the trust values at steady state are very small, that is, close to 0. Thus under the simple voting rule, it is almost impossible to establish trust even when all voting values are 1. In order to overcome this problem, *headers* are introduced that are pretrusted entities with trust value 1. An entity can be considered as a header only if it has been proven to be reliable and attack-resistant. Let us define the number of headers i trusts as y_i and the algebraic average of votes provided by those y_i headers for node 0 as b_i. Let the matrix $Y = \text{diag}[y_1,..., y_{N-1}]$ and the vector $\mathbf{B} = [b_1,..., b_{N-1}]$. Then the updating rule in Equation 8.6 changes to

$$T(\tau) = (Z + Y)^{-1}\left[CT(\tau - 1) + YB\right]. \tag{8.9}$$

If the trust graph is strongly connected, that is, C is irreducible, the trust vector converges to a unique trust vector with

$$T = (Z + Y - C)^{-1}YB. \tag{8.10}$$

The trustworthiness of node 0 is a function of the number of headers and their votes. In order to increase its trust values to be above the threshold η^+, node 0 needs to gain higher trust from the headers. More interestingly, the voting scheme can be interpreted as a Markov chain on a weighted, directed graph. Each node is a state in the Markov chain with transition matrix M; suppose we are considering the trust value t_{i0}, the Markov chain starts from state i, and chooses the next state according to the transition matrix.

Define $p_{ij}^{(\tau)}$ as the probability that the Markov chain is in node j at time τ. Then the provisional trust value represents the expected trust value at current time, which is probability that the Markov chain is in node j at time τ. Then the provisional trust value represents the expected trust value at current time, which is

$$t_{i0}(\tau) = \sum_{j \in V} p_{ij}^{(\tau)} t_{j0}(0). \tag{8.11}$$

From the transition matrix, headers are actually absorbing states in the Markov chain, which have no outgoing link. Then the previous trust graph given in Figure 8.2 can be modified to represent the evaluation of node 0 as a Markov chain, as shown in Figure 8.3. The Markov chain will eventually stop in one of the absorbing states. Assume the probability of stopping in header h starting from i is q_{ih} and the trust value of h on node 0 is v_{h0}. Then the ultimate trust value is given by

$$t_{i0} = \sum_{h \in H} q_{ih} v_{h0}. \tag{8.12}$$

Markov chain can be viewed as a random walk on the digraph. Assume that a mobile agent is moving on the digraph according to the transition matrix. It tends to choose the neighbor with high transition probability as the next hop. Transition probabilities depend on confidence values. Therefore, the mobile agent has a tendency to visit nodes that are highly connected and with high confidence values. So the expected trust values take more weights on highly connected and highly trusted nodes, which usually provide relatively accurate information.

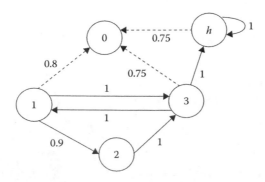

FIGURE 8.3
Markov chain for trust evaluation of node 0. The dotted lines denote direct confidence values and node h is the header which node 3 trusts.

Impact of Malicious Votes

In distributed trust establishment, the adversary also contributes to the evaluation for other nodes. Because one-step distrust is used, none will consider opinions of the adversary. There is nothing the adversary can do for trust evaluation. This is actually one advantage of the one-step distrust model. However, in a distributed scenario, an adversary may cheat to gain trust from some good nodes; then the votes provided by the adversary are also propagated into the network. The goal of adversaries is to reduce the trust value of the target as much as possible by choosing their strategies of voting and evaluation. We assume that the adversary employs the *worst-case* strategy which minimizes the trust value: it always votes −1 on good nodes and never changes its votes. In view of the Markov chain, the adversary using the worst-case strategy is also modeled as a header with $v_{m0} = -1$. Figure 8.4 shows an example, where node 1 mistakenly trusts the malicious node m with confidence value 0.2. If the trust value is computed using Equation 8.10 with

$$Z = \text{diag}[1.9, 1, 1],$$

$$Y = \text{diag}[0.2, 0, 1],$$

$$C = \begin{pmatrix} 0 & 0.9 & 1 \\ 0 & 0 & 1 \\ 1 & 0 & 0 \end{pmatrix}$$

$$B = \begin{pmatrix} -1 \\ 0 \\ 0.75 \end{pmatrix}$$

Then, we have $T = [0.45, 0.60, 0.60]'$ compared to the original $T = [0.75, 0.75, 0.75]'$ without the malicious node.

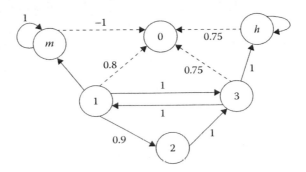

FIGURE 8.4
Markov chain for evaluation of node 0 with the presence of malicious node m. The dotted lines denote direct confidence values.

Collusion of Adversaries

Now we consider more than one, say M, adversaries are present in the network. According to the assumptions, these adversaries know each other and are able to collude in order to achieve the maximum destruction to the voting scheme. Similarly, as in the above argument, the worst-case strategy of these M adversaries is to always vote -1 for good nodes and reversely 1 for other adversaries, and they never change their votes. So adversaries are modeled as absorbing states in the digraph.

Suppose that the system is evaluating a good node, say 0. Assume that there are H headers and their votes on node 0 are denoted by v_{h0}, for an arbitrary header h. If we assume that the trust graph is ideally regular in the sense that the absorption probabilities of headers and adversaries are the same and are denoted as q, then $q(H + M) = 1$. The trust value is given by the following equation:

$$t_0 = \sum_h q v_{h0} - qM, \tag{8.13}$$

which is the same for all the evaluating nodes. Then the security impact as a function of the number of adversaries is required to be estimated. According to the threshold rule, node 0 is not trusted if $t < \eta^+$, and we have $q(\Sigma_h v_{h0} - M) < \eta^+$.

So if the number of adversaries $M > (\Sigma_h v_{h0} - \eta^+ H)/(1 + \eta^+) = M^*$, the scheme fails in assessing the trustworthiness of node 0. Therefore, M^* is the *attack-resistant threshold* for the voting scheme. In a more realistic scenario, entities with low confidence values are easy to be cheated by adversaries. Adversaries that link to low trusted entities have lower absorption probabilities, then the attack-resistant threshold is actually greater than M^*, that is, the scheme resists more adversaries. The arguments of detecting adversaries are similar with the evaluation, except that the number of adversaries as absorbing states is $(M - 1)$ instead of M, because the Mth adversary is now the evaluating target.

In this section, a trust establishment strategy based only on local interactions is formally defined. It also introduced a new notion of headers which help in distributed trust establishment. The convergence behavior of the scheme using algebraic graph theory is investigated and also interpreted as a Markov chain on a digraph. Further, the topology effects on trust spreading are discussed, which enlightens a new way of network management. Because trust management is an important component of network security, issues about trust establishment in the presence of adversaries are investigated.

Trust Credential Distribution in Autonomic Networks

Autonomic networks are networks that are self-organized, and controlled and managed in a decentralized way. Trust relations between neighbors are necessary to ensure that the transmitting messages are not leaked to the enemy. Because of the mobility and dynamics of autonomic networks, it often requires trust value to be updated. In addition, the validity of trust credentials changes with time as well as with environmental conditions. Thus, trust establishment and maintenance in autonomic networks is a much more dynamic problem than in traditional server-based networks. In autonomic networks, trust credentials are typically issued and often stored in a distributed manner. Centralized trust management where all credentials are stored in a single location is unreliable. In the case of decentralized mode, credential distribution raises many interesting questions such as what are the best places to store the credentials so that they can be easily located, well protected, and timely perceived.

After the user obtains necessary trust credentials for the target user, he applies an evaluation policy to draw conclusions about the trustworthiness of the target user. Different policies or rules have been developed and studied, such as those provided by Refs. [15,20,21]. Although choosing the right model to evaluate trust and obtaining credentials to compute trust go hand in hand, trust credential distribution is fairly independent of the specific evaluation model. Trust credentials are viewed as files to be shared for trust management. The problem of trust credential distribution derives many characteristics from distributed peer-to-peer (P2P) file sharing systems, such as those given in Refs. [22–24]. Jiang and Baras [25] proposed a trust credential distribution scheme that uses linear network coding [26] to combine credentials during transmissions. The proposed decentralized scheme used a request–response approach, which effectively handles the dynamic nature of the problem. The scheme called *network coding* has been used in P2P file sharing as well. Gkantsidis et al. [19] designed a P2P file sharing system named Avalanche, in which intermediate peers produce linear combinations of file blocks as in network coding. Acedanski et al. [27] analyzed random linear coding-based storage.

Different trust contexts require different types of credentials such as digital certificate and integrated automated fingerprint identification system (IAFIS). *Digital certificate* is issued by a certificate authority or entity and verifies that a public key is owned by a particular entity. Digital certificates are used to thwart attempts to substitute one person's key for another. Digital certificates are widely used in PGP and X.509. IAFIS uses human fingerprints as identities.

Trust value denotes the degree of trust the issuer of the credentials has on the target user. The value can be binary valued, either trust or distrust, or multivalued, such as four levels of trust in PGP [28], or even continuous in an interval, say [−1, 1]. Credentials change over time and space. Every user has confidence values on the credentials he stores. The confidence value depends on several factors such as time elapsed since the credential is issued, or communication distance taken by the credential to reach the user. When the confidence value is below certain threshold, the corresponding credential is considered to be invalid.

Network Coding-Based Scheme

In this section, we will describe the linear network coding [26] that produces efficient propagation of trust credentials and the associated operations that are decentralized. Network coding is a recent field in information theory proposed to improve the throughput utilization of a given network topology. Ahlswede et al. [29] were the first to introduce network coding. Let us introduce *trustors* that are interested in the trustworthiness of a particular user called *trustee*. The trust credentials about the trustee are initially stored in the nodes scattered throughout the network. These nodes issue the trust credentials, or retrieve these credentials from others. Trust credentials will only point to that of a single trustee, and all the operations are applied on those credentials about the trustee. Each trust credential, in the form of a digital document, has a unique ID, which is obtained by applying hash function to the document. In autonomic networks, nodes only communicate directly with a small subset of other nodes called *neighborhood* and nodes frequently check with their neighbors for new credentials. The new credentials are forwarded to the node, once it discovers them in its neighborhood. Whenever a node forwards trust credentials, it produces a linear combination of all the credentials it currently has in its storage and the combined documents it has received from its neighbors.

Consider a network as a graph $G = (V, E)$, where V is the set of nodes and E is the set of edges. Edges are denoted by $e = (v, v') \in E$, in which $v = \text{head}(e)$ and $v' = \text{tail}(e)$. The set of incoming edges is denoted as $I(v) = \{e \in E : \text{head}(e) = v\}$, and the in-degree of v is $\deg_I(v) = |I(v)|$. Similarly, the set of outgoing edges is denoted as $O(v) = \{e \in E : \text{tail}(e) = v\}$, and the out-degree of v is $\deg_O(v) = |O(v)|$. Let us denote by X_l^v, $l = 1,\dots,h$ one of the h credentials the node v stores and by $W_{e'}^v$ the combined document transmitted to v via edge e'. Consider one outgoing edge of v, edge e. The combined document transmitted via link e is defined as follows:

$$W_e = \sum_{l=1}^{h} a_{l,e} X_l^v + \sum_{e':\text{head}(e')=v} b_{e',e} W_{e'}^v. \tag{8.14}$$

All these operations take place in the finite field $\mathbf{F}_q$. The coefficient $a_{l,e}$ and $b_{e',e}$ are randomly picked from $\mathbf{F}_q$ and the data content of each trust document is represented by d elements from $\mathbf{F}_q$. Coefficient vector of a combined document W_e is defined as $C_e = [c_{e,1}, c_{e,2}, ..., c_{e,m}]$, where m is the total number of trust credentials available. According to Equation 8.14, the coefficient of credential X_v^l is $a_{l,e}$. Suppose the coefficient of a credential $X_{l'}$ in the combined document $W_v^{e'}$ is $c_{e',l'}$. Then we have that

$$c_{e,l'} = \sum_{e':\text{head}(e')=\text{tail}(e)} b_{e',e} c_{e',l'}. \qquad (8.15)$$

Figure 8.5 depicts the flow of documents from edge e' to node v and further to edge e. It is observed that given m distinct trust documents, a node can recover them after receiving m combined documents for which the associated coefficient vectors are *linearly independent* of each other. The reconstruction process is similar to solving linear equations.

Network coding-based scheme involves only local interactions. Nodes need not be aware of the existence of any trust credential and its location. They only contact their neighbors to check whether there are new credentials or new combined documents. This is an advantage as compared to any request–response scheme, such as the Freenet-based scheme [30]. Request–response schemes require routing tables. If the topology of the network changes when new nodes enter the network or some nodes move out, routing tables need to be updated immediately. Network coding scheme adapts quickly to the topology changes, because nodes only contact their current neighbors. In addition, request–response schemes require that nodes should know the document IDs before sending out requests, which requires global information exchange. Network coding scheme operates without knowing any document ID.

Heterogeneity

There are different types of trust credentials with various confidence values, or in other words, the trust credentials in the network are *heterogeneous*. A

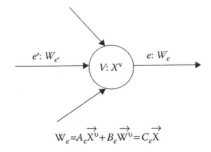

FIGURE 8.5
Flow of documents.

good trust credential distribution scheme should take such heterogeneity into consideration. Credentials with high confidence values and high importance should have high priority in transmission and they should be recovered and updated faster than ordinary credentials. In this scheme, high priority credentials are separately combined and also marked. If node v discovers either a high priority credential or a high priority credential updated at time t, it immediately sends out new combined documents to all its outgoing edges in the following ways:

1. Check the latest document node v transmitted to edge e, denoted as $W_e(t')$ at time t'.
2. Create a new combined document $W_e(t) = aX_H + bW_e(t')$, where X_H is the high priority credential. Node v transmits $W_e(t)$ to edge e immediately.

From the above, it is understood that only two linear independent coefficient vectors of $W_e(t')$ and X_H are needed to recover X_H, instead of waiting for the entire m independent coefficient vectors. Thus, high priority credentials have much higher chance to be recovered and updated. If more than one high priority credentials are discovered at node v, one combined document is created for each high priority credential. It is shown that this scheme recovers and updates high priority credentials much faster than that without this scheme.

In this section, we discuss a scheme to distribute trust documents in autonomic networks based on network coding. The scheme is highly decentralized with no global information exchange and very easy to implement in self-organized mode. Investigation of the results of the simulation study shows that the scheme is efficient and adapts quickly to network changes.

Trust Estimation in Autonomic Networks Using Statistical Mechanics Approach

Trust is broadly interpreted as a belief relationship, where an entity is confident that another peer will operate fairly, or as it is designed [31]. As pointed out in Refs. [12,14], most of the work on trust management in the literature is prevalently based on heuristics and simulation as evaluation method and almost none are implemented and tested in a real environment. Solutions are often hard to compare even on a simulative basis, because they often rely on different hypotheses and are aimed at different application scenarios. The comparison between different methods is therefore extremely difficult to accomplish, mainly because of the great simulative effort that would be required. Spin glass models such as the famous Sherrington–Kirckpatrick model [32] have already proved to be extremely versatile and valuable in understanding the global behavior of complex systems such as neural networks, whose dynamics can be modeled by the statistical mechanics of infinite-range Ising spin glasses [33]. In this section, it has been shown how statistical methods can also be used to understand the complex dynamics that arise from the local interactions of peers in decentralized networks as per the work of Ermon et al. [34].

Consider a network consisting of N nodes, represented by a directed graph $G = (V, E)$ in which each entity can communicate with a certain subset of other nodes according to an adjacency matrix A. The real trustworthiness status of each node i is represented as with a bit variable $T_i \in \{-1,1\}$, so that we collectively describe the trust status of the network with a *real trust vector* $\boldsymbol{T} \in \{-1, 1\}^N$, adopting the convention

$$T_i = \begin{cases} 1 & \text{if node } i \text{ is trustworthy} \\ -1 & \text{otherwise} \end{cases}.$$

In a real setting, the complete *real trust vector* $\boldsymbol{T}$ is unknown to the peers and nodes are usually able to judge their neighbors on the basis of the history of their previous interactions that are statistically correlated with the real trustworthiness status of the nodes. It is assumed that $\boldsymbol{T}$ does not change over time and it is related to an *opinion matrix* $C \in R^{N \times N}$ by the following equation:

$$C = f(\boldsymbol{T}, w) \, w \in \Omega, \tag{8.16}$$

where:
Ω is a sample space
$f(\cdot)$ represents the way in which opinions are formed

Each element c_{ij} of the opinion matrix C is the opinion that node i has on node j, and we assume that it is significant only if i and j are neighbors, because it is based on the history of their previous interactions. In this model, the role of a trust management algorithm is that of estimating T from C, assuming that C and the form of $f(\cdot)$ in Equation 8.16 are known.

The role of the trust estimation algorithm is to find a trust configuration $\hat{T} \in \{-1,1\}^N$ that estimates the real trust vector T. The approach used here is to search for the configuration that is more likely to have generated a certain observed opinion matrix C, or in other words the trust configuration with the highest a posteriori probability, given $\bar{C}$. The $LH(S;C)$ of any configuration S, given an opinion matrix, C is illustrated in the following equation:

$$LH(T;\bar{C}) := p(\boldsymbol{T}|\bar{C}), \tag{8.17}$$

where:
  $p(T|\bar{C})$ is the probability of T with a condition $C = \bar{C}$

According to Bayes rule,

$$p(T|\bar{C}) = \frac{p(C|T)p(T)}{p(C)},\qquad (8.18)$$

where:
  $p(T)$ is the *a priori* probability of the discrete random variable $T \in \{-1, 1\}^N$
  $p(C)$ and $p(C|T)$ are the density and conditional density of the continuous random variable $C \in R^{N \times N}$

Let us look at the symmetrical behavior of trustworthy and untrustworthy nodes. If $p(T)$ is symmetrical, the resulting likelihood function $LH(T;C)$ becomes symmetrical in T, a situation in which effectively distinguishing between the two kinds of nodes would be clearly impossible. Therefore, we will concentrate on *a priori* distributions of the real trust vector $p(T)$ that are unbalanced, that is they privilege the presence of one kind of node. Suppose that the *a priori* probability distribution is Bernoulli-distributed with parameter p, namely,

$$p := P(T_i = 1).$$

Then, assuming independence if we define $w(T) = |\{i|T = 1\}|$, then we have the following:

$$p(T) = p^{w(T)}(1 - p)^{N - w(T)} = (1 - p)^N \left(\frac{p}{1 - p}\right)^{w(T)}.\qquad (8.19)$$

Because

$$w(T) = \frac{N + \sum_i T_i}{2}.$$

Equation 8.19 becomes

$$p(T) = (1 - p)^N \left(\frac{p}{1 - p}\right)^{\frac{1}{2}(N + \Sigma_i T_i)}$$

$$= [(1 - p)p]^{N/2}\, e^{\frac{1}{2}\log\left(\frac{p}{1 - p}\right)\Sigma_i T_i}.$$

In this way we obtained that

$$p(T) = \gamma e^{-\lambda \Sigma_i T_i},$$

where:

γ normalizes to a probability distribution

$$\lambda = -\frac{1}{2}\log\left(\frac{p}{1-p}\right)$$

In case the *a priori* distribution is biased toward trustworthy nodes or the opposite for $\lambda = 0$ (or equivalently $p = .5$), we have a symmetrical case in which we cannot expect good results from the estimation.

Putting all together, we obtain the following:

$$LH(T;C) = q(C)e^{\frac{1}{\sigma^2}\Sigma_{(i,j)\in E} T_i T_j C_{ij}} \gamma e^{-\lambda \Sigma_i T_i}$$

$$= \gamma q(C)e^{\frac{1}{\sigma^2}[\Sigma_{(i,j)\in E} T_i T_j C_{ij} - \lambda\sigma^2 \Sigma_i T_i]}. \tag{8.20}$$

Therefore, from Equation 8.20, it is easy to conclude that the likelihood $LH(T;C)$ of a configuration T is proportional to a monotonic increasing function of

$$H(T) = \sum_{(i,j)\in E} T_i T_j c_{ij} - \eta\sum_i T_i, \tag{8.21}$$

where:

$\eta = \sigma^2$

$$\eta = \sigma^2 = -\frac{\sigma^2}{2}\log\left(\frac{p}{1-p}\right).$$

Therefore, we can compute a maximum likelihood estimate of the real trust vector by setting and by maximizing Equation 8.21 over all possible configurations T. Equation 8.21 is very important because it represents the energy or Hamiltonian of a configuration S in an Ising model [32] in the presence of an external magnetic field of strength η that breaks the symmetry of the system. It confirms in case *a priori* distribution of T is symmetrical, that is, $p = .5$, the magnetic field disappears and the system becomes completely symmetrical.

Trust Estimation Algorithm

A *voting rule* allows each node repeatedly to be evaluated by its neighbors. In particular, they express their opinions with a vote on its trustworthiness, and the *voting rule* takes them into consideration together with the current estimated trustworthiness status of the participants to the vote. To emulate the Metropolis–Hastings [35] algorithm, we introduce stochasticity into the rule so that we obtain the desired Markov chain structure with the proper steady-state probability distribution. At each time step, node i is chosen randomly. The trustworthiness $S_j(k+1)$ of nodes j, which are different from i, are kept constant while as in [34], the node i adopt the following voting rule for computing $S_i(k+1)$:

$$P[S_i(k+1) \mid m_i(k)] = \frac{e^{\frac{S_i(k+1)[m_i(k)-\eta]}{t(k)}}}{e^{\frac{[m_i(k)-\eta]}{t(k)}} + e^{-\frac{[m_i(k)-\eta]}{t(k)}}}, \tag{8.22}$$

where $m_i(k)$ is defined as follows:

$$m_i(k) = \sum_{j \in N_i} (c_{ij} + c_{ji}) S_j(k). \tag{8.23}$$

N_i is the set of neighbors of node i and $t(k)$ is the temperature parameter at iteration k. In this way, Markov chain with state space is obtained as $\{-1, 1\}^N$ and with transition probability $p[S,R] := P[S(k+1) = R \mid S(k) = S]$ which is equal to zero if the Hamming distance of S and R is greater than 1. If the Hamming distance of S and R is less than or equal to 1, we have that

$$p[S,R] = \frac{1}{N} \frac{e^{\frac{R_i[m_i(S)-\eta]}{t(k)}}}{e^{\frac{[m_i(S)-\eta]}{t(k)}} + e^{-\frac{[m_i(S)-\eta]}{t(k)}}}, \tag{8.24}$$

where:
  i is the index such that $S_j = R_j$ for all $j \neq i$

$$m_i(S) = \sum_{j \in N_i} (c_{ij} + c_{ji}) S_j \tag{8.25}$$

By fixing the temperature parameter t over time, the voting rule defined by Equation 8.22 is simply a modified version of the classical Metropolis–Hastings algorithm, where a Markov chain has got different transition probabilities but with the same steady-state probability distribution. The graph associated with the Markov chain is strongly connected and consists of a finite number ($2N$) of states, each one with a self-loop. Therefore, the transition matrix is primitive and the resulting chain is *regular*. Using Perron–Frobenius theorem, it is proved that a unique steady-state probability distribution exists that is reached from any initial probability distribution. Therefore, it is concluded that if $t = 1/\beta$ is fixed, then the voting rule defines a Markov chain whose steady-state probability distribution π is Boltzmann-distributed:

$$\pi T = \frac{-\beta H(T)}{Z}, \tag{8.26}$$

where:

$$Z = \sum_S e^{-\beta H(T)}$$

The choice of the *voting rule* given in Equation 8.22, however, is not fundamental because the steady-state probability distribution given in Equation 8.26 could be obtained using the standard Metropolis Algorithm, that is by choosing the following transition probability:

$$p[S, R] = \begin{cases} 1 & \text{if } \Delta U > 0 \\ e^{\beta(\Delta U)} & \text{otherwise} \end{cases}, \tag{8.27}$$

where:
$\Delta U = 2 S_i [\eta - m_i(S)]$
S and R are states with Hamming distance equal to 1
i is such that $S_j = R_j$ for all $j \neq i$ and $S_i \neq R_i$

From Equations 8.22 and 8.27, the associated Markov chain is guaranteed to converge to the steady-state probability distribution Equation 8.26. However, the convergence rate depends on the eigenvalues of the transition matrix and hence on the voting rule used.

The role of the parameter β is a coefficient that enables us to tune the difficulty of the problem, that ranges from a trivial case when $\beta = 0$ (high temperature) and a very difficult one as $\beta \to \infty$, and the system freezes concentrating all the weight on the *ground state*. In fact, even if we use algorithms such as Equations 8.22 and 8.27, to sample such a probability distribution for any value of $\beta > 0$, their convergence speed that is determined by the eigenvalues of the transition matrix to the desired steady-state probability distribution decreases as the temperature is lowered. This is due to the fact that Markov chain is ergodic for every value of β, but the degree of ergodicity decreases as β increases because downhill movement become more unlikely and therefore it is easier to get stuck in the local minima of $H(S)$. On the other hand, for small values of β there is a faster convergence, but the resulting steady-state probability distribution is noisy because of weight distribution at high temperatures.

In this section, a mathematically sound framework for trust evaluation in decentralized autonomous networks is presented. The algorithm uses the fundamental property that it is completely based on local interactions between nodes without any need for central coordination. These local interactions are characterized by several levels of randomness, both unavoidable because of uncertainty in terms of the opinions that the nodes have of their neighbors and artificially introduced by the algorithm in the voting rule. The great degree of robustness of the algorithm makes it suitable for a variety of settings in distributed networks as it ensures that to some extent the system is resilient to malfunctioning or even malicious users that try to jeopardize the system.

Summary

The autonomic computing and networking (ACN) has initiated several functional areas of an autonomic system: the autonomic system involving the autonomic properties with self* (awareness, organization, optimization, reconfiguration, regulation, healing, and protection). The current Internet is a network of interconnected and uncoordinated networks. ACN is meant to provide uninterrupted service to millions of Internet users due to its self* properties. Future Internet scenario in which services are delivered end to end in the context of value chains across interconnect enterprise, home area, access and core networks, both fixed and wireless. The essential new design paradigms have to holistically deal with the fact that the overall environmental is strongly dynamic. For example, self-healing property refers to the automatic prediction and discovery of potential failures and risks and the automatic correction to possibly avoid downtime of the computer system. In this chapter, some of the security challenges that an autonomic architecture would envisage are illustrated. The convergence behavior of the scheme using algebraic graph [12] is investigated. A scheme proposed by Jiang and Baras [25] uses distribution of trust documents in autonomic networks based on network coding, and the simulation study shows that the scheme is efficient and adapts fast to network changes. The trust model introduced by Ermon et al. [34] has elaborated about trust estimation in autonomic networks using a statistical approach.

Future networks will be complex environments of dynamic domains comprising thousands of nodes. Automaticity and trust management can provide a concrete framework toward manageability, scalability, and security of such networks. Future research works should show how these new trust frameworks are adapted to hybrid technologies and infrastructures. Performance of these frameworks under different network topologies and settings are required to be investigated and explored for the effects of mobility and

---

* Indicates the properties given in the brackets.

malicious nodes. The investigations have to reveal how resilient the new schemes are against failures and errors. Scalability is also another dimension where research will look at for hosting novel approaches and algorithms.

References

1. J. Coutaz, J. Crowley, S. Dobson, and D. Garlan. "Context is key." *Communications ACM*, 48(3): 49–53, 2005.
2. P. Gupta and P. R. Kumar. "The capacity of wireless networks." *IEEE Transactions on Information Theory*, 46(2): 388–404, 2000.
3. N. Metropolis, A. W. Rosenbluth, M. N. Rosenbluth, A. H. Teller, and E. Teller. "Equation of state calculations by fast computing machines." *The Journal of Chemical Physics*, 21(6): 1087, 1953.
4. F. Stajano and R. Anderson. "The resurrecting duckling: Security issues for ad-hoc wireless networks." *Proceedings of the 7th International Workshop on Security Protocols, Lecture Notes in Computer Science* 1796, Springer Verlag, London, pp. 172–194, April 19–21, Cambridge, 1999.
5. F. Stajano. "The resurrecting duckling—What next?" *Proceedings of the 8th International Workshop on Security Protocols, Lecture Notes in Computer Science.* April 3–6, Springer-Verlag, Berlin/Heidelberg, 2000.
6. F. Stajano and R. Anderson. "The resurrecting duckling: Security issues for ubiquitous computing." *IEEE Computer*, 35(4): 22–26, 2002.
7. T. Jiang and J. S. Baras. "Ant-based adaptive trust evidence distribution in MANET." *Proceedings of the 24th International Conference on Distributed Computing Systems*, March 23–26, Tokyo, Japan, 2004.
8. R. Ramanujan, A. Ahamad, J. Bonney, R. Hagelstrom, and K. Thurber. "Techniques for intrusion resistant ad hoc routing algorithms (TIARA)." *Proceedings of MILCOM 2000, 21st Century Military Communications Conference*, vol. 2, October 22–25, Los Angeles, CA, pp. 660–664, 2000.
9. T. Schaefer, P. Smith, M. Schoeller, A. J. Mohammad, J. P. Rohrer, D. Hutchison, and J. P. G. Sterbenz. "Towards a decision engine for self-remediating resilient networks." *Second International Workshop on Self-Organizing Systems (IWSOS)*, September 11–13, Lake District, UK, pp. 12–14, 2007.
10. L. Zhou and Z. Haas. "Securing ad hoc networks." *IEEE Network*, vol. 13, November–December, pp. 24–30, 1999.
11. S. Buchegger and J. Y. L. Boudec. "The effect of rumor spreading in reputation systems for mobile ad-hoc networks." *Proceedings of Modeling and Optimization in Mobile, Ad Hoc and Wireless Networks (WiOpt)*, March 3–5, Sophia-Antipolis, France, pp. 131–140, 2003.
12. T. Jiang and J. S. Baras. "Graph algebraic interpretation of trust establishment in autonomic networks." Downloaded from website on September 2012. http://www.isr.umd.edu/~baras/publications/papers/2005/JiangB_2005.pdf
13. S. D. Kamvar, M. T. Schlosser, and H. Garcia-Molina. "The Eigen-trust algorithm for reputation management in P2P networks." *Proceedings of the 12th International World Wide Web Conference*, May 20–24, Budapest, Hungary, pp. 640–651, 2003.

14. M. Langheinrich. "When trust does not compute-the role of trust in ubiquitous computing?" *Workshop on Privacy at UBICOMP*, October 1, ACM Press, Philadelphia, PA, 1–10, 2003.

15. G. Theodorakopoulos and J. S. Baras. "Trust evaluation in ad-hoc networks." *Proceedings of the 2004 ACM Workshop on Wireless Security (WiSe)*, ACM Press, pp. 1–10, 2004.

16. T. Beth, M. Borcherding, and B. Klein. "Valuation of trust in open networks." *Proceedings of the 3rd European Symposium on Research in Computer Security (ESORICS)*, November 7–9, Brighton, pp. 1–18, 1994.

17. J. Kong, H. Luo, K. Xu, D. L. Gu, M. Gerla, and S. Lu. "Adaptive security for multi-layer ad hoc networks." *Wireless Communications and Mobile Computing*, 2(5): 533–547, 2002.

18. R. Guha, R. Kumar, P. Raghavan, and A. Tomkins. "Propagation of trust and distrust." *Proceedings of International World Wide Web Conference*, May, New York, 2004.

19. C. Gkantsidis and P. Rodriguez. "Network coding for large scale content distribution." *Proceedings of IEEE Conference on Computer Communications Conference on Computer Communications (INFOCOM)*, vol. 4, March 13–17, Miami, FL, pp. 2235–2245, 2005.

20. M. K. Reiter and S. G. Stubblebine. "Authentication metric analysis and design." *ACM Transactions on Information and System Security*, 2(2): 138–158, 1999.

21. T. Jiang and J. S. Baras. "Trust evaluation in anarchy: A case study on autonomous networks." *Proceedings of IEEE Conference on Computer Communications Conference on Computer Communications (INFOCOM)*, April, Barcelona, Spain, 2006.

22. S. Marti and H. Garcia-Molina. "Limited reputation sharing in P2P systems." *Proceedings of the 5th ACM Conference on Electronic Commerce*, May 17–20, ACM Press, New York, pp. 91–101, 2004.

23. I. Stoica, R. Morris, D. Karger, M. F. Kaashoek, and H. Balakrishnan. "Chord: A scalable peer-to-peer lookup service for Internet applications." *Proceedings of the ACM SIGCOMM'01 Conference*, August 27–31, ACM Press, San Diego, CA, pp. 149–160, 2001.

24. L. Xiong and L. Liu. "Peer trust: Supporting reputation-based trust in peer-to-peer communities." *IEEE Transactions on Knowledge and Data Engineering*, 16(7): 843–857, 2004.

25. T. Jiang and J. S. Baras, "Trust credential distribution in autonomic networks." *Proceedings of IEEE Global Telecommunications Conference (IEEE GLOBECOM)*, November 30–December 4, New Orleans, LA, pp. 1–5, 2008.

26. S. Y. R. Li, R. W. Yeung, and N. Cai. "Linear network coding." *IEEE Transactions on Information Theory*, 49(2): 371–381, 2003.

27. S. Acedanski, S. Deb, M. Mdard, and R. Koetter. "How good is random linear coding based distributed networked storage?" *Proceedings of the 1st Workshop on Network Coding, Theory and Applications, NetCod*, April 7, Riva del Garda, Italy, 2005.

28. P. Zimmermann. *PGP User's Guide*, MIT Press, Cambridge, MA, 1994.

29. R. Ahlswede, N. Cai, S. R. Li, and R. W. Yeung. "Network information flow." *IEEE Transactions on Information Theory*, 46(4): 1204–1216, 2000.

30. I. Clarke, O. Sandberg, B. Wiley, and T. W. Hong. "Freenet: A distributed anonymous information storage and retrieval system." *Lecture Notes in Computer Science*, 2009: 46, 2001.

31. Y. Sun, Z. Han, W. Yu, and K. J. R. Liu. "A trust evaluation framework in distributed networks: Vulnerability analysis and defense against attacks." *Proceedings of IEEE International Conference on Computer Communications (INFOCOM)*, April 23–29, Barcelona, pp. 1–13, 2006.
32. D. Sherrington and S. Kirkpatrick. "Solvable model of a spin-glass." *Physical Review Letters*, 35(26): 1792–1796, 1975.
33. D. Amit, H. Gutfreund, and H. Sompolinsky. "Spin-glass models of neural networks." *Physical Review A*, 32(2): 1007–1018, 1985.
34. S. Ermon, L. Schenato, and S. Zampieri. "Trust estimation in autonomic networks: A statistical mechanics approach." *International Journal of Systems, Control and Communications*, 4(1/2): 37–54, 2012.
35. U. Maurer. "Modelling a public-key infrastructure." *Proceedings of the 4th European Symposium on Research in Computer Security (ESORICS'96)*, vol. 1146, September, pp. 325–350, 1996.

9

Security and Quality Issues in Trusting E-Government Service Delivery

Maslin Masrom, Edith AiLing Lim, and Sabariyah Din

CONTENTS

Introduction

Electronic government or e-government is defined as a method for governments to utilize the information and communications technology (ICT), for example, Internet-based applications to provide citizens, people, or users, or businesses with more easy and convenient access to government services, and to improve the quality of the services and other processes.

It encompasses government Web pages or Web sites, e-mail, service delivery over the Internet, and digital access to government information.

Security and quality issues in trusting e-government services are crucial challenges that need to be addressed for successful application of e-government either in the developed countries or in the developing countries. In the context of Asia, addressing these two issues for e-government services in public sector would require both technical and careful policy responses from policy makers, and are not easy as perceived by many people. Security forms the basis for challenge and negative implication for e-government application in public sector.

Among crucial areas of weakness in application of e-government are security program management, access controls, software development and change controls, and service continuity. It can be inferred from such alert that effective service continuity in e-government application in the public sector is not only for the availability of services delivery but also to build people trust in government institutions that data and information disclosed by citizens are secure. In addition, the quality of e-government service delivery that promotes the use of e-government, and which will lead to the sustainability of services also influences user's or citizen's trust. The trust has been seen as a motivating factor for encouraging involvement and participation of citizens in e-government services.

The quality of e-government service delivery is always related to the extent to which its Web site facilitates the efficient delivery of effective public e-services so as to assist citizens in accomplishing their governmental transactions. Therefore, the aim of this chapter is to explore the trust of citizens, people or users, security, and quality in e-government service delivery, and then propose a framework that associates security and quality with the trust of citizens toward e-government service delivery.

This chapter is organized as follows: The section "E-Government Services" provides the background of the research by presenting an overview of e-government services in Malaysia. This is followed by an explanation on trust in e-government services in the section "Trust of Citizens in E-Government Services," security in the section "Security in E-Government Services," and quality of service delivery in the section "Quality of Service Delivery." The section "Proposed Framework" highlights a proposed framework, and finally, the section "Conclusions" concludes this chapter.

E-Government Services

Due to the rapid growth of IT, particularly with the advent of the Internet, many governments, including the Malaysian government, have augmented traditional services to e-government services. The implementation of e-government in Malaysia is seen as a tool to improve government service delivery into effective and efficient delivery to the citizens, businesses, and within government agencies. Hence, with the Malaysia's Vision 2020 it focuses in depth on effectively and efficiently delivering government services to the people of Malaysia, enabling the government to become more responsive to the needs of its citizens, to tap potential synergy from the interaction between technologies, educating the population to enable an environment for knowledge-based economies, and more efficient government management through various ICT tools [1].

To accelerate the objectives of Vision 2020, a path has already been defined through the seven innovative flagship applications, which is represented by multimedia super corridor (MSC) in August 1996 management [1]. There are eight projects launched to date under the e-government flagship. The eight projects are generic office environment, electronic procurement, project monitoring system, human resource management information system, electronic services, electronic labor exchange, e-syariah, and e-land. Table 9.1 summarizes the projects and their characteristics. The e-government flagship starts with the MSC initiative and creates a multimedia haven for innovative producers and users of multimedia technology such as ICT. It also gives a chance for both local and foreign companies work with various government agencies to enhance the socioeconomic development of Malaysia.

The objectives of these flagship applications are to accelerate the growth of MSC, to enhance national competitiveness, to reduce digital divide as in reinventing the government by transforming the way it operates toward modernizing, and to attract high value in job opportunities which is guided by the program Vision 2020 in Malaysia [2,3].

Hence, ICT causes a paradigm shift such as the age of network intelligence, reinventing business, government society, and individuals if the value of the ICT can find solutions to its problems for the government [5]. This technology eventually gave an outcome of growing of e-commerce and e-business transactions [6]. With the extensive use of e-commerce and e-business on the ICT by the citizens, the government gradually improved the delivery services to the people in Malaysia and improved the process quality development which enables the government to be more responsive to the needs of Malaysian citizens [2].

Due to rapid use of ICT, the Malaysian government has developed e-government for the extensive reliance of the use of IT to improve the efficiency of government services provided to citizens, employees, businesses, and agencies [7]. E-government in Malaysia serves as communication with the citizens, information access for the citizens, and provides transaction services in a friendly and efficient way all of which are the main objectives of every e-government [8]. Hence with e-government, the Malaysian government is identifying new ways of delivering information and services via electronic channels through the Internet as the service delivery alternative [9].

E-government would offer government information such as services to three main stakeholders, for example, citizens, businesses, and other government organizations on the use of ICT via the Internet [10]. E-government

TABLE 9.1

Projects under the E-Government Flagship and Implementation Agencies

Projects	Agencies	Characteristics
Generic office environment	Prime Minister's Office	Provides a new paradigm of working in a collaborative environment where governmental agencies communicate, interact, and share information.
Electronic procurement	Ministry of Finance	Links the government and suppliers in an online environment. Governmental agencies act as buyers and procure goods/services by browsing catalogues advertised by suppliers. Aimed at best value for money, timely and accurate payment.
Project monitoring system	Implementation Coordination Unit at the Prime Minister's Office	Provides a new mechanism for monitoring the implementation of development projects, incorporating operational and managerial functions, and knowledge repository.
Human resources management information system	Public Service Department	Provides a single interface for government employees to perform human resource department functions effectively and efficiently in an integrated environment.
Electronic services	Road Transport Department	Enables direct, online transactions between the public, the government, and large service providers via electronic means
Electronic labor exchange	Ministry of Human Resources	A one-stop center for labor market information, accessible to government agencies, the business sector, and the citizens
E-syariah	Islamic Justice Department at the Prime Minister's Office	Introduces administrative reforms that upgrade the quality of services in Syariah courts. To enhance the Islamic Affairs Department's effectiveness—better monitoring and coordination of its agencies and 102 Syariah courts.

Source: Multimedia Development Corporation (MDeC), Flagship Applications 2006. Available at http://www.mscmalaysia.my/topic/12073046901815

development is not only about the implementation of a new IT system, but it may also aim to improve public service delivery of government services to citizens; improve citizen empowerment through access to information and services; and increase government management, transparency, and accountability [11–16]. Hence, the characteristic of e-government is a citizen-centric approach. It has the potential to enhance the relationship between the government and its citizens by facilitating easier, smoother, more efficient, and effective interactions between citizens and government agencies [17,18]. This may lead to socioeconomic development in Malaysia and enhance citizens to participate in the process of public policy making [17] with positive impact on good public government, greater transparency, and accountability [19].

Therefore, with this technology, the government has begun to perceive citizens as customers and are more concerned about their expression of satisfaction on the use of the technology [20].

However, many governments worldwide still face the problems of low-level citizen adoption and acceptance of e-government services [21–24]. Successful e-government is based on the technology and also contingent upon the citizen's willingness to adopt it [7]. Nevertheless, the introduction of e-government has encountered many problems, even in the developed countries, where a better environment is available for such development [25]. These problems include the social, cultural, and organizational perspectives of e-government. Without understanding what motivates the public to adopt e-government services, the government will not be able to take strategic actions to increase e-government adoption or participation in e-government [26]. On the contrary, preferences and interests of the citizens are not similar to one another. In fact, not all citizens can and should do one thing in the same way as it varies considerably [27]. Therefore, a minimum level of service requirements and satisfaction of the system should be analyzed in order to make citizens' participation meaningful and helpful [27].

Malaysia ranks 25th out of 32 countries in terms of growth in e-government use [2]. The UN e-Government report shows that Malaysia has improved its ranking from the 42nd position in 2004 to the 34th position in 2008 [28]. Likewise, the Web measure index of the country has improved from 41 in 2005 to 17 in 2008 [28]. However, this position is still below the expectation of the government to fully achieve 100% usage of the e-government. The aim to achieve 100% is to reinvent how the government works on improving the quality of its interactions with citizens and business through improved connectivity, better access to information, and services delivery of the e-government. Not only does the government need to improve the delivery, but it also needs a collaborative environment where the government, businesses, and citizens work together for the benefit of the nation as a whole [29].

There are a few factors that may lead to e-government strategies toward success, which would incorporate users better in the development of e-government services, and these are given as follows:

1. The level of interest and commitment of the government project how government services are offered, enhance transparency, openness, accountability, participation, and other benefits for the citizens [30].
2. The government needs to get continuous feedback or evaluation from the citizens on the e-government services. It has to be an ongoing evaluation practice to continue improving and enhancing its services, especially in improving the program on a regular basis and monitoring government activities and goals. This ongoing evaluation would also eventually deploy e-government strategies, invoking major changes, especially in policies and legislations [31].

3. The government must make communications technology available to its citizens. In order to develop a Web-based e-government services, user needs to have a broadband connection, high-end computer, and advanced technology. These, however, can exclude a segment of people who do not own a computer and do not have access of Internet and complex system, which would deter users from using such services. Therefore, by understanding the technology access and capabilities of the various segments of users, government can develop systems that not only meet the needs of the users but also develop the types of training and support that users may need to be successfully engaged in the e-government objectives.

4. The government must establish a good strategic plan to meet the needs of the citizens as they expect better governmental services for the adoption of more efficient pubic services [32,33]. A good strategic planning can be by looking and understanding how the information is to be used, and how users seek information and identify specific types of problems that the users intend to address with the information. These observations would help the government improve and understand user's information behavior. Without this strategic planning, it would result in poor quality services with limited capacity to meet user information needs.

Trust of Citizens in E-Government Services

Trust is one essential aspect in the implementation of e-government services. People or citizens need to trust the e-government process in making the objectives of e-government achieved. Without trust, people will not participate in e-government, and much effort and strategies have to be done by government to get the trust. According to Ref. [34],

> Trust involves the belief that others will, so far as they can, look after our interests, that they will not take advantage or harm us. Therefore, trust involves personal vulnerability caused by uncertainty about the future behaviour of others, we cannot be sure, but we belief that they will be benign, or at least not malign, and act accordingly in a way which may possibly put us at risk.

Trust is defined as the subjective assessment of one party (trustor) that another party (trustee) will perform a particular transaction according to his or her confident expectations, in an environment characterized by uncertainty [35]. In this context, the degree to which a trustor trusts a trustee is dictated by the trustworthiness of the trustee. According to Ref. [35], this trustworthiness is rooted in the trustor's perceived attributes of ability, benevolence, and integrity implicit to the trustee.

Trust is an important component of e-government acceptance and adoption because it could ease citizens' anxiety and fear of being compromised by e-government service providers or managers. The major concern here is that the e-government services' Web site should display ability, benevolence, and integrity; thereby, it can persuade trust among citizens. Good quality e-government web sites that deliver services to the citizens, and conform to citizens' service expectations on a consistent basis should foster trust among citizens. In other words, designing high-quality e-government services Web site not only translates to functional efficacies but also serves the dual purpose of building citizen trust toward e-government services [36].

In order to trust e-government services, citizens need to believe that government owns the managerial and technical resources to implement and secure the system, and have confidence in the ability of service providers or managers. The citizens should also have the intention of participating in e-government. Thus, citizens' trust can be categorized into two dimensions: (1) trust on the Internet and (2) trust on the government.

Security in E-Government Services

Security is understood in the context of visibility of transaction data by other parties on the Web site. Security is one of the issues in e-government that needs a careful examination. Example of security issues are the following [37]:

1. Identification of security requirements
2. Attribute certificates
3. Public key infrastructure
4. Certification or authentication
5. Risk analysis and metrics for e-government
6. Database security

People or citizens who are keen to ensure that the e-government services should provide a sense of declaration that any personal information retrieved or submitted through the services will be kept safe. The lack of security could affect citizens' trust negatively [38]. In this regard, if e-government services were not secure enough, the citizens' data and information would be under threat and could be misused or changed by others, for instance, the hackers. The appropriate and adequate safeguards should be embedded in the system in order to protect the information from unauthorized dissemination and use [39]. The study done by Jaruwachirathanakul and Fink [40] had shown that citizens' trust in e-government services is associated with security provided to them.

Quality of Service Delivery

Quality of service delivery for the public has become an issue of great concern as manifold problems related to quality of e-service still exist. A good quality of service would eventually increase citizen's participation in e-government. Therefore, an emerging attempt to explore or focus on overall customer satisfaction on the quality of service delivery concerns here are based on the portal of information quality (IQ) and system quality (SQ) for e-government. The customer satisfaction of IQ and SQ is based on perceived citizens' quality and their expectation on the service provided by the e-government. Both quality aspects are based on information, process, and service of the e-government. Whereas the cause and effect of these IQ and SQ enable government agencies to predict how to increase satisfaction through Web enhancements of the provided information and system as illustrated in Figure 9.1.

It is predicted that the satisfaction affects behavioral intention to use e-government services that are based on future behaviors of site visitors such as return visit or referrals to the site for information. If the predictions are not up to the level of satisfaction, then the agency should prioritize improvements based on the lowest impact area on citizen satisfaction to enhance the overall satisfaction.

User satisfaction and perceived quality on the online service is measured as follows [41]:

1. Usability dimension (i.e., whether users have experienced any problems using the service)
2. Benefit experiences by the users (i.e., save time and gain flexibility)
3. Overall evaluation (i.e., overall satisfaction with the service such as whether the user expectations are met or not).

Information Quality for Different Key Players

IQ is defined as the degree to which quality information is provided to the users to fulfill their needs of getting the appropriate information [42]. There are a few key players that the government should take into consideration to shape the IQ of service delivery. The first key player is the citizens' responses toward the IQ [43]. In this regard citizens based

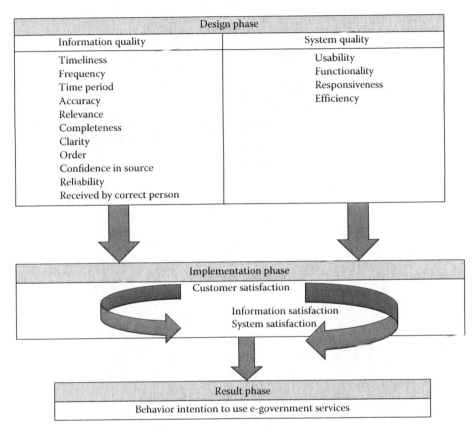

FIGURE 9.1
E-government success measurement framework.

their IQ on their value perception and the nature of interaction with the information system provided by the e-government. A good information flow is based on conveying the information to the appropriate agencies, at the place, on time, and in detail, as well as conveying accurate information to the citizens for a vital element of service-based environment [39]. Hence, the fundamental IQ criteria have been taken up such as meeting citizens' expectation of information which was initially fit for use as in the case of data quality research.

There are also other key players that the government should take into consideration in order to shape the view of understanding the IQ. The other key players besides citizens are the administration and the IT developers [43]. Administration here is regarded as the solution providers, and the IT developers are the technical support required to accomplish the integration. On the other hand, the desires and needs of the players are rooted in a different perception of the problem. The perception and vision are illustrated in Table 9.2.

TABLE 9.2

Key Players' Perception and Vision on the E-Government

Key Player(s)	Problem Perception	Vision/Hoping For
Citizen	• How to find appropriate information in due time? • How to relate with administration on getting the information on where, when, with whom, and how	• Customer-centric approach • Integrated services • Both personal human service (traditional method) and online technical support
Administrations	• Unknown who has what kind of information • Limited access to current information • Lack of reliable interoperability framework	• Integration of information and processes • Leadership/accepted standards in achieving interoperability and integration
Information technology	• Incompatible formats and interfaces • Lack of semantic interoperability and process ownership	• Provision of and adherence to standards • Agreement on technical, data formats, process control, and interfaces

Source: Klischewski, R. and Scholl, H., *Proceedings of the 39th Hawaii International Conference*, 2006.

Information Quality in Internet-Based Service Environment

The Internet is used to deliver and transfer information, and is knowledge-intensive to the end users [44]. Currently, it is classified as the most powerful medium used to gain knowledge-intensive view of information processes under service-based environment. This is completely different from the olden days when information was obtained from traditional database view of information. Therefore, with the Internet, government offers its services to citizens through online portals. Several researchers have explored the issues related to the information search under the Internet-based environment as illustrated in Table 9.3.

Developing Information Quality Conceptual Model

There are three levels of IQ assessments [39]. This is illustrated in Table 9.4. These assessments are based on the following:

- *Representation level*: This level is more on Web interface design. Web interface is based on the design and format of information that has been presented at the Web site. The criteria for this are based on the flexibility, consistency, and information in terms of form.

- *Process level*: This level is to have convenient interaction with the system interface and information consumer through effective media

TABLE 9.3

Research under Internet-Based Environment

Researchers	Issues
Knight and Burn [45]	Explore issues related to information search in the Internet and retrieval through search engines and user queries
Naumann and Rolker [46]	There are three main sources of IQ: 1. User of information 2. Information 3. Process of information search retrieval
Chae and Kim [47]	IQ from mobile Internet: • Connection quality • Content quality • Interaction quality • Contextual quality

TABLE 9.4

Measure of IQ for E-Government Services

Level of IQ		
Representation Level	**Process Level**	**Application Level**
Form, flexibility, and consistency	Accessibility, security, interactivity, and connectivity	Understandability, level of detail, currency, and applicablity

Web site design. The Web site design would assist citizens for the right information in a timely manner.

- *Application level*: This level is oriented toward utility perception of information from user's perspective, and the objective is to have appropriate and useful information that can add value in the decision making.

Representation Level

- Forms of representation are based on text, images, audio, and animations on the e-government Web site. Language in the Web site must be clear and easy to understand by the end users, whereas for images, audio and animations normally suffer from poor-quality audio and animations or have blurred images.
- Flexibility of representation is the information representation interface. This interface should display customization and content information customization, respectively.
- Consistency of representation is where information objects are represented with the same structure, format, and precision as directed by external standards or guidelines for getting the relevant information.

Process Level

- Accessibility implies the systematic manner in which the users gain access in order to obtain appropriate content on the Web portal. Systematic manner refers to the procedure to access the Web portal, for example, as login procedure to get access to the available content of information.

- Security is embedded in the system to protect information from the unauthorized dissemination.

- Interactivity implies responsive downloading and uploading of information by the user through the Web site portal. The navigation quality measure is defined as the number of clicks that a user makes to reach the appropriate information content. For uploading, there should be appropriate input fields for query and feedbacks and upload links.

- Connectivity is based on the speed and stability of interaction with the Web interface. This is because the speed of opening the content page and also the speed of the file download of information that is required by the end users. Slow speed would exasperate the user and would have negative feedbacks.

Application Level

- Understandability is the familiarity of the users to the knowledge domain to which the information belongs.

- Detail is to assess the degree of detail, completeness, comprehensiveness, and appropriate amount of information available to the end users at the Web site.

- Currency is based on the updated information at the Web site. Old information can be classified as inaccurate to the end users.

- Applicability refers to how applicable is the information to the end users. Some information at the Web site is irrelevant for use of e-government services.

Dimensions and Attributes of Information Quality

Information can have different characteristics that can form the quality. The difference between a good quality and bad quality of information can be identified by considering whether or not it has some or all the attributes of IQ as illustrated in Table 9.5. IQ is based on how the information is perceived by the users. These attributes would eventually be used as benchmarks to improve the effectiveness of information system and to develop IQ strategies for all e-government services.

TABLE 9.5

Attributes and Dimension of IQ

Attribute	Dimension	Definition
Time	Timeliness Frequency Time period	Time period that the information deals with and the frequency at which the information is received
Content	Accuracy Relevance Completeness	Describe the scope and contents of the information
Form	Clarity Order	Describe how the information is presented to the recipient
Additional characteristics	Confidence in source Reliability Received by correct person	Other attribute that would lead to IQ

Time Attributes

- Timeliness of information. Timeliness concerns which information is sufficiently up to date, current, and timely for the task at hand of the user. Here, the information depends on how quickly new information can be communicated to the appropriate user and user perceptions of the technology [48]. Timeliness can indicate the response time about how fast the information can be accessed or downloaded. The response time also depends on network traffic, server workload, etc. As with accessibility, response time can be improved by locally replicating or caching information.

- Frequency of information. Information needs to be available as and when needed by the users. This means that information must be sufficient in breadth and depth for the task of the information of the user. It must also ensure that e-government Web sites are available to the user at all times such as 24 hours a day, 7 days a week [49].

- Time period of information. Information should cover the correct time period for information accessed [50]. Information must be consistent and complete sharing among the participating government agencies and departments with the user [50]. Both departments and agencies must initiate and maintain the sharing activity over time. This is like a weekly report, weekly news, or annual report on e-government.

Content Attributes

- *Accuracy of information.* Accuracy is the measure of the correctness of the content of the information. Information should not have any claim of validity as the user should have received a true copy [51].

Therefore, government agencies should assure that information sharing is in the form of accessed, transmitted, and received accurate information.

- *Relevant information.* The information retrieval system should be relevant to meet information needs and be helpful to the user [52]. In e-government, precision will influence the user perception of usefulness, credibility, and assurance of the information sharing capacity of interoperating systems. Information must also cater to different class of users. In some cases, poor quality information may actually be a good quality because what is needed is to educate information users so that they can understand it and use it in the appropriate manner.

- *Completeness of information.* All of the information required to meet the information needs of the recipient should be provided. All information is not missing and is of sufficient depth for the user.

Form Attributes

- *Clarity of information.* The information should be presented in a user-friendly manner that is appropriate to the intended user. The user should be able to locate specific items at the Web site quickly and should be able to understand the information. Clarity in using the Web site is the most significant element that has influenced user satisfaction and behavior [53].

- *Order of information.* Information should be provided in the correct order. At the Web site, navigation should be given priority. This navigation would be able to help the visitors gain access to information from page to page easily. The navigation system should be consistent throughout the Web site that is normally put at the top or on the left of the page. The link and buttons should have understandable titles, and all hyperlinked text should be put in consistent color to differentiate the nonhyperlinked text.

Additional Characteristic Attributes

- *Confidence in source.* Here the confidence is based on the source of information received by the recipients. That is the level of acceptance and trust the information provided by the e-government Web site [54].

- *Reliability of information.* Reliability of information is the extent to which data are regarded as true and credible. The recipient should have the confidence to rely upon the information when required, and the information should be consistent in terms of quality and attributes of IQ such as accuracy and conciseness [55].

- *Received by correct person.* Some information is only applicable to the appropriate group of people. If not, that information is regarded valuable. Thus, it can be suggested that an additional attribute of IQ is that it can be verified that the information has been received and understood by that group of recipient [56].

System Quality

SQ is the measure of the actual system of the electronic information system by using the constructs of functionality, reliability, usability, and efficiency [57] as the determinant of overall user satisfaction and significant predictor of system use [58] as illustrated in Table 9.6. Hence, the measure of the SQ is on the features and performance characteristic of the e-government Web site to enhance the satisfaction of the user with the system. A better SQ and a better service are further related to user satisfaction [59].

Usability

Usability is focused on human issues on how users perceived the design objectives such as functionality, efficiency, and reliability on the Web [60]. At this level of usability, the general principles are to provide support to the user task, interface details such as menu design or use of style guide for the purpose to facilitate Web survival [61]. The usability is also a combination of

TABLE 9.6

System Quality Instrument

Construct	Factors Considered
Usability	
The Web site is easy to use	Navigation is clear
	Download speed is acceptable
	Content is well organized
Functionality	
Has the relevant mechanisms to meet the purpose on the e-government Web site	The Web site has the clear purpose for the intended audience
	Provides all the appropriate function needed by the audience and does all the things audience want to do to achieve the purpose of the visit
Responsiveness	
The Web site is reliable	Speedy and responsible response to request
Efficiency	
Benefit that user believes	Convenience
	Time saving
	Save cost

quality on how easy the user interface is operating so that users can achieve tasks easily and quickly, are able to navigate smoothly or consistently at the web site, and are able to grasp the function of the Web site after looking at the main homepage for a few seconds [61–65]. Hence, the usability key is not only how well the Web site works but also the degree to which the Web site meets user needs [66]. Majority of the users revisit the Web site because of the high-quality content, update often, minimum download time, and easy to use [61].

Functionality

Functionality provides integrative and interactive functions for the end user convenience. High-functioning government Web sites are mainly designed to search, classify, and integrate information that is linked to the existing Web site for the users [67]. High-functioning portals would give citizens one-stop shop for information that is required. Hence, functionality can be described under the terms of customization and openness.

Customization of a Web portal is mainly used to cater to the needs of the portal visitor. It means that the functioning of the Web provides content that gives users a direct need to search the information. Hence, it provides information in a specialized manner.

Responsiveness

Responsiveness is actually the response reaction by the users of the e-government. It is a customer-oriented value that provides services which fit the values and demands of the customers [42]. A speedy and suitable response to a request and timely updates are elements that impact responsiveness. E-government responsiveness can be measured in e-forums on a Web page and special Web forms for comments and suggestions. With these comments, it would help in developing government policies [68].

Efficiency

Efficiency is defined as the proficiency of the system to provide suitable performance and the relative amount of resources used. The efficiency characteristic is related to the time factor. From the user's viewpoint, time behavior is the measure used to evaluate the quality of the Web system performance. Users perceive response time to be the amount of time taken from the moment they follow a link and act on it by clicking the mouse the moment a new Web page has been fully displayed on their screen. This is called loading condition with bandwidth conditions. The main aim of efficiency should be to easily access text information that is needed without having to wait very long or wasting time even with

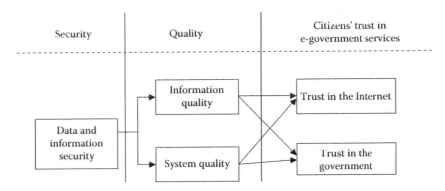

FIGURE 9.2
Proposed framework.

low bandwidth conditions, whereas uploading time limit for each Web application should be less than 30 seconds [61].

Proposed Framework

Based on the literature review, the framework for development of citizens' trust in e-government services is proposed in this research (see Figure 9.2). The constructs of this framework have been adapted from relevant past research such as those given in Refs. [35,36,38,70]. After carefully reviewing the literature, the proposed framework consists of five major constructs including data and information security, IQ, SQ, trust in the Internet, and trust in the government.

Conclusions

In this chapter, the relationships between security, quality, and citizen trust have been identified and proposed. The proposed model or framework in this chapter is based on the literature review. Generally, the success of governmental organization depends on the security and quality of e-government services provided to citizens; therefore, by understanding the constructs, concepts, and dimensions of e-government services enhances citizens' trust. The proposed framework provides an understanding of the required security and quality attributes in e-government services, and highlights what should be emphasized on among e-government service providers or managers, and it further assists them to improve e-government services performance.

References

1. Ramlah, H., Norshidah, M., Nor Shahriza, A.K., and Abdul Rahman, A. (2007). The influence of organizational factors on information systems success in e-government agencies in Malaysia. *The Electronic Journal on Information Systems in Developing Countries (EJISDC)*, 29(1): 1–17.
2. Mohsin, A. and Raha, O. (2007). Implementation of electronic government in Malaysia: The status and potential for better service to the public. *Public Sector ICT Management Review*, 1(1): 2–10.
3. Muhammad Rais, K. and Nazariah, K. (2003). *E-Government in Malaysia: Improving Responsiveness and Capacity to Serve*. Selangor: MAMPU and Pelanduk Publications.
4. Multimedia Development Corporation (MDeC) (2006). Available at http://www.mscmalaysia.my/topic/12073046901815
5. Gichoya, D. (2005). Factor affecting the successful implementation of ICT projects in government. *The Electronic Journal of e-Government*, 3(4): 175–184.
6. Moon, M. (2002). The evolution of e-government among municipalities: Rhetoric or reality. *Public Administration Review*, 62(4): 424–433.
7. Carter, L. and Belanger, F. (2005). The utilization of e-government services: Citizen trust, innovation and acceptance factors. *Information Systems Journal*, 15(1): 5–25.
8. Carenini, M., Whyte, A., Bertovrello, L., and Vonanchi, M. (2007). Improving communication in e-democracy using natural language processing. *Journal of Computer Society*, 22(1): 20–27.
9. D'Agostino, M., Schwester, R., Carizales, T., and Melitski, J. (2011). A study of e-government and e-governance: An empirical examination of municipal websites. *Public Administration Quarterly*, 35(1): 3–25.
10. Schware, R. and Deane, A. (2003). Deploying e-government programs: The strategic importance of "I" before "E." *Info*, 5(4): 10–19.
11. Cater, L. and Belanger, F. (2004). The influence of perceived characteristics of innovating on e-government adoption. *The Electronic Journal of e-Government*, 2(1): 11–20.
12. Ciborra, C. and Navarra, D. (2005). Good governance, development theory and aid policy: Risks and challenges of e-government in Jordan. *Journal of Information Technology for International Development*, 11(2): 141–159.
13. Huang, Z. and Bwoma, P.O. (2003). An overview of critical issues of e-government. *Issues of Information Systems*, 4(1): 164–170.
14. Lam, W. (2005). Barriers to e-government integration. *Journal of Enterprise Information Management*, 18(5): 511–530.
15. Ndou, V.D. (2004). E-government for developing countries: Opportunities and challenges. *The Electronic Journal on Information Systems in Developing Countries*, 18(1): 1–24.
16. Tung, L.L. and Rieck, O. (2005). Adoption of electronic government services among business organisation in Singapore. *The Journal of Strategic Information Systems*, 14(4): 417–440.
17. Lee, G.G. and Lin, H.F. (2005). Customer perceptions of eservice quality in online shopping. *International Journal of Retail & Distribution Management*, 33(2): 161–176.

18. Zeleti, F.A. (2011). Improvement of e-government in islamic republic of Iran, identify the obstacles of implementing and improving e-government. *Proceedings of the 5th Symposium on Advance and Science Technology*, Mashhad, Iran: Khavaran Higher-Education Institute, May 12–17.

19. Alhujran, O., Mahmoud, A., and Anas, A. (2011). The role of national culture and citizen adoption of e-government services: An empirical study. *The Electronic Journal of e-Government*, 9(2): 93–106.

20. Evans, D. and Yen, D. (2006). E-Government: Evolving relationship of citizens and government, domestic and international development. *Government Information Quarterly*, 23(2): 207–235.

21. Belanger, F. and Carter, L. (2008). Trust and risk in e-government adoption. *Journal of Strategic Information Systems*, 17(2): 165–176.

22. Gupta, B., Dasgupta, S., and Gupta, A. (2008). Adoption of ICT in government organization in a developing country: An empirical study. *Journal of Strategic Information Systems*, 17: 140–154.

23. Kumar, V. (2007). Factors for successful e-Government adoption: A conceptual framework. *The Electronic Journal of e-Government*, 5(1): 63–76.

24. Fu, J.R., Farn, C.K., and Chao, W.P. (2006). Acceptance of electronic tax filing: A study of taxpayer intentions. *Information & Management*, 43: 109–126.

25. Prins, J.E.J. (ed.) (2001). *Designing E-Government: On the Crossroads of Technological Innovation and Institutional Change*. Amsterdam: Kluwer.

26. Gilbert, D., Balestrini, P., and Littleboy, D. (2004). Barriers and benefits in the adoption of e-government. *International Journal of Public Sector Management*, 17(4): 286–301.

27. Gross, P. (2000). Technological support for eDemocracy: History and perspective. *11th International Workshop on Database and Expert System Applications*, London, September 4–8.

28. United Nations (2008). *World Public Sector Report: UN E-Government Survey, From E-Government to Connected Governance*. New York: United Nations Publication.

29. Karim, A.M.R. and Khairuddin, S. (1999). Electronic government: Reinventing service delivery, In Karim, A.M.R. (ed.), *Reengineering the Public Service Leadership and Change in Electronic Age*, Kuala Lumpur: Pelanduk Publication, p. 186.

30. Criado, G. and Ramilo, M. (2003). E-Government in practice: An analysis of website orientation to the citizens in Spanish municipalities. *The International Journal of Public Sector Management*, 126(3): 191–218.

31. Yoon, J. and Chae, M. (2009). Varying criticality of key critical success factors national e-strategy along the status of economic development of nations. *Government Information Quarterly*, 26: 25–34.

32. Tan, C., Pan, S., and Lim, E. (2005). Managing stakeholder interest in eGovernment implementation: Lessons learned from a Singapore e-Government project. *Journal of Global Information Management*, 13(1): 31–53.

33. Torres, L., Pina, V., and Royo, S. (2005). EGovernment and the transformation of public administration in EU countries—Beyond PHM or just a second wave of reform. *Online Information Review*, 29(5): 531–553.

34. Baier, A.C. (1986). Trust and antitrust. *Ethics*, 96: 231–260.

35. Mayer, R.C., Davis, J.H., and Schoorman, F.D. (1995). An integrative model of organizational trust. *Academy of Management Review*, 20(3): 709–734.

36. Tan, C.-W., Benbasat, I., and Cenfetelli, R.T. (2008). Building citizen trust towards e-government services: Do high quality websites matter? *Proceedings of the 41st Hawaii International Conference on System Sciences*, Big Island, HI: IEEE Computer Society Press, pp. 1–10.
37. Hwang, M.-S., Li, C.-T., Shen, J.-J., and Chu, Y.-P. (2004). Challenges in e-government and security of information. *Information & Security*, 15: 9–20.
38. Alanezi, M.A., Mahmood, A.K., and Basri, S. (2012). Conceptual model for measuring e-government service quality. *Proceedings of 2012 International Conference on Computer & Information Science*, June 12–14, Kuala Lumpur: Academy Publisher, pp. 130–135.
39. Misra, S.C.A., and Chatterjee, J. (2012). Examining information quality for e-governance services: Towards a conceptual model. *Proceedings of the 2012 International Conference on Industrial and Intelligent Information (ICIII)*, Singapore, March 17–18.
40. Jaruwachirathanakul, B. and Fink, D. (2005). Internet banking adoption strategies for a developing country: The case of Thailand. *Internet Research: Electronic Networking Applications and Policy*, 15(3): 295–311.
41. Christos, H., Xeria, P., Babis, M., and Gregonis, M. (2007). Classification and synthesis of quality approaches in e-government services. *Internet Research*, 17(4): 378–401.
42. DeLone, W.H. and McLean, E.R. (2003). The DeLone and McLean model of information systems success: A ten year update. *Journal of Management Information Systems*, 19(4): 9–30.
43. Klischewski, R. and Scholl, H. (2006). Information quality as common ground for key players in e-government integration and interoperability. *Proceedings of the 39th Hawaii International Conference* Big Island, HI: IEEE Computer Society Press, January 4–7.
44. Bevan, N. (1999). Quality in use: Meeting use needs for quality. *Journal of System and Software*, 49(1): 89–96.
45. Knight, S. and Burn, J. (2005). Developing a framework for assessing information quality on the World Wide Web. *Informing Science*, 8: 159–172.
46. Naumann, F. and Rolker, C. (2004). Assessment methods for information quality criteria. *Proceedings of the 5th Conference on Information Quality (IQ2000)*, October 20–23, Boston, MA: MIT Publication, pp. 148–162.
47. Chae, M. and Kim, J. (2002). Information quality for mobile internet services: A theoretical model with empirical validation. *Electronic Market*, 12(1): 38–46.
48. Ballou, D. (2006). Modelling data and process quality in multi-input, multi-output information system. *Management Science*, 34: 150–162.
49. Kim, J., Lee, J., and Choi, D. (2006). Designing emotionally evocative homepages: An empirical study of the quantitative relations between design factors and emotional dimensions. *International Journal of Human-Computer Studies*, 59: 899–940.
50. Marchand, D. (1990). Managing information quality. In Wormel, I. (ed.), *Information Quality: Definitions and Dimensions*. Los Angeles, CA: Taylor Graham, pp. 7–17.
51. Taylor, R. (1986). *Value Added Processes in Information Systems*. Norwood, MA: Ablex Publication.
52. Van, J. (1979). *Information Retrieval*, 2nd edn. Boston, MA: Butterworths.
53. Yoo, B. (2001). Developing a scale to measure perceived quality of an internet shopping site. *Quarterly Journal of Electronic Commerce*, 2(1): 31–46.

54. Wolfinbarger, M. (2003). Dimensionalizing, measuring and predicting e-tailing quality. *Journal of Retailing*, 79(3): 183–198.
55. Olaisen, J. (1990). Information quality factors and the cognitive authority of electronic information. In Wormel, I. (ed.), *Information Quality: Definitions and Dimensions*. Los Angeles, CA: Taylor Graham, pp. 91–121.
56. Kolsaker, A. (2006). Citizen centric e-government: A critique of the UK model. *Electronic Government: An International Journal*, 3(2): 127–138.
57. DeLone, W.H. and McLean, E.R. (2004). Measuring e-commerce success: Applying the DeLone & McLean information systems success model. *International Journal of Electronic Commerce*, 9(1): 31–47.
58. Livari, J. (2005). An empirical test of the DeLone–McLean model of information system success. *The Data Base for Advances in Information Systems*, 36(2): 8–27.
59. Teo, T., Srivastava, S., and Jiang, L. (2008). Trust and electronic government success: An empirical study. *Journal of Management Information Systems*, 25(3): 99–131.
60. Sharp, H., Rogers, Y., and Preece, J. (2007). *Interaction Design: Beyond Human-Computer Interaction*, 2nd edn. West Sussex: John Wiley & Sons.
61. Nielsen, J. and Mack, R.L. (1994). *Usability Inspection Methods*. New York: John Wiley & Sons.
62. George, H. (1996). *The Good Usability Handbook*. London: McGraw-Hill.
63. Quesenbery, W. (2008). Making personas part of your team-user friendly. *A Workshop for User Friendly*, October 24–27, Shenzen, China.
64. Roach, C. (2007). E-government: Usability of Trinidad and Tobago ministry websites, PhD Thesis, Arizona State University, Phoenix, AZ.
65. Van Welie, M., van der Veer, G.C., and Eliëns, A. (1999). Breaking down usability. *Proceedings of Interact*, August 30–September 3, Edinburgh, Scotland.
66. Thompson, K.M., McClure, C.R., and Jaeger, P.T. (2003). Evaluating federal Websites: Improving e-government for the people. In George, J.F. (ed.), *Computers in Society: Privacy, Ethics, and the Internet*. Upper Saddle River, NJ: Prentice Hall, pp. 400–412.
67. Layne, K. and Lee, J. (2001). Developing fully functional e-government: A four stage model. *Government Information Quarterly*, 18: 122–136.
68. Sims, H. (2001). *Public Confidence in Government and Government Service Delivery*. Ottawa, ON: Canadian Centre for Management Development.
69. Ba, S. and Pavlou, P.A. (2002). Evidence of the effect of trust building technology in electronic markets: Price premiums and buyer behaviour. *MIS Quarterly*, 26(3): 243–268.

10

Context-Aware E-Commerce Applications: Trust Issues and Their Solutions

Farag Azzedin, Sajjad Mahmood, and Muhammad Akhlaq

CONTENTS

Introduction

In the last decade, context-aware computing [1] has become a reality and everyday applications are becoming context-aware. Context-aware systems facilitate the use of devices available in the environment of users. A context-aware system needs to be aware of the available resources to detect changes in the environment; and adapt system functionality and behavior [2]. A context-aware system has the following three basic features [1]: *sensing, thinking,* and *acting.* Systems can vary in sophistication in each of these functionalities [2]. Context-aware applications require behavior that is highly adaptive and depends heavily on the available resources which may also be transient in nature.

Electronic commerce (e-commerce) systems have seen widespread usage and have become one of the vital facets of the Internet. E-commerce applications are distributed in nature and lately, e-commerce applications (e.g., E-Market Place [2]) have started using contextual information to provide better service to their customers. Context-aware e-commerce applications use location, customer preferences, and feedbacks as context so as to provide dynamic information to help them improve the shopping experiences of customers.

Similar to traditional e-commerce systems, context-aware e-commerce applications also face a number of issues ranging from security to trust management. Trust issue is important as it affects the willingness and confidence of customers to participate in the e-commerce system. For example, if customers refuse to engage in online activities because they are fearful that they will be cheated or overcharged for these services; then online e-commerce communities will not be able to survive. As such, a surviving and successful e-commerce system must keep an adequate level of trust to gain business in such competitive e-commerce environments by (1) keeping trustworthy interactions with its customers in terms of the delivery service and customer support and (2) presenting accurate and honest information to its customers.

In this chapter, we present a context-aware architecture for e-commerce systems. The architecture allows a developer to design context-aware e-commerce applications. We also present extensive performance evaluation experiments to illustrate the usefulness of the approach. Our results show that the approach supports the seamless integration of various context-aware trust threat solutions.

The rest of this chapter is organized in the following sections: "Related Work" presents the related work. "Context-Aware Architecture" elaborates the context-aware e-commerce architecture. "Trust Threat Solutions" discusses the trust threat solutions and "Performance Evaluation" discusses the performance analysis of the evaluations. Finally, "Conclusion" provides the gist of the context-aware architecture for e-commerce systems.

Related Work

Mark Weiser [3] introduced the term "ubiquitous computing" in 1988, which is also known as context-aware computing now. He presented an idea of invisible computers, embedded in everyday objects replacing personal computers. He highlighted the need of combining computers and humans seamlessly in an environment which would provide services to the users based on their current context. In future, nearly all applications, services, and devices will be context-aware. It is important to note that "location" is the most commonly used context in context-aware systems. This section provides a brief introduction to the location and context-aware computing, and a survey of context-aware trust models.

Location and Context-Awareness

In the "mainframe era," computing resources such as memory, secondary storage, and CPU time were conceived as valuable computing resources. Therefore, compact and memory-efficient programs were promoted. In the "portability era," users' ease and time were considered as the most valuable things. In the present "mobility era," user attention is regarded as the most valuable thing. As the capability of human attention is very restricted, we need context-aware systems to guarantee minimal user interruption. A system is called as context-aware if it can adapt to its changing context of use.

Schilit and Theimer [4] gave the definition of context as location, identities of nearby people, objects, and changes to those objects. Several researchers later on tried to define context in a better way [5–17]. From these definitions, we can conclude that any relevant information accessible at the time of interaction with a system is called as context.

Context can be used to enhance computing in several ways. Chalmers [18] has reported five main uses of context: contextual sensing (e.g., knowing the location, time, and temperature), contextual augmentation (e.g., attaching location info to geographical data), contextual resource discovery (e.g., finding the nearest printer), context triggered actions (e.g., loading a museum map when a user enters the area), and contextual mediation (e.g., describing limits and preferences over sensitive data and skipping it when there are strangers present in the meeting).

A system is called as context-aware system if it can adapt its behavior to its changing context of use. Context-aware systems use some or all of the relevant information to make better service available to their users, and are more user-friendly, less obtrusive, and very efficient. Context-aware computing offers a number of challenging issues which are the following: universal definition of context, context acquisition, context representation and exchange, context extraction and inference, context-aware actions, smart applications, contextual presentation, authoring in context, context memory, context-aware information retrieval, context-sharing and trust, context fusion, quality of context (QoC), context-aware query languages, security and privacy, user experience, and *ad hoc* location sensing.

The introduction of small and powerful devices has enabled us to use computers while on the move. The location of a mobile device is a key context, which can find out more information such as the device's capabilities, users, owners, neighboring devices, environment, and so on. A system can either detect the user's location automatically such as Wireless [19] or request input from the user such as Electronic Guidebooks [20] in order to provide relevant services. The enabling technologies for location-aware computing are wireless networks, mobile devices, and positioning technologies.

Location is the most vital and frequently used context. Location-awareness can help entities (i.e., people and things) in a number of ways: (1) social computing [21]—where people are aware of their location so as to

communicate with each other, share data, coordinate meetings, find intentions of others, and so on; (2) service discovery [22]—where people are aware
of the position of things around so as to quickly find relevant services such
as the nearest printer, ATM/cash machine, medical center, shopping mall,
and parking lot; (3) invisible computing [23]—in which devices are aware of
people's location so as to perform useful tasks without interrupting them,
such as forwarding the phone calls to the location where a user is [24], and
location-based games that can exploit players' location [25] to shape their
behavior; and (4) interaction between devices—where devices can interact
with other devices if they know their location, such as Geocasting [26], which
routes messages to the neighbor that is closest to the destination.

Context-Aware Trust Models

A context-aware system works only when contextual information is made
available to the system and shared with all users. A great amount of personal
and sensitive information is available in context-aware systems, which includes
identity, preferences, usage history, location, and so on. Context-sharing implies
that a context-aware system should show the context of one user to others. For
example, it might be relevant to show the current location of each employee in
the office building. Context-sharing involves severe privacy issues [27]. In the
absence of sufficient level of trust, a user of context-aware system may not share
contextual information with others. We need to facilitate ubiquitous access and
restricted sharing of contextual information. Trust models can play an important role in context-sharing and deciding access to the contextual information.
In return, a trust model can use the available contextual information in order to
improve its performance, which results in a context-aware trust model.

A model for supporting trust is proposed in Ref. [28]. This trust-based
model allows peers to decide which other peers are trustworthy and allows
peers to tune their understanding of another peer's recommendations. Each
peer keeps two sets: (1) set Q for peers directly trusted and (2) set R for
recommenders. One of the drawbacks of this model is the scalability issue.
The scalability of the model is not explicitly addressed in this study [29]. In
our proposed scheme, the scalability issue can be addressed by the aggregation scheme, which improves the scalability of the mechanism both in terms
of the number of resources that can be handled by the system and the number of transactions required to converge at the appropriate trust levels.

A reputation-based approach for extending Gnutella-like environments is
proposed in Ref. [30], in which an entity uses a polling protocol to query and
select target peers. Each entity maintains information on its own experience
with target peers and shares such experiences when polled by other peers.
In this approach, a peer broadcasts its request to all its neighbors regardless
of their trustworthiness. This practice is inefficient, gives constant opportunities for dishonest peers to damage and influence the reputation of the
network, and is not scalable as the number of peers grows.

A trust management in a P2P information system is proposed in Ref. [29], where the focus is on implementing a generic infrastructure to deploy any trust model. A trust model was proposed, in which peers file complaints based on bad experiences they had while interacting with other peers. One limitation of this model is that it is based on a binary trust scale (i.e., a peer is either trustworthy or not). Hence, once a complaint is filed against peer p, it is considered untrustworthy even though it has been trustworthy for all previous transactions. In addition, this approach has no mechanism for preventing a malicious peer from filing an arbitrary number of complaints and potentially causing denial of service attack.

Since the inception of the trust systems, many researchers have tackled the notion of trustworthiness at different stages. Some trust systems assume a correlation between trustworthiness and honesty [29,31–35], that is, they assume that trustworthy peers provide honest recommendation and untrustworthy peers provide dishonest recommendation. Such trust systems are vulnerable to badmouthing attacks. A trustworthy peer might provide dishonest recommendation to isolate other competing trustworthy peers and consequently increase his profit.

Other trust systems [29,35,36] assume that the majority of the peers are honest and therefore counter the effect of the dishonest ones on the recommendation network. Others such as Ref. [37], assumed that a peer is equipped with honest recommenders, and the peers are assumed to be robust and resilient to dishonest peers and risky environments. In Ref. [34], a peer's credibility is used to offset the risk of dishonest feedback. In these approaches, no mechanism is used to identify and prevent dishonest peers from polluting the recommendation network.

The trust system in Ref. [38] uses reply consistency to predict honesty. Consistent peers are assumed to be honest and vice versa. Each peer has a set of trusted allies through whom consistency check is performed. The checking is done by asking one or more of the trusted allies to send a recommendation request for the target peer to the recommender. The source peer would compare the recommendation it gets directly with the one received by the trusted allies. Assuming the requests come in relatively short time, the recommender should give answers with no or little value difference. Therefore, if the difference is more than a certain threshold, the recommender is being inconsistent. The recommender would be replaced from the source peer's recommender list and marked as dishonest so that it would not be included again in the list. However, this method cannot detect dishonest peers who provide consistent replies.

A method to filter out dishonest feedbacks for Bayesian trust systems is presented in Ref. [39]. Bayesian trust systems use the beta distribution in predicting a peer's trust level using the number of trustworthy and untrustworthy transactions as the distribution parameters. The parameters of the distribution are $\alpha = NT + 1$ and $\beta = NU + 1$ where NT and NU are the number of trustworthy and untrustworthy transactions with the target peer reported by the recommender, respectively. The expected probability of a peer's trustworthiness is calculated as: $E(p) = \alpha/(\alpha + \beta)$. The trustworthiness of a peer having parameters (1:1), (5:2), and (3:10) is 0:5, 0:714, and 0:231, respectively. It implies optimistic stranger policy because a new peer with no transactions will have a trust of 0:5. Honesty checking is performed by identifying outliers, that is, feedbacks with expected probability less than a certain quantile q and those exceeding $(1 - q)$ quantile. The recommenders providing those outliers are considered to be dishonest.

Recently, Wang et al. [40] have proposed a context-aware trust model using a Bayesian network to integrate contextual information from a set of reliable recommenders. Mohammad et al. [41] have presented an interaction-based context-aware trust model that identifies trust properties based on context and risk information. Further, they introduce a context-similar parameter in trust calculation for handling indirect interactions. Lately, Marcin Sydow [42] has reported an early stage of research aiming at improving trust prediction in social networks by using machine-learning approaches. However, these trust models use context as a static parameter and therefore, are not fit for a context-aware computing environment.

Context-Aware Architecture

An e-commerce environment consists of multiple e-commerce systems. An e-commerce system is divided into domains as shown in Figure 10.1a. Each domain (D) is represented by a broker (B). The brokering model is a peer-to-peer network of brokers (hereafter referred to as brokers). A broker is responsible for managing the resources and the clients that are within its domain. The resources within a single domain are grouped into different classes by their expected reputation by the broker. A broker's reputation will depend on how accurate and honest it is in representing the reputations of its resources. For instance, when a resource misbehaves despite being presented as a highly reputed resource by the broker, the reputation of the broker will be damaged. Conversely, if a broker conservatively estimates the reputation of its resources, then the resources within its purview can be unnecessarily shunned by other resources. Therefore, a broker is compelled to place a resource in the most appropriate class by weighing these conflicting requirements. When a resource joins a domain, it negotiates with the broker the trust level (reputation) that will be placed on it. The resource could use recommendations from prior broker associations to lay its claim for higher trust levels. Often a broker may need to know about resources that are under the purview of brokers with whom it does not have any relationships. In this case, a broker will request recommendations regarding the target broker from its broker's peers. The predicted reputation of the target resource

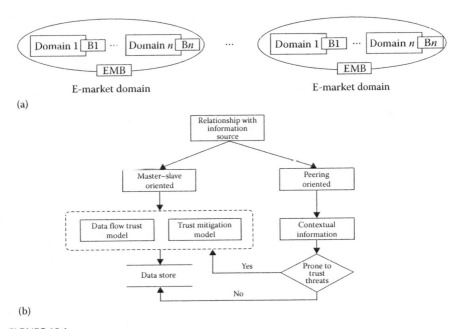

(b)

FIGURE 10.1
An e-commerce context-aware architecture: (a) overall e-market architecture and (b) context-aware data flow within an EMB.

will depend on the reputation of the broker who manages it and the reputation bestowed upon the resource by the broker. The postmortem analysis of the transactions will determine the validity of the predictions. The reputations and other trust levels of the brokers are adjusted based on the match between the predicted values and postmortem detected values.

An e-commerce system is represented by an e-commerce master broker (EMB). The EMB has a master–slave relationship with its domain brokers and a peer-to-peer relationship with other EMBs. The EMB is context-aware and uses contextual information such as relationship, location, and jurisdiction to make the e-commerce system trustworthy. An EMB analyses the received data so that it can make a context-aware informed decision. For example, as shown in Figure 10.1b, an EMB analyses the relationship contextual information to aid it in treating the received data. The relationship contextual information can be from two sources, namely, a domain broker and an EMB. The relationship contextual information from a domain broker is classified as master–slave relationship, whereas the relationship contextual information from an EMB is classified as a peer relationship. If the data has a master–slave context, then the methodology for designing trustworthy applications is applied. On the other hand, if the data has a peering context, the contextual information is analyzed by the EMB. The EMB decides whether to treat the data based on the contextual information (e.g., jurisdiction).

In an e-commerce environment, a customer can move from one domain to another. As a customer moves from one domain to another, the brokers use contextual information such as location and preferences for seamless interaction with the e-commerce environment. The information captured by these brokers is relayed to the relevant EMB. The feedback process is vulnerable to trust threats because it enables customers to provide feedback for purchased products. The feedbacks are stored in a data store.

Not all customers are honest. One of the objectives of the simulation is to model the process of uncovering the product's true values through observation of the transactions that take place among the customers and their associated broker. We model the product values that underlie among the brokers by an actual product value (APV) vector. For simplicity, we assume that the product values do not change for the duration of the simulation time. In addition to APV, we maintain a predicted product value (PPV) table to track the evolution of the customer relationships with their prospective broker. The elicitation of the PPV values is simulated by initially setting them to -1 and then updated using the following equation:

$$PPV_c^p = \begin{cases} APV^p + [0 \text{ to } \alpha], \text{ if } c \text{ is honest} \\ APV^p + [(1-\alpha) \text{ to } 1], \text{ if } c \text{ is dishonest} \end{cases}$$

where:
α is a threshold and ranges from 0 to 1 exclusively

It can be noticed that the PPV values are updated using the APV values plus a random noise. This random noise is correlated with the customer's honesty. The customer's honesty is provided as a trust service to the e-commerce environment.

Trust Threat Solutions

Euclidean Distance Method

The first solution, shown in Figure 10.2, to the dishonesty trust threat is based on Euclidean distance. The Euclidean distance between APV vector and the PPV matrix is calculated for each customer as illustrated in line 5. Then the Euclidean distances are normalized so that they range from 0 to 1 as shown in line 6. Lines 7 and 8 exclude feedbacks provided by dishonest customers. The normalized Euclidean distance for each customer is compared with a threshold (i.e., β). This comparison is necessary to identify whether a customer is honest. Finally, the mean Euclidean distance is computed for all honest customers. These steps are illustrated as follows:

Label all customers to be honest

1. $C \leftarrow$ Number of customers
2. $P \leftarrow$ Number of products
3. For $i = 1$ to C

$$d_i = \sqrt{\sum_{j=1}^{P}(\text{APV}^j - \text{CPV}_i^j)^2}$$

$$D_i = \frac{d_i}{\sqrt{P}}$$

If $D_i \geq \beta$

Label C_i to be dishonest

4. End

$$\bar{D} = \frac{\sum_{i=1}^{C} D_i}{C} \text{ over all the honest customers}$$

FIGURE 10.2
Euclidean distance method for trust threat solution.

Mean-Distance Method

The second solution, shown in Figure 10.3, to the dishonesty trust threat is based on the weighted mean distance. The weighted mean for each customer's feedbacks are calculated as illustrated in line 5. Line 7 calculates the overall weighted mean for all customers. For each customer, line 9 calculates the distance between the overall weighted mean and the weighted mean.

Label all customers to be honest

1. $C \leftarrow$ Number of customers
2. $P \leftarrow$ Number of products
3. For $i = 1$ to C

$$\bar{X}_i = \frac{\sqrt{\sum_{j=1}^{P} w_i \text{CPV}_i^j}}{\sqrt{\sum_{j=1}^{P} w_i}}$$

4. End

$$\bar{X} = \frac{\sqrt{\sum_{i=1}^{N} \bar{X}_i}}{N}$$

5. For $i = 1$ to C

$$D_i = |\bar{X} - \bar{X}_i|$$

6. End

If $D_i \geq \beta$

Label C_i to be dishonest

7. End

$$\bar{D} = \frac{\sum_{i=1}^{C} D_i}{C} \text{ over all the honest customers}$$

FIGURE 10.3
Mean-distance method for trust threat solution.

The distance for each customer is compared with a threshold (i.e., β). This comparison is necessary to identify whether a customer is honest. Finally, the mean distance is computed for all honest customers. These steps are illustrated in Figure 10.3.

Hybrid Method

The previously presented trust threat solutions consider contextual information, namely, customers' value and customers' honesty. This contextual information is valuable to a flip-flop trust threat solution where the algorithm can adapt to different solutions based on the context. For example, the presence of dishonest customers needs a filtering mechanism. The effectiveness of these filtering mechanisms depends on the percentage of dishonest customers. Therefore, we developed a context-aware trust threat adaptable solution that considers different solutions based on the contextual information such as location, dishonesty, and preference.

The following solution assumes that we have a vector of contexts available to us. Our objective is to sense contexts of interest to the e-commerce application and apply appropriate trust threat solutions that are aware of the selected contexts. Figure 10.4 shows details of the steps necessary to achieve this objective. For all available contexts, we sense T_i that is of interest to the e-commerce application. This is shown in lines 5 and 6. Lines 7–9 select other associated contexts with reference to T_i. Finally, line 12 activates appropriate trust threat solutions aware of the selected contexts available in ST.

1. $T \leftarrow$ Vector of available contexts
2. $ST \leftarrow$ Vector of available selected contexts
3. $N \leftarrow$ Number of contexts
4. For $i = 1$ to N
   If T_i is sensed
   $ST = ST + T_i$
5. For ($j = 1$ to N)
   If (T_j is associated with T_i & $i \neq j$)
6. $ST = ST + T_j$
7. End
8. End
   Activate appropriate ST-aware solution

FIGURE 10.4
Hybrid method for trust threat solution.

Performance Evaluation

We conducted a series of simulation studies to examine various properties of the context-aware e-commerce system. These experiments aid us in evaluating the performance of the system.

Performance Metrics

We use the following performance metrics to evaluate the effectiveness of the context-aware e-commerce system:

- *Satisfaction level*: One performance measure of context-aware e-commerce systems is its ability to correctly predict products' ratings. A prediction is considered successful when (1) customers are satisfied with trustworthy products and (2) customers are unsatisfied with untrustworthy products. Hence, the satisfaction level (SL) is computed as follows:

$$SL = (1 - \bar{D}) \times 100.$$

- *Detection error*: The context-aware e-commerce system ideally distinguishes honest and dishonest ratings so that only honest ones are used in predicting the products' ratings. For the detection error (DE), we measured it using (1) the percentages of dishonest customers detected as honest customers (true positive) and (2) the percentage of honest customers detected as dishonest customers (false positive). Hence the DE is computed as follows:

$$DE = a + b.$$

Simulation Setup

The requests initiating interbroker transactions are assumed to have a Poisson arrival process. The number of brokers (i.e., the number of domains) was set to 20 and the number of customers in the e-commerce systems is set to 5000. A customer can give a rating to any of the 100 products available in the e-commerce system. The number of dishonest customers was set to vary from 0% to 100% to evaluate the effect of dishonesty on the SL.

Customers can move from one domain to another and this movement was simulated using sensor proximity in which the transmission range was set to 100 m. The topology used is a Mesh meaning that a customer can be associated with more than one domain. In the simulation, we avoid bogus ratings by accepting ratings only for purchased products.

Throughout the evaluation experiments, we evaluated five trust threat solutions, namely, (1) Algorithm 1: Euclidean distance method with no filter as explained in the section "Euclidean Distance Method" without applying lines 7 and 8; (2) Algorithm 2: Euclidean distance method as explained in the section "Euclidean Distance Method"; (3) Algorithm 3: Mean-distance method as explained in the section "Mean-Distance Method." In this trust threat solution, we set the weights (w_i) to 1; (4) Algorithm 4: Weighted mean-distance method as explained in the section "Mean-Distance Method." In this trust threat solution, we set and the weights (w_i) to respective customer reputation obtained as a trust service; and (5) Algorithm 5: A hybrid method that exhibits a flip-flop pattern as explained in the section "Hybrid Method."

Results and Discussions

Figure 10.5 shows the SL as the number of dishonest customers increase for different tolerance levels.

We measured tolerance as the allowable deviation of PPV from APV. By allowing different e-commerce systems enforce a specific tolerance level, the flexibility of our context-aware architecture was proved. The acceptable difference of PPV from APV determines the slope of the curve, whereas we fixed the other parameters shown in Table 10.1. Figure 10.5 shows that for 0% tolerance, the SL is highly sensitive to the percentage of dishonest customers. This sensitivity decreases with the increase in tolerance. A 0% tolerance means that PPV must be same as APV. From the figure, it can be noticed that 20% tolerance has the advantages of low and high percentages of dishonest customers. Therefore, for the rest of the evaluation experiments, we use 20% as the basis.

Figure 10.6 shows the SL for the five trust threat solutions. However, SL is highly affected by the percentage of dishonest feedbacks by the customers. The results show that Algorithm 4 performs better when the percentage of

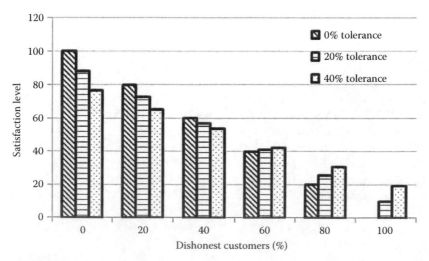

FIGURE 10.5
Tolerance of the context-aware architecture.

TABLE 10.1

Simulation Setup for an E-Commerce System

Parameter	Description
Number of domains	20
Number of customers	5000
Number of products	100
Percentage of dishonest customers	0–100
Sensors	Proximity or location
Topology	Mesh
Mobility (of customers)	Random walk with high mobility
Channel	Wireless
Transmission range	100 m

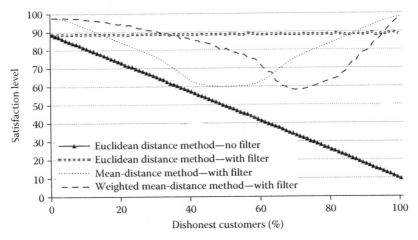

FIGURE 10.6
Satisfaction level for the trust threat solutions.

dishonest customers is lower than 35%, Algorithm 2 performs better when percent of dishonest customers is between 35% and 85%, and Algorithm 3 performs better when percent of dishonest customers is higher than 85%.

Figure 10.7 shows the hybrid method which adapts its behavior depending on the percentage of dishonest customers in the e-commerce system. We plotted the performance of the hybrid method against the performance of Algorithm 1 which we use as the base case because Algorithm 1 uses Euclidean distance without considering any weights or filters. It is clear that exhibiting a flip-flop behavior according to the contextual information results in an improved SL. The hybrid method maintains an SL ranging from 90% to 97% regardless of the increase of dishonest customers.

In Figure 10.8, we evaluated the robustness of the SL of our trust threat solutions for the duration of 10 days. During these 10 days, contextual information

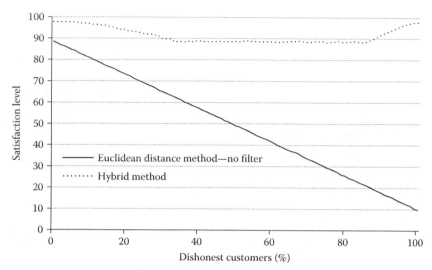

FIGURE 10.7
Satisfaction level of the hybrid method.

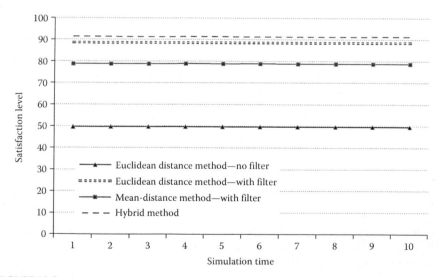

FIGURE 10.8
Robustness—The performance of the proposed algorithms in the long run.

might have changed. Such change includes location, preference, number of customers, number of products, and percentage of dishonest customers. The results indicate that our trust threat solutions withstand challenges in dynamic environments and adapts to new contextual information.

Figure 10.8 shows that as time increases, the SL achieved by the hybrid method is the highest. It is also worth noting that all the proposed trust threat solutions are robust.

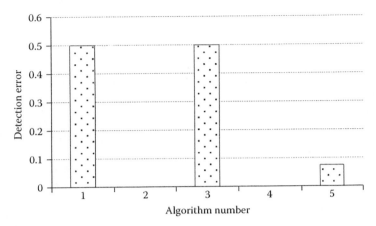

FIGURE 10.9
DE of the trust threat solutions.

Scalability shows the quality of being scalable to a number of customers or products in the e-market. All the proposed methods are highly scalable because the number of messages exchanged per customer or product remains constant.

Figure 10.9 shows the DE for the trust threat solutions. The DE for Algorithm 1: Euclidean distance method with no filter and Algorithm 3: Mean-distance method both have an error rate of 0.5, whereas Algorithm 2: Euclidean distance method and Algorithm 4: Weighted mean-distance method have error rates of 0. Algorithm 5: A hybrid method has an error rate of 0.075, that is, 7.5%.

Figures 10.10 and 10.11 show scalability as the number of customers as well as products increase. The two figures show that the computation time exhibits linear increase as the customers and as the products increase.

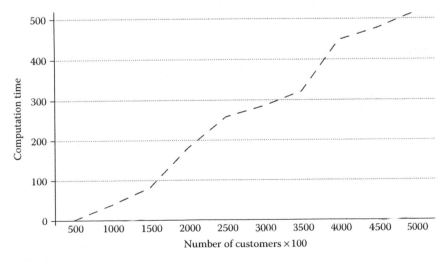

FIGURE 10.10
Computation time vs. number of customers.

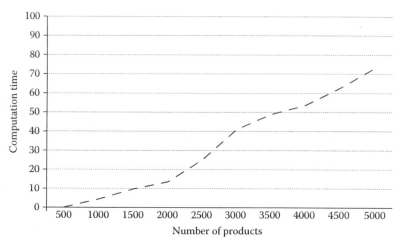

FIGURE 10.11
Computation time vs. number of products.

Conclusion

In this chapter we presented a context-aware architecture for e-commerce systems. The architecture allows a developer to design context-aware e-commerce applications. The architecture uses a methodology presented to facilitate development of context-aware e-commerce applications. We also presented extensive performance evaluation experiments to illustrate the usefulness of the approach. Our results show that the approach supports the seamless integration of various context-aware trust threat solutions.

References

1. Loke, S. *Context-Aware Pervasive Systems.* New York: Auerbach Publications, 2006.
2. Mahmood, S., Al-Barrak, A., and Al-Mulhem, M.S. "Towards requirements analysis and assessment of pervasive systems." *Proceedings of the World Congress on Engineering*, vol. II, July 4–6, London, *Lecture Notes in Engineering and Computer Science* 2198, International Association of Engineers, pp. 842–844, 2012.
3. Weiser, M. "The computer for the 21st century." *Scientific American*, 265(3): 94–104, 1991.
4. Schilit, B. and Theimer, M. "Disseminating active map information to mobile hosts." *IEEE Network*, 8(5): 22–32, 1994.
5. Brown, P.J. "The stick-e document: A framework for creating context-aware applications." *Proceedings of the Electronic Publishing*, vol. 8 (2/3), June & September, Electronic Publishing, Palo Alto, CA, pp. 259–272, 1996.

6. Hull, R., Neaves, P., and Bedford-Roberts, J. "Towards situated computing." *Proceedings of the First International Symposium on Wearable Computers (ISWC)*, Digest of Papers, IEEE Computing Society, Cambridge, October 13–14, pp. 146–153, 1997.

7. Ryan, N., Pascoe, J., and Morse, D. "Enhanced reality fieldwork: The context-aware archaeological assistant." *Computer Applications & Quantitative Methods in Archaeology*, V. Gaffney, M. van Leusen, and S. Exxon, Editors. British Archaeological Reports, Tempus Reparatum, Oxford, pp. 34–45, 1998.

8. Pascoe, J. "Adding generic contextual capabilities to wearable computers." *The 2nd International Symposium on Wearable Computers*, Pittsburgh, PA, October, IEEE Computer Society Press, Washington, DC, pp. 92–99, 1998.

9. Dey, A.K. "Context-aware computing: The CyberDesk project." *AAAI 1998 Spring Symposium on Intelligent Environments, Technical Report No. SS-98-02*, AAAI Press, pp. 51–54, 1998.

10. Schmidt, A., Beigl, M., and Gellersen, H.W. "There is more to context than location." *Proceedings of the International Workshop on Interactive Applications of Mobile Computing (IMC)*, November, Rostock, 1998.

11. Chen, G. and Kotz, D. "A survey of context-aware mobile computing research." *Technical Report No. TR2000-381*, Dartmouth College, Hanover, NH, 2000.

12. Dey, A.K., Abowd, G.D., and Salber, D. "A conceptual framework and a toolkit for supporting the rapid prototyping of context-aware applications." *Human-Computer Interaction*, 16(2–4): 97–166, 2001.

13. Patterson, C.A., Muntz, R.R., and Pancake, C.M. "Challenges in location-aware computing." *IEEE Pervasive Computing*, 2(2): 80–89, 2003.

14. Tamminen, S., Oulasvirta, A., Toiskallio, K., and Kankainen, A. "Understanding mobile contexts." *Proceedings of MobileHCI*, Udine. Springer-Verlag, London, pp. 17–31, 2003.

15. Dourish, P. "What we talk about when we talk about context?" *Personal and Ubiquitous Computing*, 8(1): 19–30, 2004.

16. de Almeida, D.R., de Souza Baptista, C., da Silva, E.R., Campelo, C.E.C., de Figueiredo, H.F., and Lacerda, Y.A. "A context-aware system based on service-oriented architecture." *Proceedings of the 20th International Conference on Advanced Information Networking and Applications (AINA)*, vol. 1, April 18–20, IEEE Computer Society, Washington, DC, pp. 205–210, 2006.

17. Korkea-aho, M. "Context-aware applications survey." http://www.cse.tkk.fi/fi/opinnot/T-110.5190/2000/applications/context-aware.html (Retrieved: January 3, 2013).

18. Chalmers, D. "Contextual mediation to support ubiquitous computing." PhD thesis, Department of Computing, Imperial College London, London, 2002.

19. Bennington, B.J. and Bartel, C.R. "Wireless Andrew: Experience building a high speed, campus-wide wireless data network." *Proceedings of the 3rd Annual ACM/IEEE international Conference on Mobile Computing and Networking (MobiCom)*, September 26–30, Budapest, ACM, New York, pp. 55–65, 1997.

20. Woodruff, A., Szymanski, M.H., Aoki, P.M., and Hurst, A. "The conversational role of electronic guidebooks." *Proceedings of the 3rd International Conference on Ubiquitous Computing, Lecture Notes in Computer Science* 2201, Atlanta, GA, Springer-Verlag, Berlin/Heidelberg, pp. 187–208, 2001.

21. Borcea, C., Gupta, A., Kalra, A., Jones, Q., and Iftode, L. "The MobiSoC middleware for mobile social computing: Challenges, design, and early experiences." *Proceedings of the 1st International Conference on Mobile Wireless Middleware,*

Operating Systems, and Applications (MOBILWARE), vol. 278, Innsbruck, February 13–15, ICST (Institute for Computer Sciences Social-Informatics and Telecommunications Engineering), Brussels, pp. 1–8, 2008.

22. Zhu, F., Mutka, M.W., and Ni, L.M. "Service discovery in pervasive computing environments." *IEEE Pervasive Computing*, 4(4): 81–90, 2005.

23. Norman, D.A. *The Invisible Computer*. Cambridge, MA: MIT Press, 1998.

24. Want, R., Hopper, A., Falcao, V., and Gibbons, J. "The active badge location system." *ACM Transactions on Information Systems*, 10(1): 91–102, 1992.

25. Drozd, A., Benford, S., Tandavanitj, N., Wright, M., and Chamberlain, A. "Hitchers: Designing for cellular positioning." *Proceedings of the 8th International Conference, UbiComp*, Dourish, P. and Friday, A., Editors, Orange County, CA, September 17–21, *Lecture Notes in Computer Science* 4206, Springer, Berlin, pp. 279–296, 2006.

26. Jiang, X. and Camp, T. "A review of geocasting protocols for a mobile ad hoc network." *Proceedings of the Grace Hopper Celebration (GHC)*, Vancouver, BC, October, 2002.

27. Salber, D. and Abowd, G.D. "The design and use of a generic context server." *Proceedings of the Perceptual User Interfaces Workshop (PUI)*, IEEE Computing Society, San Francisco, CA, November 5–6, pp. 63–66, 1998.

28. Abdul-Rahman, A. "A framework for decentralised trust reasoning." PhD thesis, University College London, London, 2005.

29. Aberer, K. and Despotovic, Z. "Managing trust in a peer-2-peer information system." *10th International Conference Information and Knowledge Management (CIKM)*, ACM, New York, pp. 310–317, 2001.

30. Damiani, E., Vimercati, S.D.C., Paraboschi, S., Samarati, P., and Violante, F. "A reputation-based approach for choosing reliable resources in peer-to-peer networks." *9th ACM Conference on Computer and Communications Security*, ACM, New York, November, pp. 207–216, 2002.

31. Kamvar, S., Schlosser, M., and Garcia-Molina, H. "The eigentrust algorithm for reputation management in P2P networks." *12th International World Wide Web Conference*, ACM, New York, pp. 640–651, 2003.

32. Papaioannou, T. and Stamoulis, G. "An incentives' mechanism promoting truthful feedback in peer-to-peer systems." *The 5th IEEE International Symposium on Cluster Computing and the Grid (CCGrid)*, IEEE Computing Society, pp. 275–283, 2005.

33. Wang, Y. and Vassileva, J. "Bayesian network-based trust model." *Proceedings of the IEEE/WIC International Conference on Web Intelligence (WI)*, Springer-Verlag, Berlin/Heidelberg, October 13–17, pp. 372–378, 2003.

34. Xiong, L. and Liu, L. "Peertrust: Supporting reputation-based trust for peer-to-peer electronic communities." *IEEE Transaction on Knowledge & Data Engineering*, 16(7): 843–857, 2004.

35. Yu, B. and Singh, M.P. "An evidential model for distributed reputation management." *1st International Joint Conference on Autonomous Agents and Multi-Agent Systems (AAMAS)*, Bologna, Italy, July 15–19, pp. 294–301, 2002.

36. Sen, S. and Sajja, N. "Robustness of reputation-based trust: Boolean case." *1st International Joint Conference on Autonomous Agents and Multi-Agent Systems (AAMAS)*, ACM, New York, pp. 288–293, 2002.

37. Abdul-Rahman, A. and Hailes, S. "Supporting trust in virtual communities." *Hawaii International Conference on System Sciences*, Hawaii, IEEE Computing Society, Washington, DC, 2000.

38. Azzedin, F., Maheswaran, M., and Mitra, A. "Trust brokering and its use for resource matchmaking in publicresource grids." *Journal of Grid Computing*, 4(3): 247–263, 2006.
39. Whitby, A., Jøsang, A., and Indulska, J. "Filtering out unfair ratings in bayesian reputation systems." *The ICFAI Journal of Management Research*, 4(2): 48–64, 2005.
40. Wang, Y., Li, M., Dillon, E., Cui, L.G., Hu, J.J., and Liao, L.J. "A context-aware computational trust model for multi-agent systems." *Proceedings of the IEEE International Conference on Networking, Sensing and Control*, Springer-Verlag, Berlin/Heidelberg, pp. 1119–1124, 2008.
41. Gias Uddin, M., Zulkernine, M., and Iqbal Ahamed, S. "CAT: A context-aware trust model for open and dynamic systems." *Proceedings of the 2008 ACM Symposium on Applied Computing*, Fortaleza, Ceara, Brazil, ACM, New York, 2008.
42. Sydow, M. "Towards context-enriched trust prediction: A proposal." *Proceedings of the International Workshop on Combining Context with Trust, Security and Privacy*, June 16, Trondheim, Norway, 2008.

11

A Walk-Through of Online Identity Management

Ravi Sankar Veerubhotla and Richa Garg

CONTENTS

Overview

Identity management (IdM) is the process of uniquely identifying and managing the entities within a local or global context. Typically, the term "identity" defines an individual or an entity. Digital identity is a combination of identity information, credentials, and a set of attributes associated with a particular entity. Access management techniques define certain rules on these identities to provide access to the resources or applications. Thus, *identity and access management* (IAM) is a union of IdM and *access management* techniques, describing a set of processes, techniques, and policies for managing digital identities and governing these identities on how to access the legitimate resources.

Entities can have multiple identities when working across diverse systems or in different roles. Today, with the proliferation of social networking websites such as Facebook, Google Plus, and LinkedIn, users can create unlimited online accounts using legitimate or fake identity information. Social networks allow these users to represent themselves in as many *avatars* as they wish. In this situation, how can one be sure of the legitimacy of the identity while dealing online? The answer to this question is *online identity management* (OIM), which provides a set of methods for generating unique identity information and managing it on the Internet. OIM is a multidisciplinary area having different facets such as technical, legal, organizational, and social aspects. This chapter focuses on technical aspects of OIM with a mild emphasis on reputation management.

IdM Life Cycle

IdM life cycle (Figure 11.1) comprises the entire life cycle of digital identities [1]. A brief description of IdM life cycle is as follows:

Provisioning. Provisioning is the process of creating a digital identity and populating it with correct attributes. The attributes could be name, e-mail, phone number, geographical location, etc.

Usage. Once the digital identity is created, it can be used across various systems or applications.

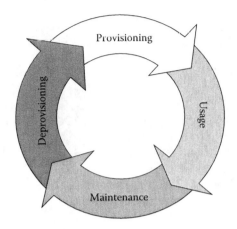

FIGURE 11.1
Life cycle of identity management.

Maintenance. This phase ensures that the attributes attached to an identity are kept updated so as to maintain its integrity.

Deprovisioning. Deprovisioning is the process of revoking the identity, when it becomes obsolete or invalid.

Traditional Identity Management

In the dawn of IdM, user management was limited to the organization's boundaries. Now, with globalization, innovative business strategies are being adopted. Consequently, new challenges in IdM have emerged. Hence, there is a need for reliable IdM solutions that can provide security as well as flexibility across the boundaries. A brief outline of the major technologies used in traditional IdM systems is as follows.

Role-Based Access Control

Role-based access control (RBAC) is based on the idea of using roles to enforce secure access inside a domain's boundary. In RBAC, access permissions are associated with the roles and users are part of appropriate roles. This greatly simplifies authorization management while giving more flexibility to implement the protection policies for an organization. Roles are defined as per the organization's structure and users are assigned these roles. If the users' role changes, they can be reassigned to another role without altering the underlying structure. These roles can be granted new permissions or the existing permissions can be revoked, as per the business need. This RBAC framework provides a great flexibility and capability to normalize user activities in a particular role.

Attribute-Based Access Control

Attribute-Based Access Control (ABAC) is a rule-based approach where attributes are used as building blocks. Attributes are the sets of labels or properties which can be used to define an entity, and must be considered for authorization determinations. Rules specify conditions under which access can be granted or denied. Each attribute consists of a key–value pair such as "role = manager." This approach might be more flexible than RBAC because it does not require separate roles for relevant sets of attributes, and rules can be applied quickly to accommodate changing needs. RBAC is often criticized for the complexity involved in setting up an initial role structure as well as for its rigidness in rapidly changing environment. In ABAC, there is a trade-off between flexibility and complexity.

Single Sign-On

Single sign-on (SSO) is an access control method for related and independent software systems or applications. User logs into the SSO only once to gain access to several systems without being prompted for credentials each time. SSO systems can be deployed within a single enterprise or across multiple enterprises and on the Web.

Industry Practices

Today organizations deal with heterogeneous applications and enormous data. Statutory laws such as Health Insurance Portability and Accountability Act (HIPAA) and the Sarbanes–Oxley (SOX) Act necessitate the use of IAM solutions for compliance. The benefits associated with IAM include granular access control to data and applications. It also provides flexibility, security, privacy, and compliance apart from the simplified user processes and reduced operational costs. Forrester, Burton, and Gartner groups are promoting the best practices for implementing IdM. According to them, the IAM solutions should be adaptable to broader IT infrastructure, such as help desk management and change and configuration management. IAM implementation should not only consider logical security, network security, and information leak prevention but also consider physical security aspects.

Clearly, the IAM infrastructure for an enterprise cannot be implemented in a single step. Instead, it may be implemented in multiple phases or with incremental functionality so as to fit them into the overall goals and objectives of the organization. For this to happen, a good coordination between various stakeholders is necessary to arrive at an agreement on the policies, procedures, and workflows to be implemented. Automated user provisioning with integrated workflow process can create identities across multiple systems. Once the identity is created, appropriate permissions are given to the user to perform the assigned tasks. Any errors in this phase can lead to grim situations where the user may either have too many privileges or lack the adequate permissions to perform the tasks. Password synchronization techniques can enhance user experience as the same credential can be used across multiple platforms. Password reset functionality provides the flexibility to the users to reset their passwords without adding any additional complexity.

Online Identity Management

OIM is the key to facilitate trust among communicating parties. OIM supports digital interactions over a public or private network, mainly in two dimensions. First dimension is the personal branding, wherein a distinguished Web presence of an individual on the Internet is preserved. Primarily, it deals with building the online reputation for an entity by maximizing the positive references, associated with an individual. The second dimension is rather technical and deals with the challenging issues such as trust, identity sharing, and protecting user privacy. This also includes techniques on how to ensure the authenticity of the digital identity, what level of trust can be attributed to a digital identity, and how the identities can be used across organizations' boundaries.

Introduction

The Internet is highly susceptible and it is very essential to identify with whom we are dealing online, before getting involved in any kind of transactions. OIM is a collection of administrative technologies for generating, maintaining, and controlling individual's unique digital identity on the Internet. The key participants in OIM are as follows:

- Subject is an individual or a group who seeks a digital identity.
- Identity provider (IdP) is the entity that creates, issues, and maintains credentials of the subject, and provides subject's identity-related information to third parties or to other subjects.
- Relying party (RP) relies on IdPs for identity-related information.

Goals and Requirements

The major objectives of OIM are as follows:

- Generating and maintaining the digital identity of the users
- Supporting authentication of digital identities
- Providing access rights to individuals at the right time

- Establishing trust for third parties who are dealing with the online entities
- Preventing identity misuse or theft
- Addressing online reputation problems
- Using digital identity in a scalable and cost-effective manner across networks

How Online Identity Works

OIM fundamentally involves three processes:

1. The process of identifying and issuing an identity credential to reflect the subject's identity, known as identification
2. The process of verifying the credentials presented by the subject named as authentication
3. The process of determining what rights and privileges are permitted to such person, after authentication, known as authorization

A detailed look at the foundations of IdM is as follows:

1. *Identification.* It is the act through which the subjects present themselves. It involves capturing a set of attributes of the subject in order to identify and define the identity to the level sufficient for the intended purpose. It typically involves the collection of personal attributes about the subject. The attributes can be permanent, temporary, acquired, or inherited. Permanent attributes include date of birth and social security number, whereas temporary attributes include current employer name. Educational degrees fall under acquired attribute category, but attributes such as DNA and user hierarchy are inherited.
2. *Scope and accuracy of identity.* Identification is usually measured across two different magnitudes, namely, scope and accuracy.
    a. Scope identifies the purpose of identification and considers relevant attributes for identification. It determines the type and amount of identity information needed. For example, some Web sites collect only basic information such as salutation, name, and e-mail id of the user to grant access to the content, whereas sensitive Web sites do require ample user information such as age and address before granting access.
    b. Accuracy aspect focuses on the precision of the identity attributes. It involves third parties to verify the attributes of the subject and establish trust. Public key infrastructure (PKI)-based digital certificates play a vital role in establishing trust among the communicating parties.

3. *Issuance of credential.* Post identification, the subject's identity is typically embodied in an electronic format, called identity credential. PINs, passwords, and digital certificates take the digital form of identification, whereas identity cards and hardware tokens provide physical identification for the given subject.

4. *Authentication.* It is the process of establishing trust and confidence in the subject's claimed identity. This is a critical step in establishing the digital identity. Authentication can be carried through any of the following factors or their combination:

   - What the subject has, for example, hardware tokens
   - What the subject knows, for example, PIN, password
   - What the subject is, for example, biometrics information

   In order to keep attackers at the bay, multifactor authentication is widely being used. A multifactor authentication requires the presentation of two or more factors stated above. When there is an access request to sensitive data, strong authentication methods using hardware tokens or biometric authentication is used for verification. Biometrics is a powerful technology that aims to uniquely identify an individual by capturing and analyzing their physiological and behavioral characteristics. Fingerprints, face, hand geometry, and iris recognition capture physiological characteristics, whereas signature, voice, and keystroke dynamics use behavioral characteristics.

5. *Authorization.* Once a user is successfully authenticated, an authorization process decides what set of actions and activities the subject is allowed to perform. Thus, authorization determines rights or privileges associated with the subject's identity, to facilitate a transaction or decision. This is a critical defense against identity theft.

6. *Assurance levels.* Identification and authentication are the most important for the functioning of any access control. But there are some possibilities by which an attacker can forge the identity. One way is when the attacker masquerades the identity of the subject. The alternate way is when the authentication information (e.g., PIN or password) is compromised, allowing the attacker to successfully complete the authentication process. Because of these risks, the relying parties must consider the degree of confidence in the subject's identity. Different types of transactions will require different assurance levels. The US federal government has defined four levels of assurance as follows:

   1. Little or no confidence
   2. Some confidence
   3. High confidence
   4. Very high confidence

If the risks associated with a transaction are high, the assurance level during authentication should also be high. The assurance will be naturally high when the concerned parties agree on a common set of policies and abide by the rules.

Motivation and Challenges

Today, an increasing number of populations use the Internet as a first-hand instrument to vouch for someone's identity. However, phishing and malware attacks are commonly used by intruders to steal the credentials of online users and impersonate those individuals. So, the authenticity of the online information and verification of digital identity on the Internet have become essential. Social networking sites facilitate good collaboration, better productivity, and more business. Hence, organizations are embracing them. But the level of assurance provided by social networking sites is low for the relying parties. It is partly due to privacy concerns among users, such as what identity information is to be revealed to the RP. Does the basic information—such as name and age sufficient, or comprehensive information including date of birth, bank account, or financial history—need to be shared? These concerns are genuine and intended to protect individual's sensitive information from the hands of wrongdoers. IdM shall address these problems and strike a balance between privacy, security, and usability. In specific, the major challenges in OIM include the following:

- Ensuring that identity information is accessible to appropriate parties
- Establishing trust among the parties involved in identity transactions
- Making use of identity in a scalable, usable, and cost-effective manner
- Building online reputation management system
- Conducting security and identity audits to avoid identity misuse

Trust Management

Trust management comprises identification and analysis of trust relations. In traditional transactions, the notion of *seeing is believing* played a vital role. However, this is no longer valid in today's virtual world. The challenge indeed is to determine how trust relationships can be created and measured when dealing online.

The Who, What, and How of Trust Management

International Telecommunications Union (ITU) defines trust between two entities A and B as; entity A is said to trust entity B when entity B behaves exactly in the same way as entity A expects. This "expected" behavior is determined by the rules established between the two entities. Trust has to be quantifiable and verifiable. Trust can be captured based on the following IAM parameters as described in Ref. [2]:

- *Identification trust level (ITL)*. It represents the level of assurance in the authenticity and integrity of an individual's claimed identity.
- *Authentication trust level (ATL)*. It verifies and confirms the individual's proclaimed identity. Credentials issued as a result of identification are used as proof when the individual claims the identity.
- *Reputation trust level (RTL)*. It aims to reinforce the quality for online communities. Trust signifies that the two parties will behave as expected, whereas reputation defines entity's standing in the community. Both trust and reputation are mutual and bidirectional. For example, a credit card company may conduct background verification for a prospective customer before issuing a credit card. In a similar way, the customer can also check the reputation of the credit card company before accepting the offer.

Trust and reputation management system [3,4] collects a set of facts about the service provider (SP) as shown in Figure 11.2. In this system, customer provides feedback about the SP. The collected information about the SP is aggregated and notified to relying parties as the reputation level of the SP.

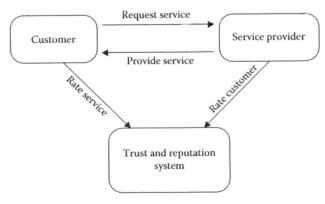

FIGURE 11.2
Trust and reputation management.

In an alternate scenario, an entity can express trust with another entity on a social networking Web site. However, this approach is highly user centric and hence subjective.

Advantages and Limitations

Trust and reputation system plays a crucial role in social networking and e-commerce Web sites. It provides a decision support system for online users who cannot check the authenticity physically. Users rely on this system to reinforce trust in an online entity. The assurance from trust management system encourages the users to use online medium to transact. Trust management system also boosts the information sharing and collaboration, thereby bringing more productivity.

However, there are some drawbacks [5] which make this decision support system weaker. A few of them are mentioned below:

- *Playbooks.* In this type of attack, an entity acts honestly and provides quality services for certain time period. Once it gains a high reputation score, it subsequently uses its reputation for making profits by providing low-quality services.

- *Unfair ratings.* In this case, the entity does not provide a genuine opinion about the other entity/services. Instead, the opinion may be exaggerated or underrated. This behavior is often considered to be unethical. However, it is particularly difficult to determine when this attack occurs.

- *Collusion.* This means that a group of entities establish certain behavior among each other, which could be running playbooks, providing unfair recommendations. It can have significant influence on rating, and hence increase the profit margins for the group entities.

- *The Sybil attack.* Here, one entity creates multiple false identities and provides multiple ratings to the same entity or service. This facilitates the attacker to get an unfair and large influence over the aggregated scores.

In order to overcome these limitations, several corrective measures have been proposed and implemented in trust management systems. E-commerce Web sites rate the registered SPs or merchants based on the customer's feedback which is aggregated at regular intervals. It helps the prospective customer to make an informed decision whether to make a purchase (from a specific merchant) or not. These Web sites also allow SPs to justify their stand, in case the customer provides unfair rating. Similarly, some novel approaches are suggested in the literature to avoid collusion and Sybil attack [6].

Web Services Trust Protocol

Web Services Trust (WS-Trust) is a framework for establishing trust between two parties in either homogeneous or heterogeneous domains. This protocol facilitates the exchange of trusted messages over simple object access protocol (SOAP), through security tokens between the two parties. A security token is typically a set of assertions made about the subject by the token issuer. In principle, WS-Trust framework [7] describes the token as a block of data that contains assertions about the user and some authentication information, needed for the application. It can use various types of tokens such as X.509 certificates, Security Assertion Markup Language (SAML), or Kerberos-based tokens as per agreed format and the requirements. The issued token is to be presented by the subject to gain access to a resource.

Figure 11.3 depicts a direct trust model. The sequence of steps in this model is as follows:

- The user requests a token from security token service (STS). This request contains the application requirements along with identity information of the user.
- STS validates the user.
- STS sends the token to the authenticated user. This token contains the assertions about the user, keys, and other information that the user can present to the resource application.
- User requests for the access by presenting the issued token to the application. Application, in turn, validates the token and grants access.

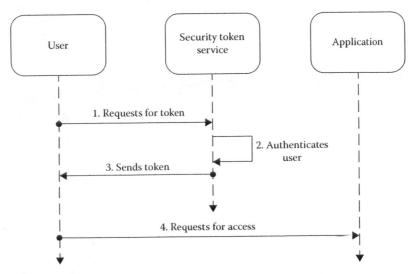

FIGURE 11.3
Direct trust model in WS-Trust.

The major responsibilities of STS are as follows:

1. *Token issuance.* STS, initially, performs the validation of user's presented identity and issues a token in case of successful validation.
2. *Token exchange and validation.* When the identity is to be used across different domains, the STS of communicating parties exchange and validate the token issued by the other party STS.

The direct trust model can be extended to other scenarios when the user needs to access the resource in different domains. This model is popularly known as trust federation or indirect trust model. In this model, there is an implicit trust between the STS of different domains. This trust can be established either by using PKI or Kerberos. The security token issued by domain A STS (STS_A) is presented as credentials to domain B STS (STS_B), to gain access to the resources in domain B. This model is efficient and facilitates interoperability, thereby avoiding the replication of identities.

Role of X.509 Certificates

Indirect trust model works well when the two heterogeneous domains agree on a predefined set of policies and decide to remain compliant with each other. But situations become complicated when the two parties cannot or do not want a federation binding. For example, users may not be comfortable signing any policy with the SP on a public network. Another limitation in token-based approach is that if the attributes are changed, a new token needs to be issued. If a real-time access control decision is to be made, tokens having incorrect/obsolete attributes may leave open doors for attackers. These issues can be addressed by the X.509-based trust management scheme shown in Figure 11.4. The steps in X.509 certificate-based trust management are as follows:

1. The user requests the service from the SP by presenting his/her X.509 certificate.
2. The SP validates the certificate and extracts certain attributes from his/her certificate such as name.
3. The SP requests more attributes about the user from the IdP.
4. The IdP validates the SP and sends the user's attributes to the SP.
5. The SP collects these dynamic attributes and then makes an informed decision about granting service to the end user.

In the above case, the user directly authenticates to the SP using standard X.509 certificates. In case the SP needs more attributes for authentication, it contacts the IdP directly, circumventing the user. This substantially reduces the authentication load from the user.

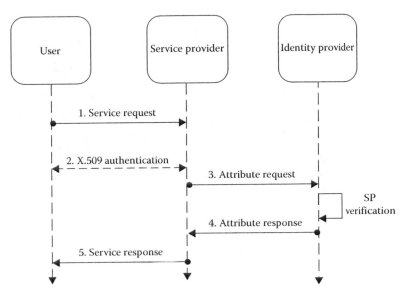

FIGURE 11.4
X.509-based trust management.

Claim-Based Identity

Claim-based identity [8,9] is a technique used by applications to obtain the user's identity information in the form of token from within or across organizations. The scheme involved here is very much similar to Kerberos, which is currently one of the most widely accepted authentication protocols.

Need for Claim-Based Identity

If an application needs to be accessed across organizations or over public network, claim-based identity can be used, wherein a user is provided with a token, which can be presented to gain access to the system. The key players are IdPs, RP, subject, and a few new participants such as claims, tokens, and STSs. A claim is a piece of information about the subject, for example, name, age, location, and so on. A token contains one or more claims, which the subjects can present to the RP to authenticate themselves. These tokens are created and digitally signed by a software application, called STS, to avoid any kind of misuse.

Claim-Based Architecture

IdPs, claims, tokens, and STSs are the building blocks of a claim-based architecture. Here, the aim is to facilitate the subject to present their digital identity to the application, and then the application decides the access

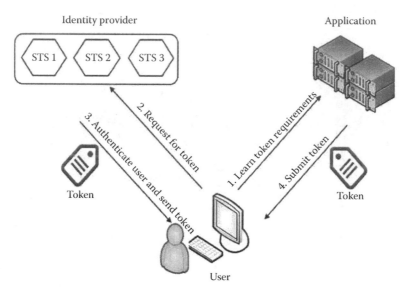

FIGURE 11.5
Claim-based architecture.

permissions, associated with the digital identity. Each application has its own requirements or criteria to filter the users for authentication. Hence, in claim-based authentication, it is important for the subject to understand the identity requirements for the application and approach appropriate STS for token issuance. The subject will be authenticated only when the application trusts the identity provided. Figure 11.5 presents the overview of claim-based architecture. The steps involved are as follows:

1. The subject understands the identity requirement of an application.
2. The subject makes a request to STS as per the identity requirements needed for the application.
3. The STS authenticates the subject and sends the token to the subject. This token is digitally signed by the STS.
4. The subject submits this token to the application, and if authorized, it gains access to the application.

This model can further be modified to suit the actual requirements of an organization. For example, in step 3, the STS authenticates the subject either by using simple passwords, Kerberos, or any alternate method. For creating tokens, the STS may capture the subject's identity by using an online repository. The token format may follow a standard or any other agreed format between the STS and the application. The issuer may also limit the life of the token for security purposes.

SSO for Web: An Example for Claim-Based Identity

SSO can be used within an organization. However, authentication becomes challenging when organization's internal applications need to be accessed remotely via the Internet. Claim-based approach provides a viable and flexible authentication solution to this problem, without modifying the underlying application. Figure 11.6 illustrates a Web-based SSO for this scenario.

Consider an example in which the employee of an organization tries to access an internal application via public Internet. Initially, the user learns the identity requirements of the application and then makes a request to STS for the token. The STS authenticates the user and returns the token. The STS could be local to the organization or may be a trusted third party. In either way, the application trusts the STS. The token is then submitted to the application for identification purpose. Once validated, the user can access the internal application via the Internet. Hence, claim-based approach simplifies the situation just like SSO scenario inside the enterprise, without having to create a virtual private network (VPN) connection.

The above process may look simple, but a closer look reveals the complexity associated with user authentication, while requesting for the token in Step 2. Kerberos is one way of authentication, but it is limited to the organization's boundaries. For Internet, the user could provide a username and password. Because the users are employees of a particular organization, they

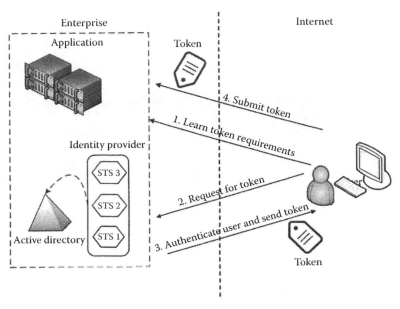

FIGURE 11.6
Web-based SSO.

would have user accounts in a repository (e.g., Active Directory), and hence they can login and authenticate themselves. Once logged-in, they can access legitimate applications without being prompted for credentials repeatedly. Although claim-based approach does not eliminate the need for credentials for Internet users, it does improve the overall situation.

Federated Identity

Federated identity management (FIM) [10] is an arrangement for IdM among multiple enterprises. It is a powerful technology that facilitates the end user to use the same credentials to gain access to the networks of all enterprises in the group. Hence, FIM is a super set of SSO and the process is referred to as identity federation. FIM offers economic advantages because of the data sharing. But in order to exploit its full potential, the organizations involved in the partnership must trust each other. Open interoperable standards such as OpenID and SAML can be used for authentication purposes. SAML 2.0 has been widely acceptable standard today, and it stimulates interoperability between different domains. It offers flexibility by abstracting the different underlying security infrastructures for different organizations.

Security Assertion Markup Language

SAML 2.0 [11], a product of OASIS Security Services Technical Committee (SSTC), is an XML-based framework for establishing a logical link between two different identities of a subject, each of which is managed by a different SP. The identity information is expressed in the form of portable SAML assertions so that applications working across the domain boundaries can trust each other. SAML 2.0 structure has four layers, namely, profile, binding, protocol, and assertion as shown in Figure 11.7.

1. An assertion is a claim or fact expressed on the subject. An SAML assertion is a package of information that is issued by IdP to the subject, containing a set of conditions, attributes, and authentication statements. This assertion can then be used by the RPs to make a decision on the subject. The subject can have multiple attributes. An authorization decision affirms that the subject is permitted to perform the desired action in the situation as specified in the statement.

2. The SAML protocol describes request–response pairs for exchanging SAML-related information. SAML protocols are independent of underlying data transmission protocols, such as HTTP/HTTPS/SOAP.

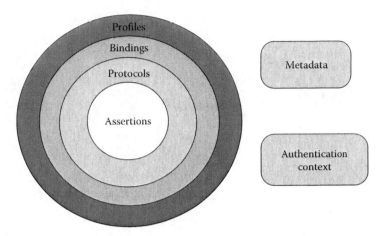

FIGURE 11.7
SAML 2.0 specification structure.

3. SAML binding describes the mapping between the SAML protocols and standard communication protocols (HTTP/SOAP). For example, the encapsulation of SAML protocol messages in the SOAP message format.
4. SAML profile is a combination of assertions, protocols, and bindings to support a defined user case. Different combinations allow different implementations of the same profile. At present, there are several profiles available, for example, SSO profile can be implemented as Web-based SSO profile, or enhanced client, or proxy profiles.

Besides these structured components, SAML 2.0 includes metadata and authentication contexts as shown in Figure 11.8. Metadata describe agreements, needed to establish trust relationships between IdPs and RPs. For example, metadata include digital certificates and bindings that can be used by IdPs and RPs. Authentication contexts provide appropriate information about authentication events.

Achieving Interoperability

The possible scenarios when the organizations usually face the interoperability issues, as described in Ref. [10] are as follows:

- When one organization is acquired and merged into another (B2B)
- When an organization outsource or take third-party service to another (B2B)
- Enhancing user experience in Web-based applications (B2C)

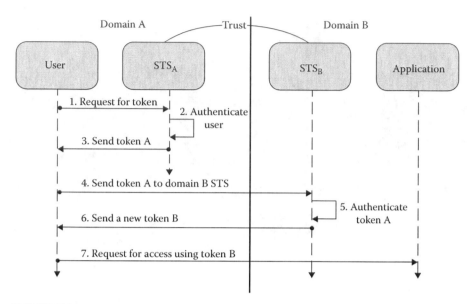

FIGURE 11.8
FIM architecture.

By considering one of the possible scenarios mentioned above, let us try to understand how FIM resolves the interoperability concerns. Figure 11.8 provides the high-level overview of the process. In this typical scenario, the end user tries to access the application that is outside its home domain A. Both domain A and domain B trust each other. The detailed steps are as follows:

- User requests a security token from STS server in its home domain, that is, STS_A. This is essential in order to access the application in domain B. The user first needs to authenticate himself/herself to STS_A by presenting its credentials.

- The STS_A verifies the user credential. Once authenticated, it issues a security token to the user, which can be used to communicate with STS_B. The issued token can include additional information such as claims for authorization purposes.

- The user presents the security token to the STS_B, issued by the STS_A, and makes a request for a new token from the STS_B to access the application in domain B.

- The STS_B validates the security token, and also verifies that the STS_A is a trusted entity in the federation group. Once authenticated, the STS_B issues a new token to the user. The STS_B may use the claims as is from the STS_A or may transform into other claims to be understood in its domain.

- Once the user gains the token from the STS_B, he/she presents this token to the application and requests for access.
- The application validates the security token and provides access to the user.

FIM facilitates the users to access resources in different domains without presenting additional credentials. This model is well adapted for Web application as well, and it can be tweaked as per the needs.

Open ID: How Google/Facebook Integrates with Other Applications?

OpenID [12] is an open standard framework in which users can create an account with an OpenID provider and subsequently use the same credentials to sign in to different Web sites that accept OpenIDs. Today, many Web sites allow their users to login with Google [13] and Facebook credentials, and support federated login. In fact, federated login has considerably simplified the login mechanism and has also taken off the load from the user to remember too many passwords. In a case when user opts to login to third-party Web sites using Google's credentials, the third-party Web site forwards the authentication request to Google. Google, in turn, returns an identifier which the third-party can use, to recognize the user. The overall scheme in OpenID is similar to that of SAML. But, OpenID provides comparatively simpler structure of identity-related data. OpenID has been widely adopted by Web 2.0 services such as blogs and social networks. Some of the provisions provided by Google are as follows:

- OpenID Attribute Exchange 1.0 facilitates different Web sites to access to different attributes of a user attributes name, e-mail, stored with Google. But this requires permission from the user.
- OpenID User Interface 1.0 provisions alternative user interface for authentication purpose. The default mechanism redirects the Web site to Google's authentication pages. Using OpenID user interface, Web developers can choose to open Google authentication page in a popup window or in a default manner.

Because OpenID is widely accepted, Facebook, the social networking giant, has also integrated itself to allow its users to login using OpenID credentials. But the reason why Facebook login has been widely accepted is because of "Facebook Connect" [14]. Facebook Connect offers the integration of third-party Web sites with itself just like OpenID. Facebook Open Graph API allows different Web sites to gain information about the individuals, after their consent. This includes photos, events, and their relationships apart from simple name and location.

Implementing OIM System

Before formulating an IdM strategy for an organization, it is necessary to conduct a survey to find out what is already in place that covers part of the field. It should also consult internal departments such as IT, physical security, HR, finance, and sales as they may have their own systems for user management. This survey shall examine the existing policies, procedures, business flows, workflows, approval process, hardware, and software. Later, it is also necessary to establish how the identity information will be accessed, be it within the organization, across organizations, or via the public network. A detailed analysis of risks associated with it should be carried out and a detailed roadmap should be in place. Finally, it is necessary to determine how the new IAM components or technologies can be integrated with the existing ones.

Identification of Scope

It is very important to define the business needs and their scope, precisely, for the solution to be designed. The major requirements of a robust IdM solution could be a few of the following:

- *Security policies, due care, due diligence, and risks.* Security policies must define appropriate level of controls. Policies should be created with due diligence and implemented with due care. Risk assessment in different categories, namely, technological, processual, and procedural challenges, and mitigation should also be considered at the initial stage.
- *Profile management.* The solution should bring the right roadmap in identification, creation, and management of unique user profiles. The solution must identify the roles and rules how accounts must be provisioned to the users.
- *Workflows.* For a robust IdM solution, workflow is a fundamental component. A workflow, in general, is a process that defines the sequence of actions to be taken in a certain order, either sequentially or in parallel. A robust solution must describe crisp workflows, which may include advanced features like timeout, escalations, and manual intervention in case of conflicts. IdM solutions should have automated workflows to reduce the administrative cost and turnaround time.
- *Single interface.* Organizations may have a complex infrastructure consisting of heterogeneous resources, such as different operating systems, servers, and databases. A good IdM solution must provide a centralized and cross-environment single-user interface to reduce the training and maintenance cost.
- *Central password management.* A good solution must have the ability for password synchronization across different platforms and services, thereby making it user-friendly and lowering the administrative cost.

- *User self-service.* A robust IdM solution should facilitate the user to reset his/her password of his/her own, based on a challenge–response scheme. The new password sharing can be done either via user's e-mail or by asking the user to enter a new password on the Web page. Additional capabilities such as generating audit records and notifications to administrative personnel may also be incorporated.
- *Centralized reporting and auditing.* Audit logs should be created in a centralized manner which would ensure timely reaction, in case of any breaches.

Requirement Analysis

A detailed requirement analysis is mandatory for a robust solution. This phase portrays what is "in" scope. The primary emphasis is on "what" not "how." One should have a clear understanding where the IdM solution will be deployed.

- *Within an enterprise.* When the solution needs to work within confined boundaries of an organization, traditional techniques will be a quick win.
- *Between enterprises.* If the boundaries are not debarring, more advanced solutions such as federated IdM should be considered.
- *Identity as a service (IDaaS) in cloud.* It refers to the management of digital identities in the cloud, public, or private. Here, the identity and access control management is provisioned as a third party. But, hybrid approaches can also be considered where identities are still maintained within the organization, but other components such as authentication can be moved to cloud. This will be conveniently effective where the initial investment cost is high and trust in the solution is less.
- *On the Internet.* If the identity has to work on public network, then trust management and claim-based identity should be considered.

Implementation Choices

When developing an IdM system, business requirements, ease of the solution, and the security aspects should be considered. A few of the best practices in implementing IdM solution are as follows:

- *Integrated system.* Consider implementing a single integrated system that provides end-to-end management of identities. This will help in mapping identities in complex systems instead of creating separate identities.
- *Automation.* User provisioning, deprovisioning, and reprovisioning are often time-consuming tasks when carried out manually. Automation of these tasks will help in reducing the manual errors and improve consistency.

- *Workflows*. A workflow process that includes approval process would be helpful in managing and tracking user access requests to perform sensitive tasks. Workflows should be chosen carefully to avoid system overheads.
- *Audit checks*. Periodic audit checks to be conducted to mitigate security risks.
- *Choose the right solution*. A careful decision should be made to choose the best IdM solutions available in the market that fit the organization's requirements after careful evaluation.

A few of the popular IAM products in the market include Oracle IdM Suit, IBM Tivoli, Novell Identity Manager, Microsoft Forefront Identity Manager, and SailPoint Identity IQ. These products offer various features ranging from user provisioning to IdM in the cloud.

Evaluation Criteria

Evaluation of IdM solution is very essential to verify its suitability to the business needs. Prior to IdM implementation, a checklist of requirements should be prepared, and each solution may be carefully evaluated and weighted. Additional care may be taken to get support from vendors for doing pilot or prototype implementations and to determine the right solution.

The right IdM solution should provide the following:

- *Ease of deployment*. The ease of deployment of an IdM solution depends on the level of openness of the system, the feature set, and the technologies used in the environment where IdM solution is to be introduced. The solution should be convenient to integrate in existing system without requiring any significant alteration. Also, all the complexities should remain abstracted.
- *Secure control*. The solution must provide the provision to control the access to resources and keeping logs of the resources accessed efficiently, regardless of the organization size.
- *Seamless integration*. The solution must have the capability to work in heterogeneous environment. The solution should work well within and across the boundaries of the organization. It may include integration with mobiles, cloud as well.
- *Cost reduction*. The solution should bring cost benefits to the organization. For example, reducing administrative costs by automating manual processes; self-help services, and so on.
- *User experience*. The solution should enrich the user's experience and abstract all the complexity from the user.

A typical example of OIM system is pan-European eIDM [15] system. The eIDM system at pan-European level works primarily on federated IdM along with legal associations and integration with private sectors.

Conclusions

IdM is an effort to make the digital identities available in a secure form. With the proliferation of online services, these digital identities gained momentum, thereby making the use of IdM solutions mandatory. In some scenarios, when the user of a particular IdM system seeks to authenticate to a different IdP, it should be able to recognize and authenticate the user. Keeping this in mind, different open standards such as OpenID, OAuth, or SAML have been designed to encourage the interoperability. Various IdM schemes such as claim-based IdM and federated IdM have also been designed to interact within or across the organization boundaries. However, the interacting parties over the Internet remain uncertain due to perceived risks such as masquerading of identities. The open challenges for an Internet-based IdM to be successful include the support of legal framework, active participation of government as well as private sectors, application integration, and trust management. Currently, a number of feature-rich and competitive identity solutions are available, which are capable of linking various heterogeneous environments. But a careful choice is to be made depending on the business requirements of the concerned entities or organizations. Last, but not the least, for a user, the usability matters. Thus, it is important to balance the business and security needs with ease of use while choosing the right IAM solution.

References

1. Bertino, E., Takahashi, K., 2011. *Identity Management: Concepts, Technologies, and Systems*. Boston, MA: Artech House.
2. Geest, G.J.V.D. and Korver, C.D.R., 2008. "Managing Identity Trust for Access Control." In Identity and Access, Journal 16, *The Architecture Journal*. http://msdn.microsoft.com/en-us/library/cc836389.aspx
3. Jøsang, A., 2010. "Trust Management in Online Communities." In *Workshop on New Forms of Collaborative Production and Innovation: Economic, Social, Legal and Technical Characteristics and Conditions*, May 5–7, SOFI, University of Göttingen.
4. Choi, H.C., Kruk, S.R., Grzonkowski, S., 2006. "Trust Models for Community Aware Identity Management." In *Proceedings of the 15th International World Wide Web (WWW)*, Edignburg, UK, May 22–26,

262   *Managing Trust in Cyberspace*

5. Jøsang, A., Golbeck, J., 2009. "Challenges for Robust Trust and Reputation Systems." In *Proceedings of Elsevier Science, 5th International Workshop on Security and Trust Management (STM)*, Saint Malo, France, September 24–25.

6. Yu, H., Kaminsky, M., Gobbons, P.B., 2006. "SybilGuard: Defending against Sybil Attacks via Social Networks." In *Proceedings of the ACM SIGCOMM Conference on Computer Communications (SIGCOMM 2006)*, Pisa, Italy, September 11–15.

7. Reeves, C., 2006. "Overview of Web Service Trust Language." CSEP 590 Practical Aspects of Modern Cryptography Project. www.cs.washington.edu/education/courses/csep590/06wi/finalprojects/reeves.doc

8. Baier, D., Bertocci, V., Brown, K., Pace, E., Woloski, M., 2010. *A Guide to Claim-Based Identity and Access Control.* Washington, DC: Microsoft Press.

9. Chappel, D., 2009. *Claim-Based Identity for Windows.* Sponsored by Microsoft Corporation. http://www.davidchappell.com/writing/white_papers.php

10. Federated Identity. http://msdn.microsoft.com/en-us/library/aa479079.aspx

11. OASIS Security Assertion Markup Language v2.0. *Technical Report.* http://docs.oasis-open.org/security/saml/Post2.0/sstc-saml-tech-overview-2.0-cd-02.html

12. OpenID: http://openid.net/get-an-openid/what-is-openid/.

13. Federated Login for Google Users: https://developers.google.com/accounts/docs/OpenID#AuthProcess

14. Facebook Platform: http://www.facebook.com/blog/blog.php?post=41735647130

15. Gracía, S.S., Oliva, A.G., Belleboni, E.P., 2012. "Is Europe Ready for Pan-European Identity Management System?" *Proceedings of the IEEE Security and Privacy Journal*, 10(4): 44–49.

12

Trust in Digital Rights Management Systems

Tony Thomas

CONTENTS

Introduction

Nowadays digital contents can be easily replicated and distributed without any loss of quality through the Internet and other portable storage devices. This has resulted in large-scale illegal replication and distribution of digital contents causing substantial loss to the content producers. Further, genuine customers may be unknowingly purchasing unauthorized or pirated digital contents available in the market. Thus, the overall trust in the digital content distribution has significantly come down due to the illegal sharing, piracy, and distribution of digital contents.

Digital rights management (DRM) technologies have been developed to protect the intellectual property rights of the producers of the content, rights of distributors, and rights of the consumers. Further, DRM enforces the duties of content producers, distributors, and consumers. Thus, DRM provides a trusted framework for digital content distribution through modern network systems. DRM uses combinations of cryptographic, digital watermarking, and licensing mechanisms to establish trust in the digital content distribution chain by protecting the rights and enforcing the duties of the producers, distributors, and consumers of digital contents. Encryption prevents unauthorized access to a digital content. However, once the content has been decrypted, it does not prevent an authorized user from illegally copying or sharing the content with others. Digital watermarking is used to complement the encryption techniques to establish and prove ownership rights and to trace copyright violators by embedding the seller's, distributor's, and buyer's information into the digital content. Thus, DRM mechanisms prevent consumers from unauthorized copying, use, and distribution of digital contents and enable the development of digital distribution mechanisms on which innovative business models can be implemented.

Various DRM architectures have been proposed for digital content and license distribution for traditional two-party scenario, where the owner and consumer (seller and buyer) are the only parties involved. However, a two-party architecture is insufficient to provide proper business scalability as it is too restrictive and may not make proper business strategies for all regions and cultures. To have more scalable business models that have the flexibility of packaging multiple contents together in a regional and culturally sensitive manner, it is necessary to have a flexible and hierarchical distribution network. Hence, multiparty, multilevel architectures involving multiple levels of distributors and subdistributors in addition to the owner and consumers have been used in many modern content distribution networks. A local distributor can better explore potentially unknown local market and can make strategies according to the market. In addition, distributors can also help in handling different price structure, quality, and sensitivity of media in different regions.

In this chapter, we first examine the trust issues in modern digital content distribution systems involving multiple parties such as content producers (owners), distributors, and consumers (clients). We then discuss how modern DRM technologies address these trust issues. The rest of this chapter is organized as follows. In the section "Digital Rights Management," a brief review of various components of DRM is given. The trust issues in modern digital content distribution systems are discussed in detail in the section "Trust Issues in Digital Content Distribution." In the section "Trust through DRM Systems," how the various components of the DRM are handling these trust issues are discussed. Finally, this chapter concludes in the section "Conclusion."

Digital Rights Management

DRM offers a trusted secure framework for distributing digital content such as music, video, images, and text data. DRM provides an electronic market-place where digital contents can be sold. DRM ensures that content providers receive adequate remuneration for the creation of the content and consumers are protected in the transactions.

DRM can be viewed as a group of rules, formats, and components used to protect digital content and the entities involved. Each content provider uses its own DRM mechanisms and formats. This makes the consumers to install different DRM systems to play/use different DRM-protected contents. Some of the current DRM-specific media players include Windows Media Player with the Windows media right manager and Real Player with the real system media commerce suite. Various DRM systems are distinguished by the difference in the cryptographic mechanisms, right expression language, and content packaging formats used. We now discuss these components in detail below.

Entities in a DRM System

Traditional DRM architecture involves two entities: a seller and a buyer. It has been observed that this architecture is not adequate to satisfactorily address the requirements of the present-day business models for content distribution [1]. Hence, multiparty, multilevel DRM architecture has been used as an alternative by many authors [1–3]. Here, the term "multiparty" refers to the multiple parties or entities such as the owner, distributors, subdistributors, and consumers; the term "multilevel" refers to the multiple levels of distributors/subdistributors involved in a digital content distribution system. The multiparty, multilevel DRM architecture used in Ref. [1] is given in Figure 12.1.

A content distribution chain is a chain of intermediate parties, where each party passes content to a party next to it in the chain before the content reaches the consumer. Depending upon the distribution chain, the parties involved can be an owner, multiple levels of distributors, and consumers. Number of levels of distributors in a distribution chain may depend upon the extent of the region and the number of consumers.

Digital Licenses

Digital license is a digital document used to establish an agreement between two entities. These entities are called the license issuer and the license requesting party. In the digital license, the license issuer allows the license requesting party to access the content regulated with the help of permissions, constraints, and content decryption keys. Permissions correspond to actions that can be performed on the content such as play, copy,

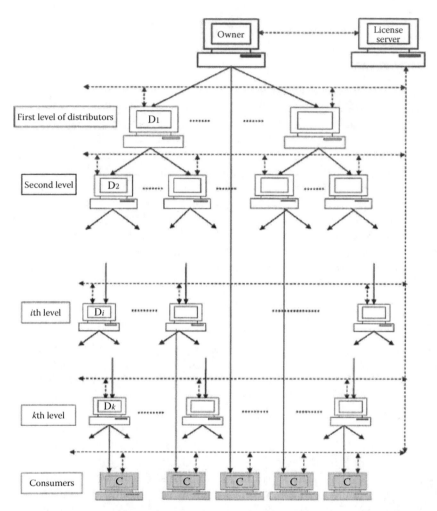

FIGURE 12.1
Multiparty, multilevel content distribution model.

edit, reuse, and redistribute. Constraints are limitations associated with the permissions in the license which can be time based, count based, location based, and so on. Content decryption keys are used to decrypt the encrypted content and can be used with a particular permission only if all the constraints associated with that permission are satisfied.

Licenses are created by the content owners and distributors for other distributors and consumers. License contains information such as identity of license issuing party, identity of the content, permissions, constraints, and decryption keys. There are mainly two types of licenses: redistribution licenses and usage licenses. Redistribution licenses are created by the owner or a distributor for redistribution of the content. Permissions given in redistribution licenses are permission for content redistribution and permission to issue new redistribution licenses. The permission for content redistribution allows distributors to create their own content package from one or more contents and upload the content on their own content server. Permission to issue redistribution license allows involving more distributors in the content distribution chain. Enforcement of redistribution license is done with the help of a license server, which keeps record of the licenses issued by the owner and distributors. Usage licenses are created jointly by the owner/distributor concerned and a license server. Consumers can use the content according to the permissions and constraints in the license with the help of the usage license. Enforcement of the usage license is done with the help of a trusted DRM agent at consumer's machine.

Trusted Platform Module

Trusted platform module (TPM), as per the specifications of the "trusted computing group" (TCG), is a tamper-resistant module [4]. TPM is a device that enables trust in computing platforms in general. The TPM contains a set of registers, called platform configuration registers (PCRs) containing various measurement digests. The values in PCR are temporal and are reset at system reboot. The only way for software to change the value of a PCR is by invoking the TPM operation. Authenticated boot, remote attestation, and secure storage are three principal operations of a TPM. A schematic diagram of the TPM module is given in Figure 12.2.

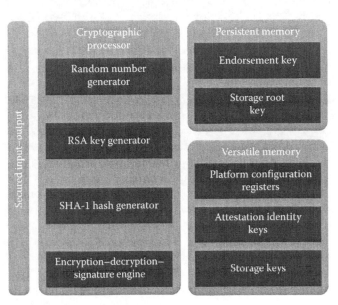

FIGURE 12.2
The components of trusted platform module.

TPM can check whether a platform is trusted. It measures the state of a platform during the boot process and stores the measurements in its registers. These measurements are typically based on the executable code involved in each stage of the boot process. In this process, a malicious code can be detected as it will cause the measurements to deviate from the expected values. This is called authenticated boot process.

Remote attestation is a process that provides proof of integrity of the host platform of the TPM to a remote party. The platform creates reports of its integrity and configuration state, which is signed by the TPM. This report can be verified by a remote verifier.

TPM protects all sensitive data by encrypting them using a nonmigratable Storage Root Key (Kroot) bound to it. The data are bound to the current platform configuration given by the PCR values.

Cryptographic Mechanisms for DRM

To protect digital contents from illegal use, various content providers encrypt their contents based on some desirable properties such as complexity, compression efficiency, perceptibility, format compliance, error resilience, scalability, and bandwidth expansion. A survey on various encryption mechanisms is given in Ref. [5]. A content provider encrypts the content using a symmetric-key encryption algorithm with the content encryption key and obtains the encrypted content.

Content scramble system (CSS) is an encryption mechanism employed on many commercially produced DVD-Video discs. CSS uses a 40-bit stream cipher algorithm for encryption of contents in the disc. CSS encryption mechanism provides a trust framework to the content distribution through discs as they cannot in principle be played on noncompliant devices; anyone wishing to build compliant devices must obtain a license, which contains the requirement that the rest of the DRM system be implemented.

CSS is getting replaced by newer encryption schemes such as content protection for recordable media (CPRM) or by block cipher encryption such as advanced encryption standard (AES) in the advanced access content system (AACS). Encryption schemes used by HD DVD and Blu-ray Disk, with 56- and 128-bit key sizes, respectively, provide a much higher level of security than the 40-bit key used in CSS.

The content encryption mechanism must be such that it must take care of the owner's concern over security of the content and distributors' concerns over illegal download of content from their content servers.

Watermarking for DRM

Digital watermarking mechanisms are used to protect digital documents against unauthorized use and distribution. It involves embedding hidden information about content owner, distributor, and consumer, which becomes

FIGURE 12.3
Effect of watermarking on an image.

inseparable with the digital content, even after copying and redistribution. Digital watermarking can be applied to digital audio, video, image, and text documents. The effect of watermarking on an image is shown in Figure 12.3. The four images in Figure 12.3 are given in the following order: first row, original image and image with one watermark (left to right); second row, image with two and ten watermarks (left to right), respectively. Figure 12.3 clearly shows the deterioration of the quality of the images with multiple watermarking [1].

For images, the hidden information can be a picture, such as a logo, carrying copyright information, or it can be a sequence of bits spread over the image according to certain algorithm. The quality of digital content such as image or video after watermark embedding can be estimated using the peak signal-to-noise ratio (PSNR) and structural similarity (SSIM) index between the original and the watermarked content. The formulas for these quantities are given below:

$$PSNR = 20\log\frac{255}{\sqrt{MSE}},$$

$$\text{MSE} = \frac{1}{mn}\sum_{i=0}^{m-1}\sum_{j=0}^{n-1}[\text{PV}_{\text{org}}(i,j)-\text{PV}_{\text{wat}}(i,j)]^2,$$

$$\text{SSIM}(x,y) = \frac{(2\mu_x\mu_y+c_1)(2\text{cov}_{xy}+c_2)}{(\mu_x^2+\mu_y^2+c_1)(\sigma_x^2+\sigma_y^2+c_2)}.$$

PSNR represents the ratio between the maximum possible power of a signal and the power of corrupting noise that affects the fidelity of its representation. SSIM is a perceptual measure used for measuring the similarity between two images. It is designed to improve on methods such as PSNR and MSE. The SSIM metric is calculated on various windows of an image.

Trust Issues in Digital Content Distribution

DRM involves not just providing a secure package containing the digital content and the accompanying metadata. DRM must also support distribution of the content package from the owner to the consumer through various distribution channels and agencies. DRM requires a trusted infrastructure that supports transport, opening, displaying, and disposing the digital contents. The major trust issues in the content distribution are discussed below.

Trust Issues for Content Owners

The content owners need to have trust in the entire digital content distribution system that there will not be any unauthorized usage such as illegal playing, copying, and distribution of their digital contents. There should be a licensing mechanism that provides different permissions and constraints for using their content. The licensing may be provided through issuance of usage and redistribution licenses through a license server. In that case, the content owners will have to the trust this license server and the licensing process. Third, a DRM agent is required at each client device that can perform actions such as playing, copying, and distribution of the content according to the licenses. This makes the owners to trust DRM agents used in the client system. The DRM agent may be coupled to TPM of the client device.

The owners will trust the content distribution system only if they can prove their ownership rights in the future in case of a dispute. Further, if there is some unauthorized use of the content due to system violation, there should be a mechanism to detect it. The unauthorized use may be detected with the help of usage logs as these can reveal the actual activities of the consumers. Therefore, usage logs should get created at the client device and reside in a trusted environment. To collect and analyze the logs, a trusted entity called log collection center should be created by the owner. The content encryption key of the owner should not be disclosed to any distributor or consumer as this may result in illegal redistribution or usage of the content. The license server could be trusted by the owner only if it stores and serves the content decryption key directly to the consumers as and when instructed by the owner or distributors. The DRM agent residing in the client devices also must prevent disclosure of keys to consumers.

Trust Issues for Content Distributors

Each content distributor in the architecture maintains a content server different from that of the owner. So, the distributors will trust the system only if they can prove their distributorship to other parties. Further, they need to get assurance regarding the protection of the contents in their content server from being downloaded by consumers of other distributors. This is a trust concern for a distributor because a malicious distributor can redirect his consumers to other distributor's content server for downloading contents without sharing profit with them. To resolve distributors' concerns and make the system trusted, a content packaging mechanism is required such that the contents in the content server of a distributor cannot be used by the owner or any other distributors. The trust framework should guarantee the distributors that their contents will not be illegally redistributed. They should be able to prove their distribution rights in the future in case of a dispute.

Trust Issues for Consumers

The consumers need to have trust on the content distribution system that the system provides them with authentic and legal contents. They should be able to prove their consumer rights in the future in case of a dispute.

In DRM systems, the distributors and users are accountable for any misuse of the contents and licenses they purchase. The content owners perform usage tracking and monitoring via license acquisitions and user authentication mechanisms so that accountability can be done. However, this process affects the user's privacy as it reveals the link between users and their usage patterns. User data gathered in this process can be used later to generate detailed profiles of the users and their purchasing history. The resulting profiles of the consumers can be misused by an untrusted content distribution system. Hence, in order for the consumers to trust any content distribution system, the privacy and anonymity of the users should be protected by the content distribution mechanism.

Trust through DRM Systems

A DRM system needs to provide sufficient trust to the content owners, the content distributors, and the consumers. We will now see how various components of a DRM system and advanced protocols address the trust issues discussed in the previous section.

Trust through Privacy Protection Mechanisms

Advances in DRM technologies have controlled the copyright violations of digital contents to a good extent. However, this has resulted in the violations of privacy of the entities involved [6,7]. The distributors and users in a digital content distribution system must be accountable for any misuse of their purchased contents and licenses. The accountability is achieved by performing usage tracking and monitoring through license acquisition transactions and user authentication mechanisms. However, these mechanisms for accountability affect the privacy of the users as it exposes the link between users, the contents they purchased and their usage patterns. These data can be used later to generate detailed profiles of the users and their activities which can be misused by the content providers.

To provide trust to the users, privacy-preserving content distribution mechanisms for DRM have been proposed by several authors [6–10]. However, accountability and privacy has not been considered together in a satisfactory way by any of the authors yet. Some of the existing content distribution mechanisms that take care of the accountability and privacy require the user to trust a third party. Whereas, other schemes avoid trusted third parties (TTPs) using complex cryptographic primitives such as zero-knowledge proofs, and they fail to satisfy many of the desirable security properties of a DRM.

A TTP is an entity that facilitates the interactions between two parties who both trust the third party [11]. In TTP-based systems, the entities use the trust to secure their interactions. However, in real life a TTP can become malicious. A malicious third-party scenario given in Ref. [11] is illustrated in Figure 12.4. Content distribution mechanisms based on TTPs must be avoided in DRM because users can never be assured that their privacy will be secured by these entities.

Privacy-preserving content distribution using TTP has been proposed in Refs. [12–15]. In Ref. [15], a mechanism using anonymity identity for providing privacy has been proposed. However, to get an anonymity ID, the users will have to trust an authentication server. However, a malicious authentication server can link all anonymity IDs to the user identities. In Refs. [12,14], the authors have addressed this problem by separating the responsibilities of certification authorities from that of the content providers. However, to revoke a user from future use, the content provider will have to collaborate with the certification authority to link the anonymity ID with the real

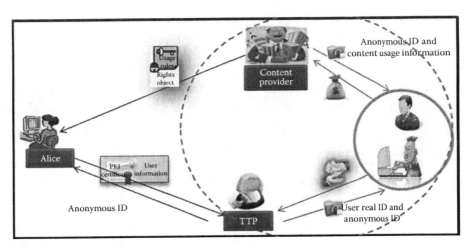

FIGURE 12.4
Malicious third-party scenario.

identity of the user. This weakens the privacy of the users as the trusted parties can collude against the innocent users. Complex cryptographic primitives such as "verifiable secret sharing," "zero-knowledge proofs," and "time capsule" have been used in Ref. [13], to design a privacy-preserving scheme for content distribution. However, this scheme requires trusting a user and two-revocation authorities.

In Refs. [6,7,9,16,17], the authors have proposed privacy-preserving content distribution mechanisms without using trusted third parties. In Ref. [17], an anonymous prepayment scheme is used to get anonymity ID. However, the real identity of the user is not getting authenticated in this scheme. Sun et al. [9] uses a mechanism called restrictive partial blind signature method for privacy-preserving consumption of digital contents. However, malicious users cannot be traced and revoked in this scheme. The problem of accountability has not been dealt in Refs. [6,7]. In certain schemes such as e-cash [18,19] and k-times anonymous authentication [20,21], the privacy of a user breaks down when a user performs the authentication operation more than a certain number of times above a threshold value. A privacy-preserving accountability mechanism for DRM using "zero-knowledge proofs" is given in Ref. [16]. However, the major limitation of this mechanism is that it requires many rounds of communications and it assumes that a user has unlimited computational power.

A privacy-preserving content distribution mechanism without using any trusted third parties is given in Ref. [11]. Their content distribution supports both accountability and privacy simultaneously. Hence, it provides trust to both content providers and consumers as the content owners can account for the usage of their contents while the privacy of the consumers is protected. Further, the authors have used simple cryptographic primitives such as blind decryption and hash chain to construct the proposed system. The authors have achieved a privacy-preserving revocation mechanism in their scheme. This preserves the anonymity of a user even after that user has been revoked for its misbehavior.

Trust through Digital Watermarking Mechanisms

Digital watermarks have been used for copy protection and copy deterrence of multimedia contents. This provides trust to the content owners on the DRM systems while distributing their contents. Copy protection and deterrence using digital watermarks are achieved by inserting a unique watermark into each copy of the digital content sold. A number of watermarking techniques have been proposed for various types of digital contents such as images, video, audio, and text files.

The watermarking techniques can be broadly classified into robust watermarking and fragile watermarking mechanisms. The purpose of robust watermarking technique is to resist several attacks such as signal processing and geometric distortion on the digital content. On the other hand, the purpose of fragile watermarking is to detect even slight changes to the digital contents. Robust watermarking prevents illegal copying and sharing of digital contents and thus provides trust to the content owners. Fragile watermarking provides authenticity protection to the content and thus provides trust to the consumers regarding the authenticity of the content they use.

The watermark information embedded in a digital content can be detected or extracted using a detection or extraction algorithm. This detection or extraction process may require a secret key. An important property of a watermark is its robustness with respect to distortions in the digital content. This means that the watermark should be detectable/extractable from the contents that underwent common signal processing operations, such as filtering, lossy compression, noise addition, histogram manipulation, and various geometrical transformations. Watermarking algorithms for copyright protection, fingerprinting, or access control must satisfy the standard security properties. This means that an attacker who knows all details of the embedding algorithm except the secret key should not be able to remove or destroy the watermark beyond detection. The watermark information embedded in a content known as capacity could be as low as one bit or several hundreds of bits. Another important aspect of watermarking is the computational complexity of the embedding and extraction algorithms. Some applications such as watermarking images in digital cameras for tamper detection require the embedding process to be as fast and as simple as possible while the extraction process can be more time consuming. In other applications, such as extracting captions from digital video, the extraction process has to be very fast.

Memon and Wong [22] described a buyer–seller watermarking mechanism using homomorphic public-key cryptosystems. Here, the seller first embeds his watermark information into the content and then embeds a transformation (permutation) of the watermark information of the buyer into the already watermarked content and passes the resultant content to the buyer. In Ref. [23], Katzenbeisser et al. proposed a buyer–seller watermarking protocol that does not use homomorphic public-key encryption mechanism and instead uses a secure watermark embedding mechanism based on partial encryption.

In a multiparty, multilevel content distribution system such as the one given in Ref. [1], the watermarking scheme needs to provide trust to the parties such as content owners, distributors, and consumers involved. This involves protecting the rights of the owner, distributors, and consumers. The owner, distributors, and consumers would trust the system only if they could prove their role, if they were part of the distribution chain of a content. On the other hand, there should be security against false framing by any party who was not part of the delivery chain of content. Further, in order to assure trust, all the parties in the distribution chain must have contributed to the watermark signal with their correct share.

The naive approach for watermarking in a multiparty, multilevel content distribution system is that each party embeds its watermark information individually into the content as and when the content reaches him. However, this approach deteriorates the quality of the resultant content to a great extent due to the presence of multiple watermark signals in the content. This also reduces the trust level of the DRM system as the security concerns of entities such as proof of their involvement, nonrepudiation of the involvement, and protection against false framing will not be properly taken care. Another approach in this case is to employ a buyer–seller watermarking protocol [22] between each pair of interacting (buyer and seller) parties. However, this is also not desirable as the quality of the resultant content will deteriorate due to the embedding of multiple watermark information into the content. In Ref. [1], the authors have proposed a mechanism to embed just one watermark signal into the content based on a watermark information jointly generated by all the parties in the content distribution chain with the help of a TTP (license server). This approach not only minimizes the amount of the watermark in the content but also takes care of the security concerns of the parties involved. This ensures trust to all the parties in the content distribution chain. This joint watermark information is generated using the Chinese Remainder Theorem. In this case, the identities of all the entities are locked in using the Chinese Remainder Theorem as a watermark and are then embedded into the content. This scheme thus takes care of the security concerns of all the parties involved and assures trust to all the parties. The identity of all the participants can be determined from the watermark signal by reverse computing the congruence relations in the Chinese Remainder Theorem. In case, the owner or distributors find an unauthorized copy, they can identify the traitors with the help of a judge. As a future research direction, the protocols may be improved by reducing the dependence on the license server.

Further, in this scheme, the watermark information is computed as digital signatures. The protocols can be made more computationally efficient if these are replaced by any other easily verifiable watermark information.

Trust through Interoperability Mechanisms

DRM technologies have focused on protecting the copyrights by binding a digital content to a device. This restricts a consumer from accessing a digital content on multiple devices seamlessly. However, a user may be interested in accessing a digital content which he has purchased on multiple devices such as laptop, TV, and mobile phone he/she possesses or in sharing that content with his friends or family members. However, this raises serious trust issues for the content owners as the contents can be misused.

Most DRM systems are neither standardized nor interoperable. In general, each content provider uses its own technologies to protect its digital content, with little or no regard for its interoperability with the DRM systems of other content owners. For example, the FairPlay DRM system of Apple has caused many iPod users to complain about their inability to play music files bought from other online services. In fact, the consumers are willing to pay a higher price for more usage rights and interoperability of the contents over multiple devices. The lack of interoperability is not only a concern for the consumers but also for the content providers, as dissatisfied customers can cause the slow growth of the digital industry. With interoperable DRM architectures, content providers can potentially reach more customers as their content will be accessible by any compliant device or application.

DRM interoperability problems have been addressed by various research-ers in different ways. MPEG-21 introduced architecture and interfaces between intellectual property management and protection (IPMP) tools [24]. IPMP is the DRM standard of MPEG. In this case, an end user needs to download and install appropriate DRM tools whenever DRM interoper-ability is required. However, this mechanism is not suitable for end–devices having less processing capability or network connectivity. In Refs. [25,26], the authors proposed mechanisms in which a local middle entity function-ally situated within a home network does the content and license translation for the home devices. However, this approach lacks a trusted framework as the content providers cannot control the content translation once the license for the content has been issued. Hence in Ref. [27], Nam et al. proposed an interoperability approach where end devices perform the DRM translation function. However, this approach also does not provide a complete trust to the content providers. This is because, in this case the end devices trans-late the content from source DRM format to a neutral format for exporting and from the neutral format to the destination DRM format for importing. Another approach to obtain interoperability is by using online third parties [28,29]. In Coral DRM [28], an online third party provides the license transla-tion mechanism.

Coral splits normal license transactions into two phases. In the first phase, the content rights encoded in DRM-independent rights token are acquired. In the second phase, the acquired rights using a native DRM technology is fulfilled. In Ref. [29], an online brokerage entity that acts as a trusted environment is used. This entity passes messages related to rights management between two different DRM agents of client devices. It requests raw digital content to be repackaged from source DRM format and sends the received raw digital content to the destination DRM format to do repackaging by the DRM agent of the destination client device. However, this mechanism requires the content providers to keep trust on the broker that it is not misusing the raw contents or messages. In Ref. [30], an interoperability mechanism that allows the content providers to designate a proxy server to perform reencryption of the content is given. However, this scheme also does not provide trust to the content providers as they cannot control the translation and redistribution actions of the proxy server. A malicious proxy server can share the delegated reencryption capability with another malicious entity or can misuse its translation rights. Thus, we can conclude that in Refs. [28,29], the existence of a TTP connected to the network is required. Moreover, a continuous online connectivity for each device is required for requesting translations. Satisfactory mechanisms to assure trustfulness of the middle entity and to achieve secure and legal distribution and sharing of different DRM-enabled contents have not been worked out yet.

In Ref. [4], the authors have proposed an interoperability mechanism using an intermediate brokerage system called local domain manager (LDM) for a multidomain architecture comprising of several authorized domains. The multidomain architecture proposed by the authors is given in Figure 12.5. In this case, the LDM need not have to be a trusted party to provide interoperable content distribution and adaptation services to multiple authorized domains. This scheme prevents any illegal content distribution and translation to other DRM supported formats by the LDM using cryptographic mechanisms, a translation and distribution license concept and secure key management mechanisms using TPM. Content providers can control translation and distribution of their contents by specifying allowable amount of translation and allowable destination DRM systems to which the contents and licenses can be translated. Thus, this mechanism provides good level of trust to the content providers.

Trust through Content Encryption Mechanisms

A content owner is concerned about the unauthorized use such as playing, copying, and distributing his contents without having the appropriate permissions. A content owner will trust a digital content distribution system only if it can assure the owner that his contents will not be misused. To resolve the concern of a content owner against unauthorized use of content, the content must be encrypted with the owner's secret key. Different content providers encrypt their contents based on the type of the media as well as some

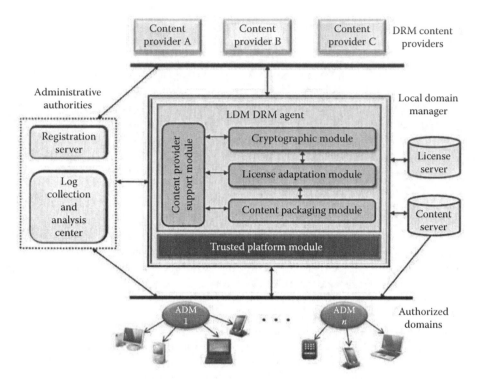

FIGURE 12.5
Multidomain interoperable content distribution.

desirable features such as complexity, compression efficiency, perceptibility, format compliance, error resilience, scalability, and bandwidth expansion. A comprehensive survey on various encryption mechanisms for digital media contents is given in Ref. [5].

A digital content can be securely delivered to a consumer in multiparty multilevel content distribution architecture [1] through simple encryption mechanisms as described in Ref. [1–35]. The encryption of content is performed using a global encryption key and a set of local encryption keys. The global encryption prevents the unauthorized use of the contents, and local encryptions prevent illegal download of the content from the content servers of the distributors. The content owner first encrypts the content with a global encryption key and then encrypts the initially encrypted content with his local encryption key using a symmetric key digital media encryption algorithm. He then uploads the resultant content on his content server. A distributor downloads the content from the content server of the owner or a higher level distributor. He then obtains the license for redistribution from a license server. The distributor is allowed to decrypt the content using of the local encryption key of the owner/distributor. Thus, the distributor gets only the content encrypted with the global encryption key. The distributor then encrypts this globally encrypted content with his local encryption key. The distributor then uploads the resultant content on his content server. A consumer can download the content of his choice from the content server of any distributor or that of the owner. The consumer then obtains a usage license for the content from the license server. The usage license will contain the global encryption key of the content owner and the local encryption key of the distributor. The consumer will have a trusted DRM agent installed in the client device. This DRM agent will decrypt the content using the local and global encryption keys. In this way, a trusted framework for content delivery can be ensured.

Conclusion

In this chapter, we discussed the trust issues in digital content delivery through modern-day networked systems. There are serious trust issues associated with content owners, distributors, and consumer in using such a system. DRM technologies have now progressed to a good extend that if properly used they can provide a trusted framework for digital content distribution.

References

1. T. Thomas, S. Emmanuel, A. V. Subramanyam, M. S. Kankanhalli, "Joint Watermarking Scheme for Multiparty Multilevel DRM Architecture," *IEEE Transactions on Information Forensics and Security*, 4(4): 758–767, 2009.
2. T. Thomas, S. Emmanuel, A. Das, M. S. Kankanhalli, "Secure Multimedia Content Delivery with Multiparty Multilevel DRM Architecture," in *The 19th International Workshop on Network and Operating Systems Support for Digital Audio and Video (NOSSDAV)*, Williamsburg, VA, June 3–5, 2009.
3. T. Thomas, S. Emmanuel, A. Das, M. S. Kankanhalli, "A CRT Based Watermark for Multiparty Multilevel DRM Architecture," in *IEEE International Conference on Multimedia and Expo (ICME)*, New York, June 28–July 2, 2009.
4. L. L. Win, T. Thomas, S. Emmanuel, "Secure Interoperable Content Distribution Mechanisms Using a Multi-Domain Architecture," *Multimedia Tools and Applications*, 60(1): 97–128, 2012.
5. N. S. Kulkarni, B. Raman, I. Gupta, "Multimedia Encryption: A Brief Overview," in *Recent Advances in Multimedia Signal Processing and Communications*, M. Grgic, K. Delac, M. Ghanbari (eds.), Studies in Computational Intelligence, Vol. 231, Berlin/Heidelberg: Springer, pp. 417–449, 2009.
6. D. J. T. Chong, R. H. Deng, "Privacy-Enhanced Superdistribution of Layered Content with Trusted Access Control," in *Proceedings of the ACM Workshop Digital Rights Management*, Alexandria, VA, ACM, October 30, pp. 37–44, 2006.

7. M. Feng, B. Zhu, "A DRM System Protecting Consumer Privacy," in *Proceedings of the CCNC*, Las Vegas, NV, January 10–12, IEEE, pp. 1075–1079, 2008.

8. R. Perlman, C. Kaufman, R. Perlner, "Privacy-Preserving DRM," in *Proceedings of the 9th Symposium on Identity and Trust on the Internet*, Gaithersburg, MD, April 13–15, ACM, pp. 69–83, 2010.

9. M. K. Sun, C. S. Laih, H. Y. Yen, J. R. Kuo, "A Ticket Based Digital Rights Management Model," in *Proceedings of the CCNC*, Las Vegas, NV, January, IEEE, pp. 1–5, 2009.

10. J. Yao, S. Lee, S. Nam, "Privacy Preserving DRM Solution with Content Classification and Superdistribution," in *Proceedings of the CCNC*, Las Vegas, NV, January 10–13, IEEE, pp. 1–5, 2009.

11. L. L. Win, T. Thomas, S. Emmanuel, "Privacy Enabled Digital Rights Management without Trusted Third Party Assumption," in *IEEE Transactions on Multimedia*, 14(3): 2012.

12 A. O. Durahim, E. Savas, "A-MAKE: An Efficient, Anonymous and Accountable Authentication Framework for WMNs," in *Proceedings of the ICIMP*, Barcelona, Spain, May 9–15, IEEE, pp. 54–59, 2010.

13. Y. S. Kim, S. H. Kim, S. H. Jin, "Accountable Privacy Based on Publicly Verifiable Secret Sharing," in *Proceedings of the ICACT*, Gangwon-Do, South Korea, February 7–10, IEEE, pp. 1583–1586, 2010.

14. L. Wenjing, R. Kui, "Security, Privacy, and Accountability in Wireless Access Networks," *IEEE Wireless Communications*, 16(4): 80–87, 2009.

15. J. Yao, S. Lee, S. Nam, "Privacy Preserving DRM Solution with Content Classification and Superdistribution," in *Proceedings of the CCNC*, Las Vegas, NV, January 10–13, IEEE, pp. 1–5, 2009.

16. P. P. Tsang, M. H. Au, A. Kapadia, S. W. Smith, "PEREA: Towards Practical TTP-Free Revocation in Anonymous Authentication," in *Proceedings of the CCS*, Alexandria, VA, October 27–31, ACM, pp. 333–344, 2008.

17. J. Zhang, B. Li, L. Zhao, S. Yang, "License Management Scheme with Anonymous Trust for Digital Rights Management," in *Proceedings of the ICME*, Amsterdam, Netherlands, July 6–8, IEEE, pp. 257–260, 2005.

18. M. H. Au, S. S. M. Chow, W. Susilo, "Short E-Cash," in *Indocrypt*, Bangalore, India, December 10–12, Springer, *LNCS* 3797, pp. 332–346, 2005.

19. J. Camenisch, S. Hohenberger, A. Lysyanskaya, "Balancing Accountability and Privacy Using E-Cash," in *Security and Cryptography for Networks*, Maiori, Italy, September 6–8, *LNCS* 4116, Berlin/Heidelberg: Springer, pp. 141–155, 2006.

20. M. H. Au, W. Susilo, Y. Mu, "Constant-Size Dynamic k-TAA," in *Security and Cryptography for Networks*, Maiori, Italy, September 6–8, *LNCS* 4116, Berlin/Heidelberg: Springer, pp. 111–125, 2006.

21. I. Teranishi, K. Sako, "K-Times Anonymous Authentication with a Constant Proving Cost," in *Public Key Cryptography*, New York, April 24–26, *LNCS* 3958, Berlin/Heidelberg: Springer, pp. 525–542, 2006.

22. N. Memon, P. W. Wong, "A Buyer Seller Watermarking Protocol," *IEEE Transactions on Image Processing*, 10(4): 643–649, 2001.

23. S. Katzenbeisser, A. Lemma, M. U. Celik, M. van der Veen, M. Maas, "A Buyer–Seller Watermarking Protocol Based on Secure Embedding," *IEEE Transactions on Information Forensics and Security*, 3(4): 783–786, 2008.

24. R. H. Koenen, J. Lacy, M. Mackay, S. Mitchell, "The Long March to Interoperable Digital Rights Management," *Proceedings of the IEEE*, 92(6): 883–897, 2004.

25. G. Taban, A. A. Cardenas, V. D. Gligor, "Towards a Secure and Interoperable DRM Architecture," in *Proceedings of the ACM Workshop on Digital Rights Management*, Alexandria, VA, October 30, ACM, pp. 69–78.

26. C. Serrao, M. Dias, J. Delgado, "Bringing DRM Interoperability to Digital Content Rendering Applications," in *The International Joint Conferences on Computer, Information, and System Sciences, and Engineering (CISSE)*, University of Bridgeport, Bridgeport, CT, 2005.

27. D. W. Nam, J. S. Lee, J. H. Kim, "Interlock System for DRM Interoperability of Streaming Contents," in *Proceedings of the IEEE International Symposium on Consumer Electronics (ISCE)*, Irving, TX, June 20–23, 2007.

28. Coral Consortium Whitepaper, Technical Report, 2006. http://www.coral-interop.org

29. C. Serrao, E. Rodriguez, J. Delgado, "Approaching the Rights Management Interoperability Problem Using Intelligent Brokerage Mechanisms," *Computer and Communications*, 34(2): 129–139, 2010.

30. S. Lee, P. Heejin, K. Jong, "A Secure and Mutual-Profitable DRM Interoperability Scheme," in *IEEE Symposium on Computers and Communications (ISCC)*, Riccione, Italy, June 22–25, IEEE, pp. 75–80, 2010.

31. A. Sachan, S. Emmanuel, A. Das, M. Kankanhalli, "Privacy Preserving Multiparty Multilevel DRM Architecture," in *Proceedings of the Consumer Communications and Networking Conference*, Las Vegas, NV, January 10–13, IEEE, pp. 1–5, 2009.

32. K. S. Gayathri, T. Thomas, J. S. Jayasudha, "Prevention of Copyright Issues of Media in Social Network," *International Journal of Computer Applications*, (Special Issue on *Advanced Computing and Communication Technologies for HPC Applications*), 2: 43–48, 2012.

33. L. L. Win, T. Thomas, S. Emmanuel, "A Privacy Preserving Content Distribution Mechanism for DRM without Trusted Third Parties," in *The 2011 IEEE International Conference on Multimedia and Expo (ICME)*, Barcelona, Spain, July 11–15, 2011.

34. L. L. Win, T. Thomas, S. Emmanuel, M. S. Kankanhalli, "Secure Domain Architecture for Interoperable Content Distribution," in *IEEE Pacific-Rim Conference on Multimedia (PCM)*, Bangkok, Thailand, December 15–18, 2009.

35. B. C. Popescu, B. Crispo, A. S. Tanenbaum, F. L. A. J. Kamperman, "A DRM Security Architecture for Home Networks," in *Proceedings of the 4th ACM Workshop on Digital Rights Management*, Washington, DC, October 25, pp. 1–10, ACM, 2004.

13

Privacy and Trust Management in Safety-Related C2X Communication

Axel Sikora

CONTENTS

Introduction

Cyber physical systems (CPSs) can be used potentially in many different applications, including traffic. With the help of CPS, future traffic can be seen as a cooperative system, in which devices, vehicles, and infrastructures are context aware and capable of providing the needed information seamlessly and on time. Intelligent transport systems (ITSs) may potentially help to make better use of the existing infrastructure (more efficient), to conserve the environment (cleaner), to provide new and better mobility services (smarter), and to increase traffic safety (safer). In Europe, a new legal framework (Directive 2010/40/EU) was published in 2010 to accelerate the deployment of these innovative transport technologies across Europe [1]. This Directive is an important instrument for the coordinated implementation of ITS in Europe. It aims to establish interoperable and seamless ITS services while leaving member states free to decide which systems to invest in.

Basically, all ITS installations are based on a combination of sensors, actuators, intelligent decision processes, and communication, where this contribution concentrates on the communication aspects. Wireless communication can significantly reduce the effort for understanding a traffic situation and can potentially enhance the precision in predictions. Among the various application services enabled by enhanced wireless communication technologies, vehicular *ad hoc* networks (VANETs) recently represent the most popular ones. A VANET is a special case of a mobile *ad hoc* network (MANET), which consists of mobile nodes that connect themselves in a decentralized, self-organizing manner and may also establish multihop routes.

For road traffic, the following communication classes can be distinguished:

- *Car-to-car (C2C) communication*, that is, the communications between cars. Sometimes, the term VANET is used for this kind of application.
- *Car-to-infrastructure (C2I) communication*, that is, the communication between cars and some road-side units (RSUs), such as traffic signs, signal lights, or measurement units.

- *Car-to-vulnerable-road-user (C2VRU) communication*, that is, the communication between cars and other nonmotorized road users, such as pedestrians and cyclists as well as motorcyclists and persons with disabilities or reduced mobility and orientation.

All these versions can be grouped under the umbrella term car-to-X (C2X) communication.

Vehicular scenarios have been classified into safety-related and nonsafety-related applications. For example, safety-related services include collision avoidance, emergency vehicle signal preemption, intersection collision avoidance, and traffic management. Other applications include payment services and infotainment (e.g., Internet access) [2]. The author in Ref. [3] gives a practical overview on how ITS and communication systems can improve traffic safety. However, it is clear that C2X networking is not only a matter of technology, but is also very closely related to policy-making about deployment. These aspects are discussed in Ref. [4].

With regard to safety-related issues, VRUs play a special role. Although the absolute number of accidents and fatalities in traffic is continuously decreasing (at least in Europe), the relative risk for vulnerable road users (VRUs) increases, that is, for pedestrians, cyclists, powered two wheelers, etc. [5]. Today's advanced driver assistance systems (ADASs) are based on on-board environment perception sensors. C2X communication technology enables a number of new use cases in order to improve driving safety or traffic efficiency and provide information or entertainment to the driver [6].

It is only for a short time that predictive pedestrian protection systems (PPPSs) can be used. Today's predictive systems use image-based perception sensors systems for the detection of a pedestrian in the vicinity. These systems are significantly limited by uncertainties in target classification. Moreover, they do not offer any benefit in case of fully or partially hidden pedestrians such as, for example, children hidden by cars parked at the roadside. Cooperative pedestrian protection systems (CPPS) that follow the communication model of secondary surveillance radar from air traffic control overcome these weaknesses. Cooperative systems may be used alone or in conjunction with image-based systems. They may be seamlessly integrated into other upcoming C2X communication systems.

Though some people have doubted the added value of security requirements, like in Ref. [7], but mostly undocumented from industry, it is now common understanding that security is closely linked with the robustness of the systems (system safety). Thus, IT security is a major and indispensable element especially for eSafety applications [8] (cf. Figure 13.1).

In addition, it is a basic assumption of this research that eSafety services will only be successful and acceptable to customers if a high level of reliability and security can be provided. The most crucial security service for VANETs is the introduction of trust and the provisioning of trustworthy service [9] preconditions for a broad acceptance of new traffic in the population.

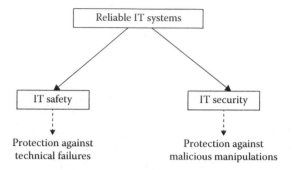

FIGURE 13.1
The relationship between IT safety and IT security. (From Wolf, M., Weimerskirch, A., Wollinger, T., *EURASIP Journal on Embedded Systems*, 74706, 2007.)

Therefore, the remainder of this chapter is dedicated to the security of safety-related C2X communication systems.

Requirements

General Security Requirements

This chapter presents the requirements on security for CP-systems in vehicular applications. In subchapter *B*, various general aspects on security are presented, whereas subchapter *C* discusses the additional requirements, which must be regarded for in C2X-applications.

Technical Scope of Security

With regard to the technical scope of the security, attacks on in-vehicular system infrastructure must be distinguished from attacks on external communication. It must be ensured that in-vehicular system infrastructure is not illegally tampered with, so that critical safety systems are not influenced. Ref. [10] gives an interesting overview of attacks in this field. The security of internal communication is not considered in this work but assumed to be granted.

Attacks on external communication must be prevented or at least detected, so that fake messages are properly identified and eliminated before influencing applications. For example, an attacker could inject messages with false information, or collect vehicle messages to track their locations and pilfer sensitive user data [7]. With regard to external communication, some examples illustrate the criticality of these attacks and the necessity of counter measures. Passive eavesdropping attacks might lead to significant risks shown as follows [11]:

- The police use broadcast messages to calculate driving behavior and issue speeding tickets.
- An employer eavesdrops on the communications from cars on the company parking lot. After distinguishing which car belongs to which employee he automatically gets the arrival and departure dates.
- A private investigator easily follows a car, without being noticed, by extracting position information from messages and broadcast frames.
- Insurance companies gather detailed statistics about the movement patterns of cars. In some cases, individual persons may be charged for traffic accidents based on the gathered movement patterns.
- A criminal organization has access to stationary communication boxes and uses the accumulated information to track law enforcement vehicles. The same technique could be used by a foreign secret service to track VIPs.

The attacks become even more critical in the case of active attacks that are shown as follows:

- Erroneous information can simulate deviations. Thus traffic jams can be stimulated.
- Alarm messages can cause cars to stop, which might lead to major disruption on highways.
- In case of eSafety applications, erroneous information can even cause accidents, that is, if one car makes an emergency operation (braking, swerving, or similar activity).

Consequently, with regard to the portfolio of security parameters (confidentiality, authentication, authorization, privacy, integrity, nonrepudiation/accountability, availability) [12–16], basically all elements must be supported, ever more as they depend on each other. Ref. [17] provides an in-depth discussion on the security parameters and their relevance to C2X-communication. In addition, Ref. [14] contains a detailed requirements analysis, which is repeated here in Table 13.1. In addition, Ref. [18] contains a well-structured listing of requirements with regard to different C2X services.

All these requirements make it clear that technical solutions from legacy securing embedded Internet devices [19] must be enhanced to meet these requirements.

Confidentiality is also required where applicable. In general cases, the gathered information can be openly shared to improve traffic efficiency and road safety. Depending on the business models, one might also think of closed group communication [20].

TABLE 13.1

Security Objectives for C2X Communication

Confidentiality	
Co1	Information sent to or from an authorized ITS user should not be revealed to any party not authorized to receive the information
Co2	Information held within the ITS-S should be protected from unauthorized access
Co3	Details relating to the identity and service capabilities of an ITS user should not be revealed to any unauthorized third party
Co4	Management information sent to or from an ITS-S should be protected from unauthorized access
Co5	Management information held within an ITS-S should be protected from unauthorized access
Co6	It should not be possible for an unauthorized party to deduce the location or identity of an ITS user by analyzing communications traffic flows to and from the ITS user's vehicle
Co7	It should not be possible for an unauthorized party to deduce the route taken by an ITS end-user by analyzing communications traffic flows to and from the ITS end-user's vehicle Integrity
In1	Information held within an ITS-S should be protected from unauthorized modification and deletion
In2	Information sent to or from a registered ITS user should be protected against unauthorized or malicious modification or manipulation during transmission
In3	Management information held within a ITS-S should be protected from unauthorized modification and deletion
In4	Management information sent to or from an ITS-S should be protected against unauthorized or malicious modification or manipulation during transmission
Availability	
Av1	Access to and the operation of ITS services by authorized users should not be prevented by malicious activity within the ITS-S environment
Accountability	
Ac1	It should be possible to audit all changes to security parameters and applications (updates, additions, and deletions)
Authenticity	
Au1	It should not be possible for an unauthorized user to pose as an ITS-S when communicating with another
ITS-S	
Au2	It should not be possible for an ITS-S to receive and process management and configuration information from an unauthorized user
Au3	Restricted ITS services should be available only to authorized users of the ITS

Source: ETSI TR 102 893 V1.1.1 (2010-03), http://www.etsi.org/deliver/etsi_tr/102800_102899/ 102893/01.01.01_60/tr_102893v010101p.pdf

Note: ITS-S, ITS-station.

Authentication is another major issue, as the emission of fake or erroneous messages must be avoided. The prevention here—in the general case—cannot come from cryptographic security protocols alone, but plausibility checks must also be applied [20].

In traffic, we have not only vehicles and objects, but also persons and personal equipment. Thus, we are not only talking about an Internet-of-things (IoT) application. For these, privacy is to be supported with the highest priority [11]. Broadly speaking, "privacy" encompasses the concept that persons own the data relating to them, and unauthorized parties should not be able to use of personal data [18].

The objective of privacy is the protection against typical privacy-infringing malicious profiling or accidental eavesdropping [20]. Much can be inferred about the driver's privacy if the whereabouts and the driving pattern of a car can be tracked (location privacy) [21]. However, it is also possible for attackers to trace vehicles by using cameras or physical tracking. But such physical attacks can only trace specific targets and are much more expensive than monitoring the communication [22]. In order to support this primary goal of privacy, message authentication, integrity, and nonrepudiation, must be guaranteed [7].

This concept can be enhanced toward conditional privacy, in which authorized parties should be able to make use of personal data only with its owner's knowledge, and people should be able to choose which part of their data they reveal to which authorized party [23].

However, it also should be mentioned that privacy is not relevant for all elements. For example, RSUs normally do not need privacy [20].

Ref. [24] shows the relationship between authorization levels and access rights (cf. Figure 13.2).

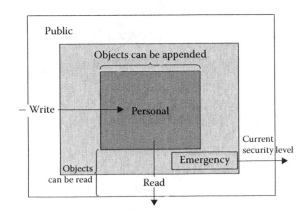

FIGURE 13.2
Relationship between authorization levels and access rights. (From Sernal, J., Luna, J., Medina, M., *Journal of Information Assurance and Security*, 4, 432–439, 2009.)

Availability guarantees that the access to and the operation of ITS services by authorized users should not be prevented by malicious activity within the environment of the ITS station [14]. Availability is an important issue and might restrict the performance of the system functionality (safety). Availability might be restricted by an overload of the physical channel. It is also closely linked to prioritization or privilege [18].

Special Security Issues in C2X Applications

In addition to the requirements concerning basic security parameters, a good number of additional requirements are connected with C2X applications, which are mostly connected with the high dynamics of the network topology:

- *Dynamic topology:* Cars may be moving at absolute speeds of up to 200 km/h (around 55 m/s) and at relative speeds of 400 km/h (around 110 m/s). Thus, communication time between two partners can be quite short, that is, in the range of a few hundreds of microseconds or a few seconds (Figure 13.3). However, the time constants of the different applications can vary significantly [25].

- *Offline infrastructure:* Objects are moving, that is, can be near other communication facilities (i.e., RSUs or mobile base stations). However, they can also be outside the range of these other communication facilities and therefore must also function in an island scenario. It can be assumed that objects might have Internet connectivity from time to time [26].

- *Auxiliary information:* Furthermore, nodes in VANETs are context aware, which means they have access to additional data such as car sensor data or GPS. The usage of the so-called side-channel information can be valuable when evaluating data obtained through communication with other nodes in the VANET [26].

- *Lifetime requirements:* Cars do have a much longer lifetime than consumer electronics. For example, in Germany the average age of a passenger car is 8 years. More than 15% of the passenger cars are older than 15 years [27]. Although, the situation in other fields of automation, such as industrial process or building automation is even worse, this long lifetime makes long-time security a necessity [25].

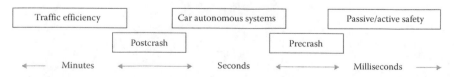

FIGURE 13.3
Variations in timing requirements. (From Schütze, T., *Embedded World Conference*, Nuremberg, 2011.)

- *Man–machine interaction:* Devices in operation must independently work without any human interaction. However, they should be capable of generating warnings and provide information to humans, for example, a driver or a pedestrian.
- *Device restrictions:* The automotive industry is very cost sensitive, which puts severe limits onto the computational power of devices. Portable devices, that is, devices being worn by VRUs, may even be more cost- and energy-efficient than their vehicle-based counterparts [28].

General Activities

Overview

The research community in industry and academia, with the endorsement of authorities, has undertaken major efforts to design for vehicular communication. Major efforts dedicated to the general development of C2X communication systems will be discussed in this chapter.

The most important activities are given as follows:

- The US Department of Transport (DOT) has conducted three major programs related to connected vehicle technologies, working primarily with the crash avoidance metrics partnership (CAMP) [29].
- The Vehicle Safety Communications Consortium (VSCC), working through the CAMP mechanism, conducted a program of research addressing a set of potential safety applications.
- The "Cooperative Intersection Collision Avoidance System Limited to Stop Sign and Traffic Signal Violations" (CICAS -V) program addressed the requirements and technology to address stop sign and traffic signal violations using dedicated short range communication (DSRC) between roadside equipment units (RSEs) and onboard equipment (OBE).
- The vehicle safety communications-applications (VSC-As) program continued many of the themes of earlier projects. The VSC-A project investigated whether DSRC-based vehicle safety applications could improve or replace autonomous vehicle-based safety systems, or enable new safety systems. This included developing crash scenarios to be targeted, system requirements for countermeasures, the development of prototype test beds, and the design and execution of objective test procedures for applications. This study also investigated the technology for absolute and relative positioning, and the potential performance levels of that positioning. The work continued earlier efforts to support technical standards development, addressed security concerns, and conducted scalability testing with multiple vehicles in close proximity.

In addition, the DOT runs the vehicle infrastructure integration (VII) initiative within its research and innovative technology administration (RITA) [30].

- In Japan, the advanced safety vehicle (ASV) project [31] and the Internet ITS consortium [32] are the main players.
- In Europe, a major driver behind the activities is the European Commission [33], which—based on their general policies—has launched various programs and projects. These efforts are backed with additional significant activities from the national governments.

This chapter describes the major activities in C2X communication, which prepare the general platform standards, whereas Chapter 14 concentrates on projects with security and privacy-enhancing solutions.

IEEE 802.11p

The IEEE 802.11p standard has been developed as the most popular basis for C2X-communication. The IEEE 802.11p task group is working on the DSRCs standard which aims at enhancing the 802.11 protocol to support wireless data communications for vehicles and the roadside infrastructure [3].

IEEE WAVE

In the United States, the IEEE 1609 working group has developed the "IEEE 1609 Family of Standards for Wireless Access in Vehicular Environment (WAVE)" [34], which defines an architecture and a complementary, standardized set of services and interfaces that secure C2C and C2I wireless communications. Figure 13.4 shows the WAVE protocol stack architecture.

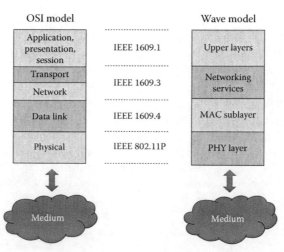

FIGURE 13.4
Protocol stack of IEEE 802.11p and IEEE 1609 WAVE.

- The IEEE 1609.1 WAVE Resource Manager defines the application read/write protocol.
- The IEEE 1609.2 WAVE Security Services [35] is the relevant standard for C2X security. It envisages anonymity, authenticity, and confidentiality. For many observers, the concrete technical standardization of security functions is most advanced in IEEE 1609.2 [25]. It was formerly described as IEEE 1556.
- The IEEE 1609.3 WAVE Networking Services provides the description and the management of the DSRC protocol stack.
- The IEEE 1609.4 WAVE Multi-Channel Operation provides the DSRC frequency band coordination and management and integrates tightly with IEEE 802.11p.

Car-to-Car Communication Consortium

The Car-to-Car Communication Consortium (C2C CC) doubles as an industrial community automotive manufacturer, supplier, and research institute. It pursues the following objectives [6]:

- The development and release of an open European standard for cooperative ITS and associated validation process with focus on intervehicle communication systems
- To be a key contributor to the development of a European standard and associated validation process for vehicle-to-roadside infrastructure communication being interoperable with the specified intervehicle communication standard
- To provide its specifications and contributions to the standardization organizations including in particular European Telecommunications Standards Institute (ETSI)'s technical committee on intelligent transport systems (TC ITSs) in order to achieve common European standards for ITS
- To push the harmonization of C2C communication standards worldwide
- To promote the allocation of a royalty-free European wide exclusive frequency band for C2C applications
- To develop realistic deployment strategies and business models to speed up market penetration
- To demonstrate the C2C system as proof of technical and commercial feasibility

The European researches are also using wireless LAN and a frequency spectrum in the 5.9 GHz range that has been allocated on a harmonized basis in Europe in line with similar allocations in the United States. The higher layers are defined by C2C CC.

A working group on security (security WG) is part of C2C CC.

ETSI

Core European activities in this field also take place at ETSI as the European regulatory body, which received the official mandate to standardize C2X. ETSI's TC ITS and is responsible for the production and maintenance of standards to support the development and implementation of ITS communications and services across the network, for transport networks, vehicles, and transport users.

- The ETSI TS 102 637 [36] defines the basic set of applications for vehicular communications. For all the different cases of use and application, there are only two ITS communication services:
  - The cooperative awareness message (CAM) service, which periodically transmits transient data on the vehicle status. CAMs contain a time stamp, car, or RSU specific fixed data (identity) as well as position, velocity, and direction. Using special flags, emergency vehicles can signal for "siren in use" or braking activities. The length of a CAM block is about 64 bytes. A CAM will be sent in message broadcast mode, or sometimes in unicast mode [25].
  - The decentralized environmental notification message (DNM, or sometimes DENM) service, which is generated on detecting an event and contain information about this event. DNM messages are typically relevant for a defined geographic area. DNMs contain geographical data: about the event area, about the target area, about the relevance area and the networking area in which resending of messages shall take place. DNMs will be transported using geonetworking or infrastructure relays. The application information in DNM are event management information, priority, generation time, validity period, etc. and of course information on the event itself: For example, the type of event, the severity of the event, the reliability of its detection and a human-readable description of the data.

    Every DNM has an action ID which is unique for each event and contains information about the originator (privacy concerns!). DNMs are typically much longer than CAMs, their length depends very much on the event type. DNMs will be conveyed in broadcast or unicast mode [25].
- The ETSI TC ITS Working Group 5 deals with security aspects in ITS and develops standards focusing on securing vehicular communications, so as to prevent eavesdropping and distribution of malware

TABLE 13.2

Available Channels in the 5.9 GHz Band as Defined by EN 302 571

Channel Number	Carrier Centre Frequency (MHz)	Maximum Channel Bandwidth (MHz)
1	5860	10 (5855–5865)
2	5865	20 (5855–5875)
3	5870	10 (5865–5875)
4	5880	10 (5875–5885)
5	5890	10 (5885–5895)
6	5900	10 (5895–5905)
7	5910	10 (5905–5915)
8	5915	20 (5905–5925)
9	5920	10 (5915–5925)
10	5890	30 (5875–5905)

Source: ETSI EN 302 571 C1.2.0 (2013-05), http://www.etsi.org/deliver/etsi_en/302500_302599/302571/01.02.00_20/en_302571v010200a.pdf

to vehicles. Essential input comes from C2C CC Security WG. Other C2C CC WGs act as preparatory platforms and discussion forums.

In this context, a detailed threat, vulnerability, and risk analysis (TVRA) has also been performed, which was published in Ref. [14]. The security objectives in the section "General Security Requirements" are taken from this technical report. The analysis also includes a very readable cost–benefit analysis.

- In ETSI EN 302 571 [37], 10 channels were opened in the 70 MHz broad frequency range between 5.855 and 5.925 GHz. This allocation is listed in Table 13.2. It should be highlighted that the seven 10 MHz-channels are nonoverlapping and can be used independently. Alternatively, the bandwidth can be split into two 20 MHz and one 30 MHz channel. IEEE 802.11p uses these channels with an OFDM PHY.

eSafety

The European Union has organized their eSafety activities around the eSafety Forum, where a Security Working Group was also established [38].

Security-Centric Projects

Overview

This subchapter presents some of the most relevant publicly funded projects, which were or are performed around C2X communication and which concentrate or include security-related issues.

SeVeCom

The SeVeCom project [39] on secure vehicular communication was one of the first large European projects (2006–2008) on the topic [40] and started to define specific security mechanisms for C2X communication. The objective of the project is to define the security architecture of future vehicle communication networks and to propose a roadmap for progressive deployment of security functions in these networks.

The main areas of research in scope of the SeVeCom project were as follows:

- Threats identification: attacker's model and potential vulnerabilities
- Specification of architecture and of security mechanisms to provide the necessary level of protection. Among the topics which are planned to be fully addressed are: key and identity management, secure communication protocols, tamper-proof device, decision on crypto-system, and privacy
- Definition of cryptographic primitives

SeVeCom defines an abstract security architecture which is clearly separate from specific technologies. This abstract architecture could be mapped to concrete technology solutions later in order to support future communication standards. Within the SeVeCom project, an approach to extract security requirements in order to develop a security solution for vehicular communication was also developed.

The SeVeCom project uses a concept of pseudonyms in order to provide privacy and security of vehicles. The pseudonyms are issued by pseudonym providers (PPs) and are bound to vehicles' long-term identities. The PP stores identity pseudonyms mappings to provide accountability in case of misuse. The pseudonyms are valid for a short period of time in order to minimize a need for certificate revocation. CAs need to distribute certificate revocation lists (CRLs) just to PPs (which are the part of infrastructure, basically on RSUs). In turn, PP revokes the vehicle's long-term identity to prevent the vehicle from receiving new pseudonyms from a PP.

Thus, the concept of PPs eliminates the need for Internet connectivity for vehicles. They just use a wireless medium to communicate with RSUs. The fact that the vehicles do not deal with CRLs processing makes the requirements for computational power of onboard units (OBUs) more relaxed.

EVITA

The EVITA project on "E-safety vehicle intrusion protected applications," another project cofunded by the European Unison within the Seventh Framework Program from 2008 to 2011 [41], was focused to design, verify, and prototype an architecture for automotive on-board networks where security-relevant components are protected against tampering and sensitive data are protected against compromise. In the project, a secure automotive on-board network, that is, secure hardware and software components as well as protocols for communication inside the car were designed and developed. As a result there are three different types of security controllers (light, medium, and full HSMs corresponding to sensor, ECU and ECU C2X communication), which perform symmetric and/or asymmetric cryptography [25].

NoW

The Networks on Wheels (NoWs) project was a joint effort of German industry and academia to solve technical key questions on the communication protocols and data security for C2C communications and submit the results to the standardization activities of the C2C CC. Further, a test bed for functional tests and demonstrations was implemented, which served as a reference system for the C2C CC specifications (cited after Ref. [42], as the project Web site is no longer available).

Oversee

Oversee [43] is a European research project, which concentrated on the development of an open, standardized, and secure in-vehicle software and communication platform, which will empower (almost) everyone to develop vehicular applications and offering these applications to all vehicles that are equipped.

PATH

Established in 1986, Partners for Advanced Transportation TecHnology (PATH) is administered by the Institute of Transportation Studies (ITSs) at the University of California, Berkeley, CA, in collaboration with Caltrans. PATH is a multidisciplinary program with staff, faculty, and students from universities statewide, and cooperative projects with private industry, state and local agencies, and nonprofit institutions [44].

PATH also includes transport safety, mainly in vehicle–highway cooperation and communication, and "science of driving" investigations on driving behavior, and in implementation of prototype vehicle–highway safety systems. In the past year, new areas of research have extended to efficient means of investigating crashes, rail–highway crossings, and importantly to Caltrans, means to understand high crash concentrations in order to embark on road safety improvements. Specific project groupings include intersections and cooperative systems with crossing path vehicle crashes, safety aspects of cooperative driver-assist systems, and VII with Expedited VII and VII California.

PRECIOSA

The goal of *PRivacy Enabled Capability In co-Operative* systems and *Safety Applications* (PRECIOSA) project is to show that cooperative systems using vehicle-to-vehicle (V2V) and vehicle-to-infrastructure (V2I) communication can comply with future privacy regulations [45]. One of the most important outcomes of this project is the definition of a privacy-aware architecture for cooperative systems.

This architecture, namely, *Privacy enforcing Runtime architecture* (PeRa), is built around a concept of privacy policy enforcement, which is implemented in a three-tier approach consisting of privacy policies, mandatory privacy control (MPC), and MPC integrity protection layer (MIP).

So the overall process contains the following steps:

- Users define an immutable privacy policy.
- All created data are combined with this policy.
- MPC components make sure that applications can only perform policy-compliant operations on data.
- The MIP layer handles encrypted storage and encrypted information exchange between PeRa instances. This layer also monitors integrity of MPC components and only grants data access if all MPC components are in a trusted state.

Regarding communication privacy the PRECIOSA uses pseudonym-based approach that was developed by the SEVECOM project. Generally, the SEVECOM baseline architecture is adopted to enforce user privacy in public communications.

sim^TD

"Sichere Intelligente Mobilität Testfeld Deutschland," translated to English means "Safe and Intelligent Mobility Test Field Germany" (sim^TD) is the worldwide first field operational trial for C2X technology [46]. The project is sponsored and supported by three federal ministries, the Federal State of Hessen, the German Automobile Industry Association, and the C2C CC. The total budget accounts for an impressive €53 million with approximately €30 million of funding plus additional infrastructure investment.

In sim^TD, partners from the automotive domain, the telecommunication domain, the public authorities, and several universities and research institutes have gathered to validate technologies and applications for C2X communication in a realistic deployment scenario in order to evaluate the entire spectrum of applications with regard to the effects on traffic safety and traffic efficiency. For that purpose, about 100 controlled vehicles with hired drivers are responsible for the creation of certain traffic situations where applications may be tested and validated. An additional 300 vehicles provide a permanent base load to the C2X network which makes the field trial comparable to the later deployment scenario. This so-called free flow fleet ensures sufficient coverage of the test area, needed for a comprehensive forwarding of C2X messages inside the entire sim^{TD} network. About 100 RSUs and two central traffic centers are deployed [47]. The system architecture of sim^{TD} is shown in Figure 13.5.

The IEEE 1609.2 specification is taken as a reference for the sim^{TD} security solution. However, parts of the IEEE 1609.2 specification were adapted and extended. At the time of writing this chapter, sim^{TD} can be understood as the most advanced practical platform. It is overruling many approaches that have been discussed in earlier literature and which, therefore, are also not repeated here. Details of the most important sim^{TD} security elements are shown in Figure 13.6.

However, not everything is ideal in sim^{TD}'s security architecture: As mentioned in Ref. [25], the RSA PKCS#1 signatures come with 512 bits and small public exponent, which leads to reduced cryptographic complexity.

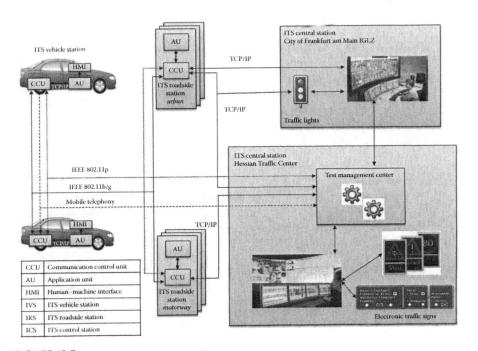

FIGURE 13.5
sim^{TD} System Architecture, original at sim^{TD} Consortium. (From Bissmeyer, N., Stübing, H., Mattheuss, M., Stotz, J., Schütte, J., Gerlach, M., Friederici, F., *7th Embedded Security in Cars Conference*, Düsseldorf, 2009; simTD Consortium, Deliverable21.2 version 3.0, 2009.)

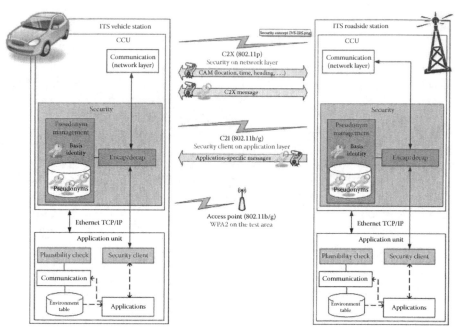

FIGURE 13.6
The detailed elements of the sim^{TD} security architecture. (From Bissmeyer, N., Stübing, H., Mattheuss, M., Stotz, J., Schütte, J., Gerlach, M., Friederici, F., *7th Embedded Security in Cars Conference*, Düsseldorf, 2009.)

General Solutions of Today

Architectures

Generally, it is assumed that each node in a VANET is equipped with a trust system, which can come to trust decisions (verify statements, be aware of trust, etc.). There are two basic options for trust establishment [26]: it can either statically rely on a security infrastructure or be built up dynamically in a self-organizing manner (cf. Figure 13.7). The former process relies on common, global, trusted, and well-known system parameters (e.g., a central CA), which can be used for message authentication. The latter process lacks this global knowledge and point of control and needs to take advantage of other trust-supporting mechanisms.

Though one important property of VANETs is that they are self-organizing and decentralized systems, consequently, from their basic understanding, successful approaches for security and privacy therefore must not rely on central services or mandatory connections to some fixed infrastructure. However, these connections to central authorities may not be completely inevitable [11].

Obviously, envisaging the large choice of alternatives, the selection of the optimum approach is a multidimensional and interdependent problem.

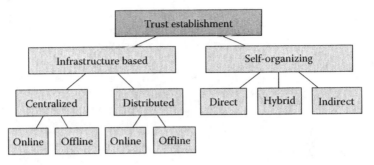

FIGURE 13.7
Classification of trust establishment approaches. (From Wex, P., Breuer, J., Held, A., Leinmüller, T., Delgrossi, L., *67th IEEE Vehicular Technology Conference*, Singapore, 2008.)

TABLE 13.3

Discussion of Technical Solutions for C2C Security Solutions

	Scalability	Signature Length	Computation Effort	Privacy
PKI + Digital Signatures	++	++	++	−
Fixed Pseudonym Pool	−	++	++	+
PKI + Dynamic Pseudonyms	−	++	+	+
PKI + Group Signatures	+	−	−	++

Source: Weyl, B., *SeVeCom Workshop*, Lausanne, 2008.
Note: +, advantageous; ++, very advantageous; −, disadvantageous.

Table 13.3 shows the results of a comparison of C2C security solutions with regard to the key parameters: scalability, signature length, computational effort, and possible privacy level [20]. From Ref. [20] and the subsequent project activities choose the option "PKI + Dynamic Pseudonyms" as the most suitable one.

The final effort, however, depends on additional parameters and functions. This is shown in the subsequent list for digital signature algorithms ([11,20] and our own extensions):

- The length of the signature (Ssig)
- The size of the public key (Spub)
- The size of a certificate (Scert)
- The computational costs for generating the signature
- The computational costs for verifying the signature

Cryptography and Encryption

First and foremost in the European groups C2X CC Security WG and SeVeCom, a purely asymmetric solution in the form of digital signatures has been discussed, that is, all parties of the system obtain digital certificates and sign all their messages with their private key. The recipient verifies the integrity of the sender and the data by signature verification [25].

Symmetric methods come with a much higher computational efficiency; however, they have the problem of key distribution and key exchange.

For this reason, also for ITS solutions, a hybrid approach is pursued, as described in IEEE 1609.2 [35]. The IEEE 1609.2 standard basically consists of three components [25]:

- Digital signatures using ECC over $\mathbb{F}_p$, specifically Elliptic Curve Digital Signature Standard (ECDSA) with NIST $\mathbb{F}_p$ curves [49]
- Asymmetric encryption with ECC, specifically Elliptic Curve Integrated Encryption Scheme (ECIES) [50,51]
- Purely symmetric scheme: authenticated encryption, specifically counter mode encryption and CBC-MAC with AES (AES-CCM) [52]

Pseudonyms

For the provisioning of an adequate privacy, pseudonymity or pseudonymous authentication is a major approach to camouflage the real identity, as it is used instead of real-world identities as identifiers [53]. It does not provide anonymity, but a higher degree of privacy [26]. Therefore, the use of the so-called source-pseudonyms [23] is an integral element for practically all proposed solutions. For these pseudonyms, a set of requirements must be met in order to use them efficiently:

- It shall be possible to change these pseudonyms, so that there is no easy mapping between the pseudonym and the real entity. The change of pseudonyms can be triggered automatically, by the user himself [26] or by some algorithm [54].
- It is common understanding that a silent period is required to avoid local tracking mechanisms. Albeit, it is not always easy to implement these silent periods, especially if we are talking about dense urban scenarios.
- In some extraordinary cases, it shall be possible to perform a mapping between the pseudonym and the real entity. However, this shall be possible only in a secure (trusted) environment, for example, in the central CA (conditional privacy). In some proposals, a pseudonym certification authority (PCA) is therefore separated from a long-term certification authority (LTCA), which holds real information about the vehicle (and possibly the drivers).
- The pseudonyms can be obtained *a priori* or be reloaded, or generated by nodes on the fly [55].

- The pseudonyms shall be unique, that is, one pseudonym shall be used by only one real entity. In case of regional communication, this uniqueness can be restricted to certain timing or spatial intervals.
- Pseudonymity must be supported on all layers. It is not sufficient to have pseudonyms on the host names, when still IP or MAC addresses can be traced [25].
- It shall be possible that one real entity has more than one pseudonyms.
- It shall be possible that the use of pseudonyms can be cryptographically bound to the use of certificates.

Important issues in conjunction with pseudonyms are the following:

- Some application protocols need long-term relationship.
- Some application protocols ask for location-based services (LBSs) and thus might endanger privacy. In these cases, a certain cloaking region might be helpful.
- If routing protocols are used, changes of pseudonyms can have a negative impact.
- The tracking of misbehaving nodes also gets more difficult.

Authentication

Public key infrastructure (PKI) promises to provide strong security features such as authentication, nonrepudiation, and confidentiality. However, with regard to their implementation, there are various questions and challenges. Some of them are discussed in the following. They especially relate to the tradeoff between computational and communication overhead in the different actions such as generation, transport, and verification [56] (Figure 13.8).

Creation and Verification Processes

Due to its highly dynamic character, C2X-communication differs significantly. Ref. [56] shows that there might occur the need to verify between 400 and

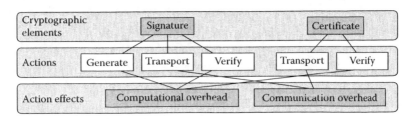

FIGURE 13.8
Overhead due to signing. (From Schoch, E., Kargl, F., *Proceedings of the 3rd ACM Conference on Wireless Network Security*, Hoboken, NJ, 2010.)

4000 signatures per second. This high requirement on the verification side leads to a shift from generation to verification in comparison with normal PKI systems.

Distribution Processes

A performance bottleneck can occur when verifying a mass of signatures within a rigorously required interval, even under adverse scenarios with bogus messages. Tree-based signature schemes like in Ref. [57] can help to alleviate this bottleneck. Another proposal for a tree-oriented signature verification scheme comes from [58]. These tree-based approaches can also be used for the construction and verification of revocation lists.

Optimized Signatures

In Ref. [59] a method is proposed to shorten the signature to less than 22 bytes (171 bits). It is also shown that it is possible to combine two distinct signatures (from different messages) to only one signature. The overall resulting overhead can therefore be reduced to about 40 bytes; approximately 20 bytes for total signature and 20 bytes for a public key. The price to pay for such short signature is a more complex verification process (about three times of RSA) [11].

ECC vs. RSA

The discussion about the best cryptographic algorithm under these circumstances shows this tradeoff clearly. For authentication purposes, a variety of asymmetric algorithms, such as NTRU, XTR, ECC, and the so-called MQ algorithms exist to name just a few [60].

Most approaches suggest the use of elliptic curve cryptography (ECC) in vehicular networks, as ECC signatures, keys, and certificates are smaller than their RSA counterparts [61]. A 224-bit ECDSA signature excluding certificate requires 56 bytes compared to 256 bytes for an equivalent RSA-2048 signature. Including certificates and additional management information, a 224-bit ECDSA signature and certificate as proposed by IEEE 1609.2 [35] needs 181 bytes, a corresponding RSA-2048 certificate about three to four times that size. Whereas ECC signature generation is fast, verification is comparatively costly with approximately double the effort for verification in comparison to generation [25].

PKI

In all existing security frameworks, the private/public keys of VANET nodes are assigned by the key generation center (KGC), which causes an inherent weakness such as the key escrow problem because the KGC issues their private keys using the master key of the KGC [2]. This does not guarantee strong nonrepudiation and private communication because the KGC can sign and decrypt any messages and abuse its access ability.

However, it should be clear that this role is inherent to KGCs [12].

The public key of the CA should be preloaded by all vehicles [62]. Although this is easy to be said, this preloading should also be supported under roaming conditions, if the objects move between different domains [63].

A certificate $\text{Cert}_V[\text{PuK}_V]$ of public key of a vehicle V should include at least the following [62]:

$$\text{Cert}_V\left[\text{PuK}_V\right] = \text{PuK}_V \mid \text{Sig}_{\text{PrK}_{CA}}[\text{PuK}_V \mid \text{ID}_{CA}].$$

In the certificate, PrK_{CA} is CA's private key and ID_{CA} is the unique identifier of CA.

The centralized certificate update process in the classical PKI may be impractical in large-scale VANETs because [63]

- Each CA encounters a large number of certificate update requests, which can render the CA with a bottleneck.
- The certificate-update delay is long in comparison to the short V2I communication duration between the immobile RSUs and the highly mobile OBUs, during which a new certificate should be delivered to the requesting OBU.

The long certificate update delay is because a request submitted by an OBU to an RSU must be forwarded to the CA, and the CA has to send the new certificate to that RSU, which, in turn, forwards the new certificate to the requesting OBU. Accordingly, the classical PKI should be pruned or optimized to satisfy the certificate service requirement in volatile vehicular communication scenarios. To provide a practical certification service for VANETs, each OBU is required to efficiently update its certificate in a timely manner. The certification service should also be decentralized to enable VANETs to efficiently process the expectedly large number of certificate update requests. Moreover, to protect user privacy, the updated certificates should be anonymous and free from the key escrow issue.

Revocation Lists

Nodes holding keys and credentials, however, do not necessarily comply with the implemented protocols. They may be faulty or illegitimately obtain private keys [3]. To ensure the robustness of the VC system, it is important to remove faulty nodes and prevent the utilization of compromised keys. The distribution of CRLs is the basic approach: each CA adds registered nodes' certificates to its CRL which have not expired yet and which it deems it must revoke. It also periodically publicizes the CRL [64].

Various approaches have been proposed to reduce the bottleneck of centralized CRL architectures, which is caused by high-speed mobility and extremely large amount of network entities in VANETs [23]:

- The collaboration between regional CAs might be helpful, so that CRLs contain only regional revocation information and their size is kept low [64].
- Group certificates might allow batch verification [65].
- Ref. [66] describes an efficient, decentralized revocation (EDR) protocol based on a pairing-based threshold scheme and a probabilistic key distribution technique. Because of the decentralized nature of the EDR protocol, it enables a group of legitimate vehicles to perform fast revocation of a nearby misbehaving vehicle.
- Ref. [67] proposes three certificate revocation protocols for VANETs: revocation using compressed CRLs (RC²RLs), revocation of the tamper-proof device (RTPD), and distributed revocation protocol (DRP). RC²RL uses a compression technique to reduce the overhead of distribution of the CRL. Instead of checking the status of a certificate, RTPD removes revoked certificates from their corresponding vehicles' certificate stores by introducing a tamper-proof device as a vehicle key and certificate management tool. In this case the vehicle possessing the revoked certificates is informed of the certification revocation incident after which the tamper-proof device automatically removes those revoked certificates. Different from RC²RL and RTPD, a distributed certificate revocation mechanism is implemented in DRP to determine the status of a certificate. In DRP each vehicle is equipped with an attacker detection system, which enables a vehicle to identify any compromised peer (cited after Ref. [23]).
- In addition to other proposals, it is also described in Ref. [68] to only distribute delta CRLs, that is, CRLs which contain only those certificate revocations that are not yet distributed.
- Ref. [69] proposes a binary authentication tree (BAT) scheme, which promises to effectively eliminate the performance bottleneck when verifying a mass of signatures within a rigorously required interval, even under adverse scenarios with bogus messages.
- The VSC project proposes to maintain a list of short-lived anonymous certificates for the purpose of keeping the privacy of drivers, where the short-lived certificates are discarded after being used. The scheme can provide a higher security assurance [23].
- Also, the sim^{TD} project decided to use short-lived certificates without any revocation list.

In addition, a special compact form of certificates is defined, the so-called WAVE certificate. The secured messages are of unsecured, signed, or encrypted type.

The security level of IEEE 1609.2 WAVE secured messages is 112 bit for short-time messages and 128 bit for others. This leads to the following requirements: ECDSA with NIST P-224 curve and SHA-224, ECDSA with NIST P-256 curve and SHA-256, ECIES with NIST P-256 curve, and AES-128 encryption. Modern hybrid encryption schemes consist of a data encapsulation mechanism (DEM) and a key encapsulation mechanism (KEM). For ECIES, ElGamal-based encryption is used as KEM. Encryption and MAC keys, are derived for the DEM which uses symmetric schemes [25].

Apart from the CRL discussion [70], the possible algorithms for the fast exclusion of errant devices from vehicular networks are additionally discussed.

Hardware Security Device

It is a common assumption that each vehicle has a trusted tamper-resistant hardware security device (HSD) or HSM on board that has computation and storage capabilities to calculate private/public keys [71]. It can be implemented as trusted computing module (TCM) [72] or trusted platform module (TPM) [73], as customized security controller or as field-programmable gate array (FPGA) [8] (cf. Table 13.4). Such a device is also the cornerstone of in-vehicle security.

The HSD is preferably reconfigurable such that it can be reconfigured by downloading data from the central CA [71].

The integration of an HSD is gaining increasing importance as a precondition. This not only holds true for C2X, but also for other applications such as smart metering communication [74], in which the additional hardware cost saves muchcomputational and communication overhead, as well as reduces the process complexity, and is able to prevent runtime software and hardware attacks at least [8].

TABLE 13.4

Hardware Security Module

	TPM	Customized Security Controller	FPGA
Standardized	Yes	No	No
Flexibility	Very limited	Yes, until release	Yes, even after release
Cost	Medium	Low (high volumes)	High
Security level	High	Adaptable	Medium–high

Source: Wolf, M., Weimerskirch, A., Wollinger, T. *EURASIP Journal on Embedded Systems*, 74706, 2007.

Extension Toward Localization

The information about the locality of the objects offers additional possibilities, but also challenges. These are as follows:

- The information can help to offer additional LBSs.
- The information about the locality and the potential tracking of objects and persons can be an additional threat to privacy (location privacy) [21].
- The information about the geographical area can also improve the trust level with regard to Certification Authorities and Attributes Authorities [24].
- The change of localization can also help to generate pseudonyms [75].

Extension Toward Car-to-VRU Communication

VRUs play a major role in current and future safety systems, as they are a key parameter for a further reduction of traffic-related fatalities. However, VRU-based communication units may suffer from even more stringent restrictions with regard to energy (battery lifetime), cost, and Internet connectivity [76]. Therefore, VRU units are handled as a special subsystem in past and present projects [75,77,78]. Ref. [79] discusses the special requirements, which derive from C2VRU communication.

In Ref. [80], an extension of the basic architecture is enhanced also with regard to security. The proposal includes backend infrastructure that comprises two main modules: certificate authority and PP. A concept of pseudonyms can be applied to conceal vehicle and VRU identities. Because RSU do not require privacy, the PP is included in the backend infrastructure and uses an Internet connection to update pseudonyms of a vehicle or of a VRU (via a vehicle). The PP maintains mapping between long-term identities and the corresponding pseudonyms. A CA is used to sign basic identities and pseudonyms for vehicles and VRUs.

Summary and Outlook

This chapter discusses the general and specific requirements for security solutions for C2X communication. It has shown different proposals and described state-of-the-art proposals. For the time being, it seems that— after several years of very extensive and broad discussions of hundreds of proposals from the worldwide community—there is a good consolidation in process. More and more it seems that the practical issues of implementation will be the upcoming topic, as C2X communication will be the cornerstone of automated driving.

Acknowledgments

The author's team has been and still is active in various projects. The author is especially grateful for the involvement in the WATCHOVER-project [77], which has been cofunded by the European Commission Information Society Technologies with the strategic objective "eSafety Co-operative Systems for Road Transport," and in the Ko-FAS initiative for cooperative ADASs [75], which is cofunded by the German Ministry of Economics and Technology.

In addition, the author thanks his team for various kinds of dedicated support and the project partners for excellent collaboration.

References

1. European Commission, "Road Transport: A Change of Gear," 2012, available at: http://ec.europa.eu/transport/modes/road/doc/broch-road-transport_en.pdf
2. J. Choi, S. Jung, "A Security Framework with Strong Non-repudiation and Privacy in VANETs," *6th IEEE Consumer Communications and Networking Conference (CCNC)*, Las Vegas, NV, January 10–13, 2009.
3. U.S. Department of Transportation, National Highway Traffic Safety Administration, "Vehicle Safety Communications Project: Task 3 Final Report—Identify Intelligent Vehicle Safety Applications Enabled by DSRC," March 2005, available at: http://www.its.dot.gov/research_docs/pdf/59vehicle-safety.pdf (last accessed August 1, 2013).
4. T. Kosch, C. Schroth, M. Strassberger, M. Bechler, *Automotive Inter-networking (Intelligent Transport Systems)*, Chichester: John Wiley & Sons, 2012.
5. European Commission, "Towards a European Road Safety Area: Policy Orientations on Road Safety 2011–2020," SEC(2010)903, July 20, 2010, available at: ec.europa.eu/transport/road_safety/pdf/com_20072010_en.pdf
6. R. Baldessari, et al. "Car2Car Communication Consortium—Manifesto Overview of the C2C-CC System," August 28, 2007, Version 1.1, available at: http://elib.dlr.de/48380/ (last accessed August 1, 2013).
7. P. Papadimitratos, G. Calandriello, J.-P. Hubaux, A. Lioy, "Impact of Vehicular Communications Security on Transportation Safety," *IEEE Conference on Computer Communications (IEEE INFOCOM) Workshop on Mobile Networking for Vehicular Environments*, Phoenix, AZ, April 2008.

8. M. Wolf, A. Weimerskirch, T. Wollinger, "State of the Art: Embedding Security in Vehicles," *EURASIP Journal on Embedded Systems*, Article ID 74706, 2007.
9. S. Eichler, C. Schroth, J. Eberspächer, "Car-to-Car Communication," *VDE-Kongress*, Aachen, Germany, October 23–25, 2006.
10. K. Koscher, et al., "Experimental Security Analysis of a Modern Automobile," *31st IEEE Symposium on Security and Privacy*, Berkeley, CA, May 16–19, vol. 31, 2010.
11. F. Doetzer, "Privacy Issues in Vehicular Ad Hoc Networks," in: G. Danezis, D. Martin (eds.), *Privacy Enhancing Technologies*, Heidelberg: Springer, *Lecture Notes in Computer Science* 3856, pp. 197–209, 2006.
12. C. Eckert, *IT-Sicherheit: Konzepte, Verfahren, Protokolle*, 7th edn., Munich: Oldenbourg-Verlag, 2012.
13. A. Sikora, *Technische Grundlagen der Rechnerkommunikation: Internet-Protokolle und Anwendungen*, Munich: Leipzig, 2003.
14. Intelligent Transport Systems (ITS); Security; Threat, Vulnerability and Risk Analysis (TVRA), ETSI TR 102 893 V1.1.1 (2010-03), available at: http://www.etsi.org/deliver/etsi_tr/102800_102899/102893/01.01.01_60/tr_102893v010101p.pdf (last accessed August 1, 2013).
15. Intelligent Transport Systems (ITS); Security; ITS communications security architecture and security management, ETSI TS 102 940 V1.1.1 (2012-06), available at http://www.etsi.org/deliver/etsi_ts/102900_102999/102940/01.01.01_60/ts_102940v010101p.pdf (accessed on August 8, 2013).
16. Intelligent Transport Systems (ITS); Security; Trust and Privacy Management, ETSI TS 102 941 V1.1.1 (2012-06), available at http://www.etsi.org/deliver/etsi_ts/102900_102999/102941/01.01.01_60/ts_102941v010101p.pdf (accessed on August 8, 2013).
17. K. Plobl, T. Nowey, C. Mletzko, "Towards a Security Architecture for Vehicular Ad Hoc Networks," *Proceedings of the 1st International Conference on Availability, Reliability and Security (ARES)*, Vienna, April 20–22, pp. 374–381, 2006.
18. W. Whyte, "Security, Privacy, Identifications," in: A. Eskandarian (ed.), *Handbook of Intelligent Vehicles*, London: Springer, pp. 1219–1267, 2012.
19. S. Jaeckel, N. Braun, A. Sikora, "Design Strategies for Secure Embedded Networking," in: Workshop "Long-term Security," in: A.U. Schmidt, M. Kreutzer, R. Accorsi (eds.), *Long-Term and Dynamical Aspects of Information Security: Emerging Trends in Information and Communication Security*, Hauppauge, NY: Nova Science Publisher, 2007.
20. B. Weyl, "Secure Vehicular Communication: Results and Challenges Ahead," *SeVeCom Workshop*, February 20–21, Lausanne, Switzerland, 2008.
21. C. Zhang, R. Lu, P.-H. Ho, A. Chen, "A Location Privacy Preserving Authentication Scheme in Vehicular Networks," *IEEE Wireless Communications & Networking Conference (WCNC)*, Las Vegas, NV, March 31–April 3, 2008.
22. J. Domingo-Ferrer, Q. Wu, "Safety and Privacy in Vehicular Communications," in: C. Bettini, S. Jajodia, P. Samarati, S. Wang (eds.), *Privacy in Location-Based Applications*, Heidelberg: Springer, *Lecture Notes in Computer Science* 5599, pp. 173–189, 2009.
23. X. Lin, R. Lu, C. Zhang, H. Zhu, P.-H. Ho, X. Shen, "Security in Vehicular Ad Hoc Networks," *IEEE Communications Magazine*, April 2008.
24. J. Serna1, J. Luna, M. Medina, "Geolocation-based Trust for Vanet's Privacy," *Journal of Information Assurance and Security*, 4: 432–439, 2009.
25. T. Schütze, "Automotive Security: Cryptography for Car2X Communication," *Embedded World Conference*, Nuremberg, Germany, March 1–3, 2011.

26. P. Wex, J. Breuer, A. Held, T. Leinmüller, L. Delgrossi, "Trust Issues for Vehicular Ad Hoc Networks," *IEEE 67th Vehicular Technology Conference (VTC-Spring)*, Marina Bay, Singapore, May 11–14, 2008.

27. Federal Motor Transport Authority (KBA), "Fahrzeugalter," version 15.04.2011, available at: http://www.kba.de/cln_031/nn_125398/DE/Statistik/Fahrzeuge/Bestand/Fahrzeugalter/fahrzeugalter__node.html?__nnn=true (last accessed August 1, 2013).

28. D. Lill, M. Schappacher, A. Gutjahr, A. Sikora, "Development of a Wireless Communication and Localization System for VRU eSafety," *2nd International Workshop on Communication Technologies for Vehicles (Nets4Cars) at 7th IEEE I'l Symposium on Communication Systems, Networks and DSP (CSNDSP)*, Newcastle, UK, July 21–23, 2010.

29. National Highway Traffic Safety Administration (NHTSA), "Interoperability Issues for Commercial Vehicle Safety Applications," DOT HS 811 674, September 2012. http://www.nhtsa.gov/DOT/NHTSA/NVS/Crash%20Avoidance/Technical%20Publications/2012/811674.pdf

30. U.S. Department of Transportation, Research and Innovative Technology Administration, "Reports from the Vehicle Infrastructure Integration Proof of Concept Project," available at: http://www.its.dot.gov/vii/ (last accessed August 1, 2013).

31. National Agency for Automotive Safety & Victim's Aid, "Advanced Safety Vehicle (ASV) Technology Put to Practical Use," available at: http://www.nasva.go.jp/mamoru/en/assessment_car/asv.html (last accessed August 1, 2013).

32. Internet ITS Consortium, available at: http://www.internetits.org/en/top.html (last accessed August 1, 2013).

33. The European Commission, "M/453 EN Standardisation Mandate Addressed to CEN, CENELEC and ETSI in the Field of Information and Communication Technologies to Support the Interoperability of Co-operative Systems for Intelligent Transport in the European Community," October 2009, available at: http://www.etsi.org/WebSite/document/aboutETSI/EC_Mandates/m453%20EN.pdf (last accessed August 1, 2013).

34. U.S. Department of Transportation, "IEEE 1609—Family of Standards for Wireless Access in Vehicular Environments (WAVE), Intelligent Transportation Systems Standards Fact Sheet," September 25, 2009, available at: http://www.standards.its.dot.gov/Factsheets/Factsheet/80 (last accessed August 1, 2013).

35. IEEE P1609.2, "Trial Use Standard for Wireless Access in Vehicular Environments (WAVE)—Security Services for Applications and Management Messages," 2006, available at: http://www.standards.its.dot.gov/Factsheets/Factsheet/80 (last accessed August 1, 2013).

36. Intelligent Transport Systems (ITS); Vehicular Communications; Basic Set of Applications; Part 2: Specification of Cooperative Awareness Basic Service, ETSI TS 102 637-2 V1.2.1 (2011-03), available at: http://www.etsi.org/deliver/etsi_ts/102600_102699/10263702/01.02.01_60/ts_10263702v010201p.pdf (last accessed August 1, 2013).

37. Harmonized European Standard (Telecommunication series), ETSI EN 302 571 V1.1.1: "Intelligent Transport Systems (ITS); Radiocommunications Equipment Operating in the 5855 MHz to 5925 MHz frequency band;

Harmonized EN Covering the Essential Requirements of Article 3.2 of the R&TTE Directive," available at: http://www.etsi.org/deliver/etsi_en/30250 0_302599/302571/01.02.00_20/en_302571v010200a.pdf (last accessed August 1, 2013).

38. European Commission, "The eSafety Initiative," available at: http://ec.europa. eu//information_society/activities/esafety/index_en.htm or www.esafetysup-port.org

39. T. Leinmüller, L. Buttyan, J.-P. Hubaux, F. Kargl, R. Kroh, P. Papadimitratos, M. Raya, E. Schoch, "SEVECOM—Secure Vehicle Communication," *Proceedings of the IST Mobile Summit*, Mykonos, June 4–8, 2006.

40. Secure Vehicular Communication, available at: http://www.sevecom.org (last accessed August 1, 2013).

41. E-safety Vehicle Intrusion Protected Applications, available at: http://www. evita-project.org (last accessed August 1, 2013).

42. A. Festag, G. Noecker, M. Strassberger, A. Lübke, B. Bochow, M. Torrent-Moreno, S. Schnaufer, R. Eigner, C. Catrinescu, J. Kunisch, " 'NoW—Network on Wheels': Project Objectives, Technology and Achievements," *Proceedings of the 5th International Workshop on Intelligent Transportation (WIT 2008)*, Hamburg, Germany, March 18–19, 2008.

43. Open Vehicular Secure Platform, available at: https://www.oversee-project. com/ (last accessed August 1, 2013).

44. "California Partners for Advanced Transit and Highways (PATH)," available at: http://www.path.berkeley.edu/.

45. Privacy Enabled Capability in Co-operative Systems and Safety Applications, available at: http://www.preciosa-project.org/ (last accessed August 1, 2013).

46. "Sichere Intelligente Mobilität Testfeld Deutschland," available at: http:// www.simtd.org/ (last accessed August 1, 2013).

47. N. Bissmeyer, H. Stübing, M. Mattheuss, J. Stotz, J. Schütte, M. Gerlach, F. Friederici, "simTD Security Architecture: Deployment of a Security and Privacy Architecture in Field Operational Tests," *7th Embedded Security in Cars Conference (escar 2009)*, Düsseldorf, Germany, November 24–25, 2009.

48. simTD Consortium, "Konsolidierter Systemarchitekturentwurf," Deliverable 21.2 version 3.0, October 2009, available at: http://www.simtd.de/index. dhtml/0951f9ef59624750942t/object.media/deDE/6459/CS/-/backup_pub-lications/Projektergebnisse/simTD-Deliverable-D21.2_Konsolidierter_ Systemarchitekturentwurf.pdf (last accessed August 1, 2013).

49. National Institute of Standards and Technology, Federal Information Processing Standards (FIPS) 186–3, "Digital Signature Standard (DSS)," June 2009, avail-able at: http://csrc.nist.gov/publications/fips/fips186-3/fips_186-3.pdf (last accessed August 1, 2013).

50. ISO/IEC 18033-2-2006, "Information Technology—Security Techniques— Encryption Algorithms—Part 2: Asymmetric Ciphers," ISO/IEC, 2006, available at: http://standardsdevelopment.bsigroup.com/Home/Committee/50001780 (last accessed August 1, 2013).

51. IEEE P1363, "Standard Specifications for Public Key Cryptography," 2000, available at: grouper.ieee.org/groups/1363/ (last accessed August 1, 2013).

52. M. Dworkin, "Recommendation for Block Cipher Modes of Operation: The CCM Mode for Authentication and Confidentiality," NIST Special Publication 800-38c, National Institute of Standards and Technology (NIST), May 2004, available at: http://csrc.nist.gov/publications/nistpubs/800-38C/SP800-38C.pdf (last accessed August 1, 2013).

53. P. Ardelean, P. Papadimitratos, "Secure and Privacy-Enhancing Vehicular Communication," *IEEE 67th Vehicular Technology Conference: VTC2008-Spring*, Marina Bay, Singapore, May 11–14, 2008.

54. J.-H. Song, V.W.S. Wong, V.C.M. Leung, "Wireless Location Privacy Protection in Vehicular Ad-Hoc Networks," *Mobile Networks & Applications*, 15: 160–171, 2010.

55. G. Calandriello, P. Papadimitratos, A. Lioy, J.-P. Hubaux, "Efficient and Robust Pseudonymous Authentication in VANET," *The 4th ACM International Workshop on Vehicular Ad Hoc Networks (VANET), in conjunction with ACM MobiCom 2007*, Montréal, QC, September 9–14, 2007.

56. E. Schoch, F. Kargl, "On the Efficiency of Secure Beaconing in VANETs," *Proceedings of the 3rd ACM Conference on Wireless Network Security, (ACM WiSec 2010)*, Hoboken, NJ, March 22–24, 2010.

57. Y. Jiang, M. Shi, X. Shen, C. Lin, "A Tree-Based Signature Scheme for VANETs," *IEEE Global Telecommunications Conference GLOBECOM*, New Orleans, LA, November 30–December 4, 2008.

58. J. Forne, J.L. Munoz, O. Esparza, F. Hinaerejos, "Certificate Status Validation in Mobile Ad Hoc Networks," *IEEE Wireless Communications*, 16(1): 52–62, 2009.

59. D. Boneh, H. Shacham, B. Lynn, "Short Signatures from the Weil Pairing," *Advances in Cryptology—ASIACRYPT 2001, 7th International Conference on the Theory and Application of Cryptology and Information Security*, Gold Coast, QLD, December 9–13, 2001.

60. B. Driessen, A. Poschmann, C. Paar, "Comparison of Innovative Signature Algorithms for WSNs," *1st ACM Conference on Wireless Network Security (WiSec 2008)*, Alexandria, VA, March 31–April 2, 2008.

61. M. Raya, J.-P. Hubaux, "The Security of Vehicular Ad Hoc Networks," *The 3rd ACM Workshop on Security of Ad Hoc and Sensor Networks (SASN 2005)*, Alexandria, VA, November 7, 2005.

62. N.-W. Wang, Y.-M. Huang, W.-M. Chen, "A Novel Secure Communication Scheme in Vehicular Ad Hoc Networks," *Computer Communications*, 31: 2827–2837, 2008.

63. A. Wasef, Y. Jiang, X. Shen, "DCS: An Efficient Distributed-Certificate-Service Scheme for Vehicular Networks," *IEEE Transactions on Vehicular Technology*, 59(2): 2010.

64. P. Papadimitratos, G. Mezzour, J.-P. Hubaux, "Certificate Revocation List Distribution in Vehicular Communication Systems," *Proceedings of the 5th ACM International Workshop on VehiculAr InterNETworking (VANET 2008) at MobiCom*, San Francisco, CA, September 14–19, 2008.

65. A. Wasef, X. Shen, "Efficient Group Signature Scheme Supporting Batch Verification For Securing Vehicular Networks," *IEEE ICC 2010—IEEE International Conference on Communications*, Capetown, South Africa, May 23–27, 2010.

66. A. Wasef, X. Shen, "EDR: Efficient Decentralized Revocation Protocol for Vehicular Ad Hoc Networks," *IEEE Transactions on Vehicular Technology*, 58(9): 5214–5224, 2009.

67. M. Raya, J.-P. Hubaux, "Securing Vehicular Ad Hoc Networks," *Journal of Computer Security, Special Issue on Security of Ad Hoc and Sensor Networks*, 15(1): 39–68, 2007.
68. J. J. Haas, Y. C. Hu, K. P. Laberteaux, "Design and Analysis of a Lightweight Certificate Revocation Mechanism for VANET," *6th ACM International Workshop on VehiculAr Inter-NETworking (VANET)*, Beijing, China, September 25, 2009.
69. Y. Jiang, M. Shi, X. Shen, C. Lin, "BAT: A Robust Signature Scheme for Vehicular Networks Using Binary Authentication Tree," *IEEE Transactions on Wireless Communications*, 8(4): 2009.
70. T. Moore, M. Raya, J. Clulow, P. Papadimitratos, R. Anderson, J.-P. Hubaux, "Fast Exclusion of Errant Devices from Vehicular Networks," *5th Annual IEEE Communications Society Conference on Sensor, Mesh and Ad Hoc Communications and Networks (SECON 2008)*, San Fransisco, CA, June 16–20, 2008.
71. R. Peplow, D.S. Dawoud, J. Merwe, "Ensuring Privacy in Vehicular Communication," *Proceedings of the 1st International Conference on Wireless Communication, Vehicular Technology, Information Theory and Aerospace & Electronic Systems Technology*, Aalborg, Denmark, May 17–20, pp. 610–614, 2009.
72. Trusted Computing Group, available at: http://www.trustedcomputinggroup. org/ (last accessed August 1, 2013).
73. Trusted Computing Group—Information for Developers, available at: https:// www.trustedcomputinggroup.org/developers/ (last accessed August 1, 2013).
74. P. Digeser, A. Sikora, "Management of Routed Wireless M-Bus Networks for Sparsely Populated Large-Scale Smart-Metering Installations," *International Conference on Security in Computer Networks and Distributed Systems (SNDS)*, Trivandrum, India, October 11–12, 2012.
75. Forschungsinitiative Kooperative Fahrerassistenzsysteme, available at: http:// www.kofas.de/ (last accessed August 1, 2013).
76. A. Sikora, R.H. Rasshofer, "Cooperative Sensors Using Localization and Communication for Increased VRU Safety," *4th International Conference on Sensing Technology (ICST)*, June 3–5, Lecce, Italy, 2010.
77. Vehicle-to-Vulnerable roAd user cooperaTive communication and sensing teCHnologies to imprOVE transpoRt safety, available at: http://www.watcho-ver-eu.org/ (last accessed August 1, 2013).
78. Aktive mobile Unfallvermeidung und Unfallfolgenminderung durch koopera-tive Erfassungs- und Trackingtechnologie, available at: http://www.projekt-amulett.de/ (last accessed August 1, 2013).
79. A. Sikora, "Security Solutions for Highly Dynamic Car2X Networks in the KoFAS Initiative," *4th International Workshop Nets4Cars/Nets4Trains, Lecture Notes in Computer Science 7266*, Vilnius, Lithuania, April 25–27, 2012.
80. P. Stepanov, "Security Algorithms and Protocols for Highly Dynamic Distributed Wireless C2X-Networks," Masters Thesis 2012, University of Applied Sciences, Offenburg, Germany.

14

Using Trust and Argumentation in Multiagent Recommender Systems

Punam Bedi and Pooja Vashisth

CONTENTS

Introduction

In the recent past, considerable research has been devoted to trust mechanisms to simplify complex transactions for open environments in social networking, e-commerce, and recommender systems (RSs) [1,2]. Such open environments are largely developed using multiagent systems, which are vulnerable to malicious agents and hence pose a big challenge: the detection and prevention of undesirable behaviors. Trust in multiagent systems is used for minimizing the uncertainty in the interactions among the autonomous entities such as agents. Nevertheless, in spite of this challenge, various user support systems have been created in the last few years using multiagent systems because of the many benefits they offer. User support systems are specialized tools to assist users in a variety of computer-mediated tasks by providing guidelines or hints. RSs are a special class of user support tools that act in cooperation with users, complementing their abilities and augmenting their performance by offering proactive or on-demand, context-sensitive support. Agent-based RSs incorporate techniques such as inferring user preferences and smart reasoning based on the available data, which are key requirements in proactive, autonomous operation for achieving intelligent assistance for the users [3]. The focus is not so much on the working details of one particular agent, but on the interplay between agents; and on the communication, coordination, and cooperation required for decision-making tasks.

Though the efficiency of the existing recommenders is notable, they still have some serious restrictions [4]. On the one hand, they are unable to deal formally with the unstable manner of users preferences in complex environments. Decisions about user preferences are mostly based on statistics which depend on ranking previous user selections or collecting data from other similar users [5]. On the other hand, they are not equipped with explicit inference abilities. Hence, they cannot provide explanation which could help the user to evaluate the analysis underlying the recommendations provided. The quantitative techniques adopted by current user support systems undergo this limitation. The lack of an underlying formal model makes it difficult to provide users with an explanation of the reasons and processes that led the system to come up with some particular recommendations. As a result, severe trustworthiness issues may appear, especially in those cases where commercial interests are involved, or when outdoor influence is possible.

Logic-based approaches could aid in overcoming these issues, improving recommendation technology by providing a means to formally express constraints and thereby draw conclusions. In this setting, frameworks for argumentation [6,7] provide an exciting substitute for empowering recommendation technologies by providing suitable inference mechanisms for qualitative reasoning. In fact, the argumentation paradigm has been recognized to be effective in an increasing number of real-world applications which are based on multiagent systems, intelligent Web-based forms, semantic Web, and legal reasoning, among many others [4,8–11].

This chapter presents a generic approach to using trust and argumentation in a multiagent system, and in particular we illustrate the concept using a recommendation application, that is, an agent-based user support application in which trustworthy recommendations are provided and improved on the basis of arguments and user preferences. The approach presented in this chapter proposes to integrate existing user support technologies with appropriate trust-based inferential mechanisms for qualitative reasoning. The use of trust and argumentation allows multiagent RSs to resolve conflicts in opinions and present trustworthy recommendations with reasoned justifications. A user is then able to further investigate these arguments and accept the recommendations only if a trustworthy case can be made by the recommendation tool. We study the integration of computational trust and distrust measures on agents using argumentation for reasoning and interaction; by combining an approach for computing trust, distrust, and an argumentation system for an agent-based RS. This is done by interpretation of trust information in RSs using logic; and its integration with argument-based reasoning.

This chapter is organized as follows: The "Introduction" section presents the introduction and motivations. It highlights research challenges faced by present-day RSs as multiagent applications. The section "Traditional Recommender

System Technologies and the Challenges" presents an overview of current RS technologies and briefly describes the importance of trust and argumentation in the RSs. The section "Basic Concepts: Trust and Argumentation" briefs on the basic argumentation and trust concepts. These concepts are needed to fully grasp the rationale behind the trust-integrated argumentation-based recommender techniques that are discussed in the central part of this chapter, which focuses on the application of trust and argument-based reasoning in RSs. The section "Modeling Trust for an Argumentation-Based Recommender System" presents a novel trust–distrust model suitable for argumentation-based RSs. Its formulation takes into account the elements we use in our recommendation approach (accepted and refused arguments, satisfied and violated preferences, liked and disliked recommendations). We later compare (in subsequent sections) the above-said computational trust model with other computational trust mechanisms applied to RSs. The section "Unifying Trust and Argumentation with Agent Reasoning" discusses our approach to empowering recommendation technologies through trust and argumentation. It deals with the integration of the argumentation-based trust with an agent's practical reasoning for allowing agents to initiate, evaluate information, reason, decide, and propagate trust values. This integration enables agents to argue about trust or a statement having a weak trust support structure. Hence, the agents are able to reason with and about trust using logic and argument-based analysis. In the section "An Illustration from a Book Recommendation Scenario," we demonstrate the benefit of the proposed approach using an example in which agents reason with and about trust; at the same time this may result in argumentative attacks and conflicts, which are then resolved using weighted arguments. The above-said example is based on a particular application which emerged as an instance of this approach, oriented toward providing suitable decision support in the context of a book recommendation system. Finally, the section "State-of-the-Art Recommender Systems and Possible Extensions" discusses some state-of-the-art systems, recent developments, and open challenges, such as visualizing trust relationships and argument attacks in an RS, alleviating the cold start problem in a trust network of an RS by using argumentation, studying the effect of involving argumentation in the recommendation process, and investigating the potential of social influences on the agents in a multiagent system. The "Conclusion" section presents the conclusion and future directions.

Traditional Recommender System Technologies and the Challenges

RSs became an important research area due to information overload on the Internet and increase in online e-commerce. Much work has been done both in the industry and academia on developing new approaches to RSs over the last decade. The interest in this area still remains high because it constitutes a problem-ridden research area because of the abundance of practical applications that help users in dealing with information overload and provides personalized recommendations, content, and services to them. Examples of such applications include recommending books, CDs, and other products at Amazon.com, movies by MovieLens, etc. Moreover, some of the vendors have incorporated recommendation capabilities into their commerce servers as well [5,12].

In its most common formulation, the recommendation problem is reduced to the problem of estimating ratings for the items that have not been seen by a user. Intuitively, this estimation is usually based on the ratings given by this user to other items. Once we can estimate ratings for the yet unrated items, we can recommend to the user the item(s) with the highest estimated rating(s). RSs are usually classified according to their approach to rating estimation. Moreover, RSs are usually classified into the following categories, based on how recommendations are made (see Ref. [5], Chapters 2 through 5):

- *Content-based recommendations:* The user is recommended items similar to the ones the user preferred in the past.
- *Collaborative recommendations:* The user is recommended items that people with similar tastes and preferences liked in the past.
- *Hybrid approaches:* These methods combine collaborative and content-based methods.

In addition to RSs that predict the absolute values of ratings that individual users would give to the yet unseen items (as discussed above), work has been done on preference-based filtering, that is, predicting the relative preferences of users [13]. For example, in a movie recommendation application preference-based filtering techniques would focus on predicting the correct relative order of the movies, rather than their individual ratings. However, despite all these advances, the current generation of RSs still requires further improvements to make recommendation methods more effective, persuasive, and applicable to an even broader range of real-life applications which consider the trust issue as well [14,15]. These improvements include better methods of representing user behavior and information about the items to be recommended, more advanced recommendation modeling methods, incorporation of various background information into the recommendation process, utilization of multicriteria ratings, and development of less-intrusive and more trustworthy recommendation methods. In this chapter, we describe various ways to extend the capabilities of RSs.

A solution to some of the research problems faced by current day RSs can be provided by integrating existing user support technologies with appropriate trust-based inferential mechanisms for qualitative reasoning [7,16]. The use of trust and argumentation allows multiagent RSs to resolve conflicts in opinions and present trustworthy recommendations with reasoned justifications. A user is then able to further investigate these arguments and accept the recommendations only if a trustworthy case can be made by the recommendation tool.

Basic Concepts: Trust and Argumentation

Trust is a mechanism for managing the uncertainty about autonomous entities and the information they deal with. Formally trust is defined to be *a subjective expectation a partner has about another's future behavior based on the history of their encounters.* As computer systems have become increasingly distributed, and control in those systems has become more decentralized, trust has become an increasingly more important concept in computer science [17]. Much of the work on trust in computer science has concentrated on dealing with specific scenarios in which trust has to be established or handled in some fashion. There have been studies on the development of trust in ecommerce through the use of reputation systems and studies on how such systems perform [18]. Another area of concern is the reliability of sources of information on the Web, like the one provided by the RSs [19]. Bedi and Vashisth [7] rate the individuals who provide information by looking at the history of the arguments they have provided. Trust is also an important issue from the perspective of autonomous agents and multiagent systems. As a result, we find much work on trust in agent-based systems [9,18,20–22] and in RSs as well [1,15,23].

The argumentation approach can be used to improve performance of the trust mechanisms. This can happen when trust is computed by an agent in isolation (relying on past interactions with a target) or when agents can exchange and share information about the trustworthiness of possible targets (or one another). Computing trust is a problem of reasoning under uncertainty, requiring the prediction and anticipation by an agent (the evaluator) of the future behavior of another agent (the target). Despite the acknowledged ability of argumentation to support reasoning under uncertainty [8,24], only Prade [25], Bedi and Vashisth [16], and Parsons et al. [26] have considered the use of arguments for computing trust in a local trust rating setting. Our recent work [16] proposed a fuzzy trust model based on all accepted and unaccepted arguments generated by the agents in an RS. These agents can argue about each other's beliefs, goals, and plans under an argumentation system. Argumentation system is simply a set of arguments and a binary relation representing the attack relation between the arguments. The following definitions, describe formally an argument and attack relations in their most basic form. Here, BB indicates a possibly inconsistent *Belief Base*. Let $\vdash$ stand for inference and $\equiv$ for logical equivalence. The symbol "$\oplus$" defines a generic operation that can be used to combine trust, belief, or similarity values.

Definition 1 (Argument). *An argument is a pair* (H, h) *where* h *is a formula of a logical language and* H *a subset of BB such that (1)* H *is consistent, (2)* H ⊢ h, *and (3)* H *is minimal, so no subset of* H *satisfying both (1) and (2) exists.* H *is called the support of the argument and* h *its conclusion.*

Definition 2 (Attack Relation). *Let* (H_1, h_1), (H_2, h_2) *be two arguments.* (H_1, h_1) *attacks* (H_2, h_2) *either by rebut or undercut.* (H_1, h_1) *rebuts* (H_2, h_2) *iff* $h_1 \equiv \sim h_2$. (H_1, h_1) *undercuts* (H_2, h_2) *if* $h_1 \equiv \sim h_2'$ *where* $h_2' \in H_2$.

Before making a suggestion h, the speaker agent must use its argumentation system to build an *argument* (H, h). The idea is to be able to persuade the addressee agent about h, if he decides to refuse a suggestion. On the other side, the addressee agent must use his own argumentation system to select the answer he will give. To be able to communicate and argue, the agents use a set of logic rules based on the facts stored in their BB.

Modeling Trust for an Argumentation-Based Recommender System

Trust systems are rating systems where each individual is asked to give his opinion after completion of each interaction in the form of ratings (it can be implicit or explicit). In our work, trust and distrust values are inferred from the rating database of the RS and thereafter these values are used to enhance the accuracy of the recommendation process. These interactions consist of the arguments and recommendations generated by agents. Arguments are important for giving the explanation behind any kind of interaction between the agents. This improves the quality and utility of recommendation iteratively, over several cycles [6]. During argumentation, the agents may agree or disagree over certain issues, before; finally, the user either accepts or rejects a recommendation. We believe that whatever may be the eventual result of a recommendation process, always various arguments (in support or against) are responsible for it. This is because these arguments form the basis for the generated recommendations. Therefore, it is vital to determine the agents' responses on such arguments besides recommendations. Agreement and acceptance would determine a more accurate trust value for agents in the system whereas disagreement and rejection can determine the distrust in the system at the same time.

From a research perspective, too, it is generally acknowledged that distrust plays an important role [27–29], but much ground remains to be covered in this domain. As mentioned above, we calculated the trust and distrust values separately based on the agents' interactions in the system. But as discussed in Ref. [15], all the approaches [28,29] that treat trust and distrust separately, so far, have not been able to reach at a consensus on how to propagate distrust.

Different operators yield different results depending on the interpretation, therefore choosing an appropriate propagation scheme for the application at hand is very important. In our case, because we propose a generalized trust–distrust model for any recommendation application where we propagate and aggregate all trust–distrust values using agent reasoning, therefore we obtained the final suggested trust value by subtracting a weighted fuzzy distrust value from a weighted fuzzy trust value. Guha et al. [29] have also supported merging of trust and distrust values due to similar issues related to distrust propagation, but they did not consider fuzzy measures to improve the accuracy further.

More formally, let $A = \{a_1, a_2, \ldots, a_M\}$ be the set of all agents (user as well as recommender agents), where M is the number of agents in the system. We assume each user agent will rate a recommender agent after completing the recommendation process. An interaction $i \in I$, where $r_{xy}(i_k)$ is the rating, an agent x has given to agent y for an interaction i_k. The rating scale or grade for interactions is defined as $G = \{-2, -1, 0, +1, +2\}$. The set of ratings an agent x has given to an agent y is $S_{xy} = \{r_{xy}(i_k) \mid i_k \in I\}$ and the whole past history of the agent x is $H_x = \{S_{xy} \mid \forall y(\neq x) \in A\}$. The agent's rating for a recommendation or an argument "i_k" is estimated implicitly from the number of matches the argument parameters strike with the agent's preference list. The rating can also be given explicitly by the user on scale "G" as mentioned above.

Given that A is the set of agents. We define an agent's trustworthiness as follows: Trusts : $A \times A \times F \to [0,1]$. This function associates to each recommender agent a fuzzy measure representing its trustworthiness according to other user agents. To compute the trustworthiness of an agent y (denoted as y), an agent x (denoted as x) uses the history of its interactions (both arguments and recommendations) with y.

$$\text{Trusts}_{xy} = \frac{\sum_I \text{FM\_trust}_{xy} - \sum_I \text{FM\_distrust}_{xy}}{\text{T\_N\_Arg}_{xy} + \text{T\_N\_R}_{xy}}, \tag{14.1a}$$

$$= \frac{\left[\sum_{\text{Arg} \in I} \text{FM\_agree(Arg}_{xy}) + \sum_{R \in I} \text{FM}(R_{xy}) * v_{\text{satisfied}} - \left(\sum_{\text{Arg} \in I} \text{FM\_disagree(Arg}_{xy}) + \sum_{R \in I} \text{FM}(R_{xy}) * v_{\text{unsatisfied}} \right) \right]}{\text{T\_N\_Arg}_{xy} + \text{T\_N\_R}_{xy}}. \tag{14.1b}$$

Here, Trusts_{xy} denotes the final trustworthiness of y according to x's viewpoint.

T_N_Arg_{xy} is the total number of arguments made by y towards x, a count maintained by the agent's persistent belief base in the system.

T_N_R$_{xy}$ is the total number of recommendations made by y for x, a count maintained by the agent's persistent belief base in the system.

$\sum_{\text{Arg}\in I}$ FM_agree(Arg$_{xy}$) is the summation of the fuzzy measure of the degree of x's agreement over y's arguments that are acceptable to x, obtained from Equation 14.7.

$\sum_{\text{Arg}\in I}$ FM_disagree(Arg$_{xy}$) is the summation of the fuzzy measure of the degree of x's disagreement over y's arguments that are unacceptable to x, obtained from Equation 14.8.

$\sum_{\text{R}\in I}$ FM(R$_{xy}$)$*v_{\text{satisfied}}$ is the summation of the fuzzy measure of the degree of x's satisfaction over y's recommendations obtained from Equations 14.4 and 14.5. Hence, these recommendations are acceptable to x.

$\sum_{\text{R}\in I}$ FM(R$_{xy}$)$*v_{\text{unsatisfied}}$ is the summation of the fuzzy measure of the degree of x's dissatisfaction over y's recommendations obtained from Equation 14.6. Hence, these recommendations are unacceptable to x. Here, $v_{\text{satisfied}}$ and $v_{\text{unsatisfied}}$ are the values (weights) attached to the acceptable and unacceptable recommendation arguments, respectively.

We now need to find the user satisfaction (satisfied to what extent) and dissatisfaction (unsatisfied to what extent) for a recommendation or an argument generated by an agent. To do so, we can define two fuzzy subsets on each agent's ratings, say satisfied and unsatisfied. This is because the fuzzy sets can clearly capture the concept of finding the extent (membership value) to which an agent is satisfied or unsatisfied with an interaction. The satisfied and unsatisfied fuzzy subsets for x are defined as below:

$$\text{Satisfied}(x) = \{\text{sat}_x(i_k) \mid i_k \in H_x\} \tag{14.2}$$

$$\text{Unsatisfied}(x) = \{\text{unsat}_x(i_k) \mid i_k \in H_x\} \tag{14.3}$$

where:
sat$_x(i_k)$ and unsat$_x(i_k)$ give us crisp membership values for x's ratings over an interaction i_k (arguments and recommendations) in the fuzzy subsets satisfied(x) and unsatisfied(x), respectively (see Figure 14.1)

We then determine the type-2 fuzzy membership values for x's ratings over an interaction i_k (see Figure 14.2) using Equations 14.5 and 14.6.

Now, we give a simple triangular membership function for satisfied(x) and unsatisfied(x) type-1 fuzzy subsets that are defined by the following popular equations [30]:

$$\text{Sat}_x(i_k) = \begin{cases} 0 & r_{xy}(i_k) = g_{\min} \\ \dfrac{r_{xy}(i_k) - g_{\min}}{g_{\max} - g_{\min}} & g_{\min} < r_{xy}(i_k) < g_{\max}, \\ 1 & r_{xy}(i_k) = g_{\max} \end{cases} \tag{14.4}$$

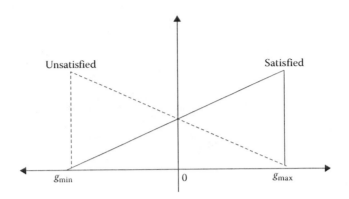

FIGURE 14.1
Membership function to determine values of x's ratings for an interaction i_k in the fuzzy sub-sets satisfied(x) and unsatisfied(x).

where:
g_{min} and g_{max} are the minimum and the maximum ratings for a given system with $G = \{g_{min}, ..., 0, ..., g_{max}\}$

To fuzzify the membership function value as obtained from Equation 14.4, we use the interval type-2 fuzzy sets to generalize type-1 fuzzy sets. This way more uncertainty can be handled. Hence, now $\text{sat}_x(i_k) \in \{\text{sat}_x(i_k)^2, \sqrt{\text{sat}_x(i_k)}\}$ for instead of being equal to a numeric value $r_{xy}(i_k) - g_{min}/g_{max} - g_{min}$. This now gives us the flexibility to define the user satisfaction level (extent) using interval type-2 fuzzy sets. We believe that different users (say old-frequent and new user) may have different levels of satisfaction for a given output. A frequent user expects more relevant recommendations from the system than a new user or a user who makes use of the system less frequently. Say, we have three groups of satisfaction level (less satisfied, moderately satisfied, and highly satisfied). So, the range of satisfaction level may vary differently for new user and frequent user given the same set of recommendations.

Further, the degree of user satisfaction (taking experience in system usage under consideration) can be determined as:

$$\text{Sat}_x(i_k) = \frac{\alpha\left[\text{sat}_x(i_k)^2\right] + \beta\left[\sqrt{\text{sat}_x(i_k)}\right]}{\alpha + \beta}, \tag{14.5}$$

where:
α and β are the weights attached to the lower and upper bounds for the interval type-2 fuzzy membership set $\{\text{sat}_x(i_k)^2, \sqrt{\text{sat}_x(i_k)}\}$ for $\text{sat}_x(i_k)$, respectively

Here, α and β will take values during execution, according to the type of user logged into the system and in need of recommendations (Figure 14.2).

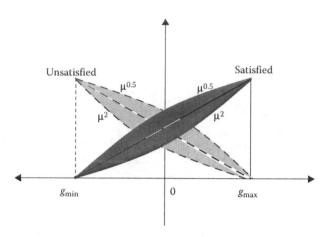

FIGURE 14.2
Type-2 fuzzy membership function to determine values of x's ratings for an interaction i_k in the fuzzy subsets satisfied(x) and unsatisfied(x).

Using Equation 14.5, we get

$$\text{Unsat}_x(i_k) = 1 - \text{sat}_x(i_k). \tag{14.6}$$

Type-2 fuzzy sets and systems generalize (type-1) fuzzy sets and systems so more uncertainty can be handled. Membership function of a type-1 fuzzy set has no uncertainty associated with it. A type-2 fuzzy set lets us incorporate uncertainty about the membership function into fuzzy set theory, and is a way to address the above criticism of type-1 fuzzy sets. And, if there is no uncertainty, then a type-2 fuzzy set reduces to a type-1 fuzzy set, which is analogous to probability reducing to determinism when unpredictability vanishes. For example, in the case study under consideration for this chapter, we represent membership of cost preference of a product by the interval type-2 fuzzy set, that is, membership of cost for each product in every category, that is, low, medium, or high is an interval rather than a crisp value. Thus, more uncertainty can be handled using type-2 fuzzy sets than by using type-1 fuzzy sets. User satisfaction level can be defined using type-2 fuzzy sets as well. As explained above that, different users (old-frequent and new user) may have different levels of satisfaction with the generated recommendations. A frequent user expects more relevant recommendation from the system than a new user or a user who makes use of system less frequently. Therefore, the range of satisfaction levels will be different for new users and frequent users.

Next, for the argumentation between agents in an RS, we define two combinations of satisfied and unsatisfied fuzzy subsets, that is, satisfied–satisfied written as $SS(x, y)$, and unsatisfied–satisfied written as $US(x, y)$. We assume that x represents a user whereas y represents a recommender. Therefore, for a rating fuzzy trust system, the agreement and disagreement values over an argument between any two agents x and y are given by

$$FM\_agree(Arg_{xy}) \equiv SS(x, y) = \frac{|satisfied(x) \cap satisfied(y)|}{|satisfied(x) \cup satisfied(y)|}. \qquad (14.7)$$

$$FM\_disagree(Arg_{xy}) \equiv US(x, y) = \frac{|unsatisfied(x) \cap satisfied(y)|}{|unsatisfied(x) \cup satisfied(y)|}. \qquad (14.8)$$

Fuzzy sets literature describes many alternatives for union and intersection of crisp sets. The popular one is minimum for intersection and maximum for union. The min–max alternative and the definition of fuzzy set's cardinality yield the following [30]:

$$|Satisfied(x) \cap satisfied(y)| = \sum_{i_k \in H_x \cap H_y} min[sat_x(i_k), sat_y(i_k)]. \qquad (14.9)$$

$$|Satisfied(x) \cup satisfied(y)| = \sum_{i_k \in H_x \cap H_y} max[sat_x(i_k), sat_y(i_k)]. \qquad (14.10)$$

All these arguments and recommendations are related to a particular domain. The basic idea is that, the trust and distrust degree of an agent can be estimated according to how much information acquired from him has been accepted or rejected as belief in the past, respectively.

Unifying Trust and Argumentation with Agent Reasoning

We are now interested in a finite set of agents A, how these agents trust one another and use this representation in logic. If $Trusts(Ag_i, Ag_j)$, where $Ag_i, Ag_j \in A$, then Ag_i trusts Ag_j. This is not a symmetric relation, so it is not necessarily the case that $Trusts(Ag_i, Ag_j) \Rightarrow Trusts(Ag_j, Ag_i)$. It is usual in work on trust to consider performing inference about trust by assuming that trust relations are transitive. The process of inference is what [29] calls "direct propagation." Another type of trust relation is known as indirect trust. This type of trust can be computed using trust value of an agent collected from its neighbors or other similar agents. Our aim here is to show the use of logic and argumentation, to propagate trust values between agents in the RS. In other words, we want an argumentation-based approach that an agent can use to determine that he has a reason to trust another agent, and then combine this trust with the other knowledge he has, to make decisions about recommendations. In the system, every agent Ag_i has a BB, that is, some collection of information about the world, which we will call BB_i, and it is expressed in logic. BB_i is made up of a number of partitions as follows:

$$BB_i = BB_i^{trust} \cup BB_i^{bel} \cup BB_i^{similar} \cup \bigcup_j BB_i^j. \qquad (14.11)$$

Here, BB_i^{trust} holds information about the degree of trust Ag_i has in other agents it knows, BB_i^{bel} is a set of beliefs or facts (in the form of predicates) of Ag_i about the world (which we assume come with some measure of belief), $BB_i^{similar}$ contains measure of similarity of Ag_i with other agents (similarity between two agents is computed by applying cosine similarity [31] to their likes and dislikes for preferences), and some information BB_i^j that Ag_i is provided by each of its neighbors Ag_j. The elements of BB_i are triplets as described in the work by Parsons [32]. Each element has the form: (*type_id*; *data*; *value*). The first parameter *type_id* is a means of referring to the element of the particular type BB_i, the second parameter is a *formula*, and the third one is a quantitative *measure* of *trust/belief/similarity* depending upon which partition of BB_i the element is referring to.

All arguments take the form: (*conclusion*; *grounds*; *rules*; *value*). The *conclusion* is inferred from the *grounds* using the rules of inference that is *rules* (as specified below) and the *conclusion* is inferred with a degree having a *value*.

$$Arg^{trust} \frac{(type\_id; \text{trusts}(x, y); v) \in BB_i^{trust}}{BB_i^{trust} \vdash (\text{trusts}(x, y); \{type\_id\}; \{Arg^{trust}\}; v)}. \tag{14.12}$$

The rule Arg^{trust} says that if some agent Ag_i has a triplet: $(t1; \text{trusts}(x, y); 0.4)$ in its BB_i^{trust} then it can construct an argument for $\text{trusts}(x, y)$ where the grounds are $t1$, the degree of trust is 0.4, and which records that the Arg^{trust} rule was used in its derivation.

$$Arg^{similar} \frac{(type\_id; \text{sim}(x, y); v) \in BB_i^{similar}}{BB_i^{similar} \vdash (\text{sim}(x, y); G; \{Arg^{similar}\}; v)}. \tag{14.13}$$

Similarly, the rule $Arg^{similar}$ says that if some agent Ag_i has a triplet: $(s1; \text{sim}(x, y); 0.9)$ in its $BB_i^{similar}$ then it can construct an argument for $\text{sim}(x, y)$ where the grounds are given by a set G (will combine all the facts used to determine the similarity), the degree of similarity is 0.9, and which records that the $Arg^{similar}$ rule was used in its derivation.

$$Arg^{gdt} \frac{\left[\begin{array}{c} BB_i^{trust} \vdash (\text{trusts}(x, y_n); G_n; R_n; v_n) \text{ and} \\ BB_i^{trust} \vdash (\text{trusts}(y_n, z); H_n; S_n; w_n) \end{array}\right]^p}{\begin{array}{c} BB_i^{trust} \vdash (\text{trusts}(x, z); G_n \cup H_n; R_n \cup S_n \cup \{Arg^{gdt}\}; \\ v_n \oplus^{agg(trust)} w_n) \end{array}} \tag{14.14}$$

where:

$n = 1$ to p

p denotes the number of agents directly known to x (i.e., neighbors of x)

The numerator in the above expression can be repeated at most p number of times as p determines the maximum number of agents known to x who also know z.

The rule Arg^{gdt} captures *group direct propagation of trust* values for z obtained from a group of neighbors of x. It says that if we can show that $\text{trusts}(x, y_n)$ holds with degree v_n and we can show that $\text{trusts}(y_n, z)$ holds with degree w_n for some agent y_n, then we are allowed to conclude $\text{trusts}(x, z)$ with a degree $v_n \oplus^{\text{agg(trust)}} w_n$, and that the conclusion is based on the union of the information that supported the premises, and is computed using all the rules used by both the premises. Here, $v_n \oplus^{\text{agg(trust)}} w_n$ is interpreted to be an aggregation over various trust values (combined individually by the operation $\oplus^{\text{trust}}$) obtained from different sources trusted by agent i. Therefore, we expand the combination as follows:

$$v_n \oplus^{\text{agg(trust)}} w_n = (v_i \oplus^{\text{trust}} w_i) \oplus^{\text{agg(trust)}} (v_{i+1} \oplus^{\text{trust}} w_{i+1}) \cdots \oplus^{\text{agg(trust)}} (v_p \oplus^{\text{trust}} w_p).$$

The reason we are interested in using argumentation to handle trust, is that we want to record, in the form of the argument for some proposition, the reasons that why it should be believed. Because information on the source of some piece of data, and the trust that an agent has in the source, is relevant, then it should be recorded in the argument. This is easier to achieve if we encode data about who trusts whom in logic.

So far, we have explained how agent Ag_i can reason about the trustworthiness of other agents. The reason for doing this is so Ag_i can use its trust information to decide how to use information that it gets from those agents. Consider now, the following set of *belief inference rules*:

$$\text{Arg}^{\text{belief}} \frac{(type\_id; \alpha; v) \in \text{BB}_i^{\text{belief}}}{\text{BB}_i \vdash (\alpha; G; \{\text{Arg}^{\text{belief}}\}; v)}. \tag{14.15}$$

$$\text{Trust} \frac{\begin{bmatrix} \text{BB}_i^{\text{trust}} \vdash (\text{trusts}(i, j); G; R; a) \text{ and } \text{BB}_i^{\text{similar}} \vdash \\ (\text{sim}(i, j); H; S; b) \text{ and } \text{BB}_i^j \vdash (\alpha; K; T; c) \end{bmatrix}}{\begin{array}{c} \text{BB}_i \vdash (\alpha; G \cup H \cup K; R \cup S \cup T \cup \{\text{Trust}\}; \\ a \oplus^{\text{sim}} b \oplus^{\text{bel}} c) \end{array}}. \tag{14.16}$$

$$\text{IC} \frac{\text{BB}_i \vdash (\alpha; G; R; v) \text{ and } \text{BB}_i \vdash (\beta; H; S; w)}{\text{BB}_i \vdash (\alpha \wedge \beta; G \cup H; R \cup S \cup \{\text{IC}\}; v \oplus^{\text{bel}} w)}. \tag{14.17}$$

$$\text{MP} \frac{\text{BB}_i \vdash (\alpha; G; R; v) \text{ and } \text{BB}_i \vdash (\alpha \rightarrow \beta; H; S; w)}{\text{BB}_i \vdash (\beta; G \cup H; R \cup S \cup \{\text{MP}\}; v \oplus^{\text{bel}} w)}. \tag{14.18}$$

All these trust information, that is, the rules in Equations 14.11 through 14.14 can then be used, along with the above rules given in Equations 14.15 through 14.18, to construct arguments that combine trust and beliefs of the agents. The rule Arg^{belief}, extracts an argument from a single item of information, while the rules introducing a conjunction (IC) and modus ponens (MP) are typical natural deduction rules. The rules for IC and eliminating implication or MP are augmented with the combination of degrees of belief, and the collection of information consisting of data and proof rules. The key rule is the rule named Trust. This says that if it is possible to construct an argument for α from some BB_i^j, indicating that the information comes from Ag_j and Ag_i trusts Ag_j and Ag_i similar to Ag_j, then Ag_i has an argument for α. The grounds of this argument combine all the data that was used from BB_i^j, all the information about similarity used to determine that Ag_i is similar to Ag_j, and all the information about trust used to determine that Ag_i trusts Ag_j, and the set of rules in the argument record all the inferences needed to build this combined argument. Finally, the belief that Ag_i has in the argument is the belief in α as it was derived from BB_i^j combined with the similarity and trust Ag_i has in Ag_j. Hence, this rule sanctions the use of information from an agent's acquaintances, provided that the degree of belief in that piece of information is modified by the agent's similarity and trust in that acquaintance. Thus, one agent can only import information from another agent if the first agent can construct a trust argument that determines that it should trust the other (second) agent, using the *Trust* rule.

An Illustration from a Book Recommendation Scenario

As an example, consider Figure 14.3 which shows a persuasion dialogue between the RS's seller agent and the buyer agent. Both the agents have their own belief bases having values of their trust in various information sources,

Buyer 1: Your book is not popular. *Claim*

Seller 1: Why is my book not popular? *Challenge*

Buyer 2: Because the newspapers recently reported negative on its popularity. *Argument*

Seller 2: Yes, that is what the newspapers say, *concession*, but that does not prove anything, because newspapers are unreliable sources of information. *Undercutter*

Buyer 3: Never mind, *concession*, but your book is still not popular. *Counterclaim*

Seller 3: Is the book popular due to the customer survey? *Rebuttal*

Buyer 4: Because the author quality is not good, the book is not popular. *Rebuttal*

Seller 4: That is not true, because the author is a *Booker* prize winner. *Undercutter*

Buyer 5: OK, I did not know that, I was wrong that book is not popular. *Retraction*

FIGURE 14.3
An example persuasion dialogue between buyer and seller agent over a book recommendation.

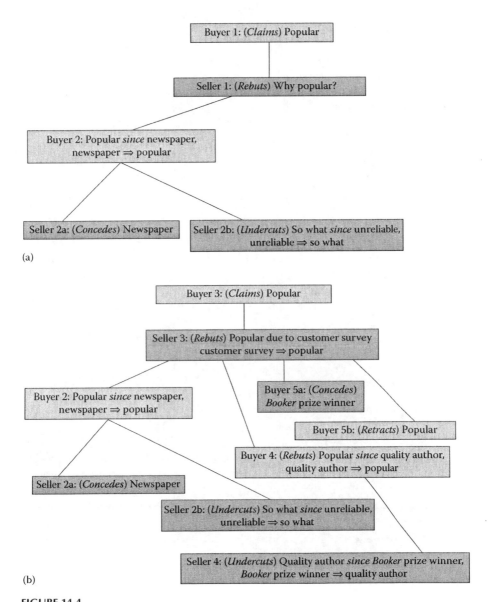

FIGURE 14.4
A persuasion dialogue between buyer and seller over a book recommendation represented as arguments (a and b).

similarity with other agents, information communicated by other agents and their own beliefs. The trust values lie between 0 and 1. In this example, the buyer agent needs to take a decision about a book recommended to it by the *RS*s seller agent. For example, our buyer agent say *buyer* might have the following collection of information:

So, BB_{buyer}^{trust} contains:

($t1$; trusts(buyer, *newspaper*); 0.8)

($t2$; trusts(buyer, seller); 0.6)

Also, BB_{buyer}^{bel} contains:

($b1$; Author(a, b); 1)

($b2$; ~goodquality(a); 0.7)

($b3$; ~goodquality(x) → ~popular(x); 0.8)

($b4$; Bookerprize(x) → goodquality(x); 1)

Buyer also has some information from *newspaper*'s connections: $BB_{buyer}^{newspaper} \vdash$ ($bn1$; ~popular (a); 1)

Each argument can then be used with *belief inference rules* given as Equations 14.15 through 14.18 to construct arguments by the *buyer* about the question of whether the recommended book is popular or not. Using information from its own bases and *newspaper*, the *buyer* can determine:

$$BB_{buyer} \vdash (\sim popular\ (a);\ \{t1, bn1\};\ rules\ 1;\ t_1)$$

where:

rules1 = {Arg^{trust}, Trust}

This shows that after the communication of information from *newspaper* about ~*popular* (a), its proof requires the application of *Trust* to establish a degree of belief in *newspaper*'s information. So, an application of *Trust* to import $bn1$ from *newspaper* is required. We interpret the degree of trust in an information source to be a degree of belief that what the source says is true. To compute the degrees of trust, we follow in taking operation $\oplus^{trust}$ to be *minimum* [33]. So, $t_1 = 0.8 \oplus^{trust} 1 = \min \{0.8, 1\} = 0.8$. Similarly, the *buyer* can also construct other arguments as depicted in Figure 14.4(a), Figure 14.4(b), and Figure 14.5.

The *RSs seller* agent also has its own belief bases containing some of the following information, which is then used to construct arguments using *belief inference rules* given as Equations 14.15 through 14.18.

So, BB_{seller}^{trust} contains:

($t1$; trusts(seller, *newspaper*); 0.2)

($t2$; trusts(seller, customersurvey); 0.9)

Also, BB_{seller}^{bel} contains:

($s1$; Author(a, b); 1)

($s2$; Bookerprize(a); 1)

($s3$; Bookerprize(x) → goodquality(x); 1)

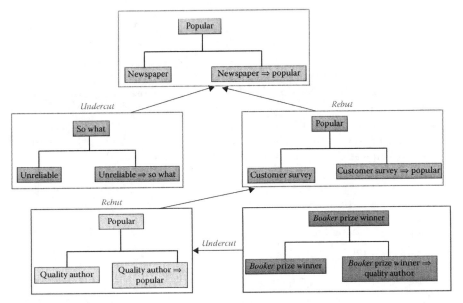

FIGURE 14.5
An argument graph for the above example to show attack relations.

Seller also has some information from *customersurvey*'s connections:

$$BB_{seller}^{customersurvey} \vdash (sc1; \text{popular } (a); 0.9)$$

The *seller* agent uses these arguments (constructed from earlier information) during the persuasion dialogue as shown later. As a result, new information, such as "The author (a) is a Booker prize winner," is revealed by the *seller* to the *buyer* and this helps in convincing the *buyer* about the good popularity of the book under question. Hence, the *buyer* accepts the recommendation for the particular book as good.

State-of-the-Art Recommender Systems and Possible Extensions

In recent years, several models of trust have been developed in general and in the context of MAS [9,15,19,20,23,34]. However, these models are not designed to trust argumentation-based agents in an RS. Their formulations do not take into account the elements we use in our recommendation approach (accepted and rejected arguments, satisfaction and dissatisfaction for interactions and recommendations). In addition, these models have some limitations regarding the user's acceptance of the collected information from other agents and their trustworthiness, which we elaborate further.

Much of the work on trust in computer science has concentrated on dealing with specific scenarios in which trust has to be established or handled in some fashion. There have been studies on the development of trust in ecommerce through the use of reputation systems and studies on how such systems perform [18]. Another area of concern is the reliability of sources of information on the Web, like the one provided by the RSs. For example, Dong et al. [12] investigated mechanisms to determine which sources to trust when faced with multiple, conflicting information. Bedi and Vashisth [7,16] extend this idea to rate the individuals who provide information by looking at the history of the arguments they have provided. Trust is an especially important issue from the perspective of autonomous agents and multiagent systems. The idea behind the multiagent systems is developing software agents that will work in the interest of their owners, carrying out their owners' wishes while interacting with other entities [3]. In such interactions, agents will have to reason about the degree to which they should trust those other entities, whether they trust those entities to carry out some task, or whether they trust those entities for not misusing crucial information. As a result, we find much work on trust in agent-based systems [9,18,20–22] and in RSs as well [1,15,23].

The argumentation approach can be used to improve performance of the trust mechanisms. Despite the acknowledged ability of argumentation to support reasoning under uncertainty [8,24,35], only Prade [25], Bedi and Vashisth [16], and Parsons et al. [26] have considered the use of arguments for computing trust in a local trust rating setting. Prade [25] proposes an argumentation-based approach for trust evaluation that is bipolar (separating arguments for trust and for distrust) and qualitative (as arguments can support various degrees of trust/distrust). Parsons et al. [26] derive argumentation logic where arguments support measures of trust, for example, qualitative measures such as "very reliable" or "somewhat unreliable." Our recent work [16] proposed a fuzzy trust model based on all accepted and unaccepted arguments generated by the agents in an RS. These agents can argue about each other's beliefs, goals, and plans under an argumentation system.

As shown in the previous sections, one of the main contributions of the approach described in this chapter is that it allows us to integrate recommendation and argumentation with trust, within the practical reasoning of an agent. This integration allows the agent to take trustworthy decisions and reason about them at the level of beliefs and goals using argumentation. RSs can be seen as a particular instance of decision-making systems oriented to assist users in solving computer-mediated tasks. In the last few years, there have been several efforts toward integrating argumentation in generic decision-making systems. Agent technology has also been integrated within these systems to model many decision-making tasks requiring recommendation and negotiation, especially because these agents are an excellent tool to assist users or to allow them to act on behalf of users [2,4,9]. Our approach can be applied in both directions. That is, on the one hand, the agent can act on behalf of a user, taking into account his/her preferences, goals, etc.

(for instance, as can be seen in Refs. [6,16]). Unlike the work done in Ref. [4], the agent based on the proposed concept will provide and also improve recommendations autonomously by reasoning. This helps in detecting and then resolving any possible conflicts in user preferences and the recommendations. On the other hand, a personal agent can assist a user during the recommendation process by using the information extracted due to argumentation. That is, a user operating a recommendation system can receive assistance from a personal agent in decision making. Further, comparing our work by the trust model developed for an agent-based system [9], we proposed a trust model based on all accepted and unaccepted interactions generated by the agents in the system taking the user's acceptance into consideration (see Equations 14.1 through 14.10). In such interactions, agents will also have to reason about the degree to which they should trust those other entities, whether they trust those entities to carry out some task, or whether they trust those entities for not misusing crucial information. In the recent works [34,36], an argument-based approach to argue about trust is presented, in which argumentation is used to support decision task related to trust. In contrast to our approach, the authors use argumentation for trust only, whereas we also focus on using fuzzy trust and argumentation for agent's practical reasoning about beliefs. This helps in eliciting conclusions for a particular user on the basis of available information.

Some Recent Developments and Open Challenges

In the previous sections, we have covered the basics of trust modeling, argumentation, and their role in RSs. In this section, we want to give the reader a foretaste of new directions in the research area of argumentation and trust-based recommendation systems. This is certainly not meant to be a complete overview, but rather a selection of recent developments in the field. In particular, we will briefly discuss the following issues: visualizing trust relationships and argument attacks in an RS, alleviating the cold start problem in a trust network of an RS by using argumentation, studying the effect of involving argumentation in the recommendation process, and investigating the potential of social influences on the agents in a multiagent system.

Parsons et al. [32] have shown that the user cold start problem can be alleviated by including information exchange through argumentation amongst users. They demonstrated that new users can benefit from reasoning and propagation of trust values using arguments. This can be useful in order to get good recommendations from the system. However, Victor et al. [15,37] have shown that cold start users in the classical sense (who rated only a few items) are very often cold start users in the trust sense as well. Hence, new users must be encouraged to communicate with other users through agent argumentation and interaction to expand the trust network as soon as possible, but choosing who to associate with is often a difficult task. Given the impact this choice has on the delivered recommendations, it is critical to guide newcomers through this early stage connection process. In Refs. [16,37], this problem is tackled by identifying users with similar tastes. The authors show that, for a cold start user, connecting to one of the identified similar users is much more advantageous than including a randomly chosen user, with respect to coverage as well as accuracy of the generated recommendations.

O'Donovan [38] introduced PeerChooser, a new procedure to visualize a trust-based collaborative filtering RS. More specifically, PeerChooser visualizes both information coming from the traditional similarity measure Pearson Correlation Coefficient (PCC), and information coming from the underlying trust-space generated from the rating data. One of the main features of the system is its possibility to extract trust information directly from the user at recommendation time. This is done by moving specific icons (representing users in the system) on an interactive interface. In this way, the user can indicate his mood and preferences, thereby actively providing real-time trust information. This can be improvised further if users have the means of visualizing reasoning beyond trusting a given source. In this direction, the argument visualization tools can cause higher quality arguments, critical discussion, or coherent arguments. Given the fact that most critical review results by the authors in Ref. [39] point in the same direction, the authors think it is reasonable to assume that these tools have a positive effect on the users' argumentation skills and hence can enhance the persuasive power of a system. However, a lot still remains to be done, because until now experiments have not been able to provide significant evidence for the benefits of argument visualization tools in the user support systems.

There are also other ways to establish trust relations when the information is not explicitly given by the users. Several sources of social data can be consulted, such as online friend and business networks (think, for example, of Facebook or LinkedIn), e-mail communication, reputation systems, etc. Potentially all these social data sources could be incorporated into a (trust-enhanced) RS, but so far not much research has been conducted to find out which ones will be most useful [40], and whether these sources would provide similar results as the trust-based recommendation approaches discussed in this chapter. Arazy et al. [41] embark upon this problem and argue that the design of such social recommenders should be grounded in theory, rather than making *ad hoc* design choices as is often the case in current algorithms.

Another recent research direction of a completely different nature is the investigation of the potential of distrust in trust-based RSs. Whereas in the trust modeling domain only a few attempts have been made to incorporate distrust, in the recommender domain this is even less so. This is due to several reasons, the most important ones being that very few data sets containing distrust information are available, and that there is no general consensus yet about how to propagate it and to use it for recommendation purposes hence most of the works prefer to merge the two (trust and distrust). A first experimental evaluation of the effects of involving distrust in the recommendation process is reported in Ref. [37]. In work [37], three distrust strategies are investigated. The first two strategies are based on the rationale that trust can be used to select similar users (neighbors) in collaborative filtering systems, while the latter strategy suggests taking distrust as a filter for selecting neighbors and it looks more promising as well. But it is clear that much work remains to be done in this emerging research area before one can come to a more precise conclusion.

Conclusion

In this chapter, we have given an introduction to the research area of modeling trust and distrust using argumentation. We illustrated how the approach can be incorporated into agent reasoning to improve the performance of classical RSs. Recommender applications that maintain a trust network among their users can benefit from trust propagation strategies that have proven to yield a surplus value, whereas in cases where it is not immediately possible to collect explicit trust statements, methods that are able to automatically compute trust values seem to be the most ideal solution. Argumentation technologies are promising tools for the settings where autonomous agents can support humans in decision making and hence enhance automation. Agents can help their users in identifying the most profitable choice (recommendation) of all to take a decision accordingly. Especially, in an argumentation-based recommendation system, the arguments uttered to persuade each other over a product are not the result of an isolated analysis, but of an integral view of the problem that we want to agree about. In this work, we have used argumentation for handling trust and vice versa. The feature of using influence of trust on argumentation provided a useful versatility to the problem of reasoning for autonomous agents. This integration allowed the user to take well-reasoned decisions as well which were based on trustworthy recommendations.

References

1. A. Jøsang, R. Ismail, C. Boyd, A survey of trust and reputation systems for online service provision, *Decision Support Systems*, 43: 618–644, 2007.
2. J. Lang, M. Spear, S.F. Wu, Social manipulation of online recommender systems, in *Proceedings of the 2nd International Conference on Social Informatics (SocInfo'10)*, October 27–29, Laxenburg, Austria, 2010.
3. P. Maes, R. Guttman, A. Moukas, Agents that buy and sell, *Communications of the ACM*, 42(3): 81, 1999.
4. C. Chesnevar, A.G. Maguitman, M.P. Gonzalez, Empowering recommendation technologies through argumentation, in I. Rahwan, G. Simari (Eds.), *Argumentation in Artificial Intelligence*, Berlin: Springer, p. 504, 2009.

5. D. Jannach, M. Zanker, A. Felfernig, G. Friedrich, Recommender Systems—An Introduction, New York: Cambridge University Press, 2011.
6. P. Bedi, P. Vashisth, Improving recommendation by exchanging meta-information, in *Proceedings of the 2011 International Conference on Computational Intelligence and Communication Networks (CICN)*, October 7–9, MIR Labs, Gwalior, India. IEEE Computer Society, Washington, DC, pp. 448–453, 2011.
7. P. Bedi, P. Vashisth, A fuzzy trust model for argumentation-based recommender systems, in Kusum Deep et al. (Eds.), *International Conference on SocPros*, IIT Roorkee, India, December 20–22, AISC130: 493–502, Berlin/Heidelberg: Springer, 2011.
8. A. Monteserin, A. Amandi, Argumentation-based negotiation planning for autonomous agents, *Decision Support Systems*, 51(3): 532–548, 2011. doi:10.1016/j.dss.2011.02.016.
9. J. Bentahar, J.J.C. Meyer, A new quantitative trust model for negotiating agents using argumentation, *International Journal of Computer Science & Applications*, IV(II): 1–21, 2006.
10. M. Morge, An argumentation-based computational model of trust for negotiation, in *AISB 2008 Convention Communication, Interaction and Social Intelligence*, April, Aberdeen, UK, 2008.
11. P. Bedi, H. Kaur, S. Marwaha, Trust based recommender system for the semantic web, in *Proceedings of International Joint Conference on Artificial Intelligence (IJCAI)*, January 9–12, Hyderabad, India. AAAI Press, Palo Alto, CA, pp. 2677–2682, 2007.
12. J. Schafer, J. Konstan, J. Riedl, E-commerce recommendation applications, *Data Mining and Knowledge Discovery*, 5: 115–153, 2001.
13. G. Beliakov, T. Calvo, S. James, Aggregation of preferences in recommender systems, in F. Ricci, L. Rokach, B. Shapira, P.B. Kantor (Eds.), *Recommender Systems Handbook*. Berlin: Springer, pp. 705–734, 2011.
14. P. Bedi, R. Sharma, Trust based recommender system using ant colony for trust computation, *Expert Systems with Applications*, 39: 1183–1190, 2011. doi:10.1016/j.eswa.2011.07.124.
15. P. Victor, M. Cock, C. Cornelis, Trust and recommendations, in F. Ricci, L. Rokach, B. Shapira, P.B. Kantor (Eds.), *Recommender Systems Handbook*. Berlin: Springer, pp. 645–675, 2011.
16. P. Vashisth, D. Chandoliya, B. Kr. Yadav, P. Bedi, Trust enabled argumentation based recommender system, in *Proceedings ISDA 2012—12th International Conference on Intelligent Systems Design and Applications*, November 27–29, Kochi, India. IEEE Xplore, pp. 137–142, 2012.
17. D. Artz, Y. Gil, A survey of trust in computer science and the semantic web, *Journal of Web Semantics*, 5(2): 58–71, 2007.
18. J. Sabater, C. Sierra, Review on computational trust and reputation models, *Artificial Intelligence Review*, 24(1): 33–60, 2005.
19. X.L. Dong, L. Berti-Equille, D. Srivastava, Integrating conflicting data: The role of source dependence, in *Proceedings of the 35th International Conference on Very Large Databases*, August, Lyon, 2009.
20. S. Villata, G. Boella, D.M. Gabbay, L. van der Torre, Arguing about trust in multiagent systems, in *11th Symposium on Artificial Intelligence of the Italian Association for Artificial Intelligence*, Brescia, Italy, December 1–3, 2010.
21. T.D. Huynh, N.R. Jennings, N.R. Shadbolt, An integrated trust and reputation model for open multi-agent systems, *Autonomous Agents and Multi-Agent Systems*, 13: 119–154, 2006. doi:10.1007/s10458-005-6825-4.

22. T.D. Huynh, Trust and reputation in open multi-agent systems, PhD thesis, Faculty of Engineering and Applied Science, School of Electronics and CS, University of Southampton, 2006.

23. J.O. Donovan, B. Smyth, Trust in recommender systems, in *Proceedings of the 10th International Conference on Intelligent User Interfaces (IUI'05)*, New York: ACM, pp. 167–174, 2005.

24. I. Rahwan, L. Amgoud, An argumentation based approach for practical reasoning, in N. Maudet, S. Parsons, I. Rahwan (Eds.), *Argumentation in Multi-Agent Systems (ArgMAS 2007)*, LNAI 4766, Honolulu, HI, May 15, Berlin: Springer, pp. 74–90, 2007.

25. H. Prade, A qualitative bipolar argumentative view of trust, in H. Prade, V. S. Subrahmanian (Eds.), *Scalable Uncertainty Management (SUM)*. Berlin: Springer, pp. 268–276, 2007.

26. S. Parsons, Y. Tang, E. Sklar, P. McBurney, K. Cai, Argumentation-based reasoning in agents with varying degrees of trust, in *Proceedings of 10th International Conference on Autonomous Agents and Multiagent Systems (AAMAS)*, Taipei, Taiwan, May 2–6, Vols. 1–3, pp. 879–886, 2011.

27. C. Ziegler, G. Lausen, Propagation models for trust and distrust in social networks, *Information System Frontiers*, 7: 337–358, 2005.

28. P. Victor, C. Cornelis, M. De Cock, P. Pinheiro da Silva, Gradual trust and distrust in recommender systems, *Fuzzy Sets and Systems*, 160: 1367–1382, 2009.

29. R. Guha, R. Kumar, P. Raghavan, A. Tomkins, Propagation of trust and distrust, in *Proceedings of the 13th International Conference on the World Wide Web*, New York, May 17 22, 2004.

30. K.K. Bharadwaj, M.Y.H. Al-Shamri, Fuzzy computational models for trust and reputation systems, *Electronic Commerce Research and Applications*, 8(1): 37–47, 2009. doi:10.1016/j.elerap.2008.08.001.

31. F. Cacheda, V. Carneiro, D. Fern'andez, V. Formoso, Comparison of collaborative filtering algorithms: Limitations of current techniques and proposals for scalable, high-performance recommender systems, *ACM Transactions on the Web*, 5(1): 33, 2011.

32. S. Parsons, E. Sklar, P. McBurney, Using argumentation to reason with and about trust, in *Proceedings of the 8th International Workshop on Argumentation in Multiagent Systems (ArgMAS)*, Taipei, Taiwan, May 3, 2011.

33. M. Richardson, R. Agrawal, P. Domingos, Trust management for the semantic web, in *Proceedings of the 2nd International Semantic Web Conference*, Sanibel Island, FL, October 20–23, 2003.

34. R. Stranders, M. Weerdt, C. Witteveen, Fuzzy argumentation for trust, in F. Sadri, K. Satoh (Eds.), *Computational Logic in Multi-Agent Systems. 8th International Workshop, CLIMA VIII*, LNAI 5056, Porto, Portugal, September 10–11, Berlin: Springer, pp. 214–230, 2008.

35. Y. Tang, K. Cai, E. Sklar, P. McBurney, S. Parsons, A system of argumentation for reasoning about trust, in *Proceedings of the 8th European Workshop on Multi-Agent Systems (EUMAS)*, Paris, France, December 16–17, 2010.

36. S. Villata, G. Boella, D.M. Gabbay, L. Torre, Arguing about trust in multiagent systems, in *Association for the Advancement of Artificial Intelligence*, Palo Alto, CA, March 22–24, 2010.

37. P. Victor, C. Cornelis, M. De Cock, A.M. Teredesai, Trust and distrust-based recommendations for controversial reviews, *IEEE Intelligent Systems*, 26(1): 48–55, 2011.

38. J.O. Donovan, Capturing trust in social web applications, in J. Golbeck (Ed.), *Computing With Social Trust*, Human–Computer Interaction Series, London: Springer-Verlag, pp. 213–257, 2009.

39. V.D. Braak, S.W. Oostendorp, H. Prakken, G.A. Vreeswijk, A critical review of argument visualization tools: Do users become better reasoners? in *Proceedings of the ECAI-06 Workshop*, Held at the 17th European Conference on Artificial Intelligence (ECAI'06), Riva del Garda, Italy, August 28–30, 2006.

40. O. Arazy, I. Elsane, B. Shapira, N. Kumar, Social relationships in recommender systems, in *Proceedings of the 17th Workshop on Information Technologies and Systems (WITS)*, Montreal, QC, December 8–9, 2007.

41. O. Arazy, N. Kumar, B. Shapira, Improving social recommender systems, *IT Professional*, 11(4): 31–37, 2009.

15

A Multiagent Framework for Selection of Trustworthy Service Providers

Punam Bedi, Bhavna Gupta, and Harmeet Kaur

CONTENTS

Introduction

Since 2008, there has been an upsurge of interest in service-oriented environments, which seeks to integrate computational and storage resources seamlessly. A key attribute of all these systems are that providers of these systems offer services that can be invoked remotely, and users employ these services for their applications. However, despite the surge in activity, numerous advantages and interest, there are certain concerns that compromise the vision of these environments as a new IT procurement model and its adoption by the community [1–3]. One of the major concerns is the lack of effective methods for making provision for and the discovery of providers for the sensitive application of users.

There is need for selection of service providers because sensitive applications and user data are placed on the computing resources of the service providers, which can easily be tampered unless the providers give assurance for the security of the users' data. Till date, the only assurance from the providers to the users is given by signing a service-level agreement (SLA) as prepared by providers of these systems. But because of an unjustifiable interest of the service providers, it may happen that the SLA may provide incomplete, false, or even vague service descriptions, which may mislead users and make the services open, uncertain, and deceptive [4]. Moreover, on specifying the legislation in the SLA, internally or externally imposed policies that would be followed during that service have no connotation to it because there is no method to measure whether those written policies have been actually followed by the service providers or not [5]. Hence, if users are able to assure themselves about the service provider before giving their sensitive data and applications, their confidence to get the work done from that particular service provider increases, helping them to delegate the work without worries.

For selection of different providers, traditional methods such as PKI authorization fail due to the scale and *ad hoc* relationship that exist between the providers and requestors of the system. Because trust and reputation are viewed as a measureable belief that utilizes trustworthy experiences to make decisions in many fields, from the ancient fish market system to e-commerce, online service environment, wireless sensor and *ad hoc* networks for choosing relay nodes to forward packets [6]. Even trust-based management approaches have been placed for grid computing and cloud computing by selecting the resources [7,8].

Based on the human society notion in which recommendations are taken from trustworthy acquaintances to decide from among a plethora of choices, a multiagent framework for developing and evaluating providers is proposed, based on the metric of their reputation. The user agent selects a trustworthy provider's agent for his sensitive applications using a social recommendation process by sending a query to his trustworthy acquaintances who may further in turn further send it to their acquaintances to get recommendations, if $T_{timeout}$ permits, where $T_{timeout}$ is the time specified by the user agent to get the recommendations. But if the user is new to the system and has no trustworthy acquaintances, then first, the user would enter the web of trust of the agents by taking recommendations from other agents about their trustworthy acquaintances. The web of trust is a network of agents in which each agent maintains trust information about its trustworthy acquaintances and thus creates a "web." The recommendations provided to the user agent to decide about trustworthy acquaintances or the service providers are given as attributes of the requested entity based on time decay function, assuming that recent information has more weightage than past interactions. To handle the uncertainty and fuzziness from those recommended attributes when given as input to the fuzzy inference engine, the attributes give the reputation of entity as output, for which the recommendations were asked. In the case of computing the reputation of recommended service providers, the attributes supplied to the fuzzy inference engine are the combination of weighted, direct experience as well as recommendatory information.

In summary, the following items are discussed in this chapter: First, the need for selection of service providers in service-oriented environments is explained followed by the motivation for it. Basics of fuzzy inference systems (FISs) which are used to resolve uncertain recommendations are discussed in the next section. This is followed by the section on prioritizing the service providers using FIS based on reputation with updation of trust on recommenders. This chapter concludes with a discussion about two applications where the proposed framework can be employed.

Need for Selection of Service Providers

Service-oriented computing has emerged as a paradigm for delivering on demand services to users. Traditionally, organizations or users had to make a big investment for procuring IT infrastructure and skilled developers, which resulted in high cost of ownership. But a service-oriented environment offers a significant benefit to users by freeing them from these low-level tasks, thus enabling more focus on innovations. For example, a bioinformatics person may want to process protein data hosted by one institution with a remote application server for data mining run by another, and then may want to store the result using public data service [9]. Instead of investing in the infrastructure for all of these tasks, he/she can use these services being offered by different service providers. Due to these benefits provided by different service providers, many organizations and users have now started building their applications using elastic and flexible services given by service-oriented environments. But shifting to these environments is not straightforward because numerous challenges exist to leverage the full potential that these systems have to offer. These challenges are often related to the fact that corresponding to each of the application-specific requirements and characteristics, a number of competing providers exist who can fulfill those requirements with different quality of services (QoSs) and different prices. Moreover, there may be a trade-off between different functional and nonfunctional services given by different providers, which make the problem of selection even worse. Some malicious providers may provide the services at a cheap price, though maybe at the cost of security or privacy of the submitted application. Thus, it is certain that evaluation or selection of providers must be done on a measure such that reliability and security of an application is preserved.

Motivation

The key idea behind this work is to base the selection on the notion of trust of society, that is, instead of service consumers working individually, which would limit their effectiveness in arbitrating only those service providers with whom they have directly interacted, it is helpful for them to share their knowledge about service providers with their trustworthy acquaintances. The notion of taking recommendations from trustworthy acquaintances for making a decision from among a plethora of choices is common in human society because decision in real life is often made through satisficing as opposed to aiming for optimal selection [10]. Usually, this is because human beings are considered as cognitive misers [11] who take advantage from trust due to their limited capacity to process incomplete and uncertain information. Trust helps them to eliminate the choices that are not trusted from consideration. In a way, trust can be taken as a tool for making balanced predication [12].

From trustworthy acquaintances' knowledge, not only will the user get the information about those service providers with whom he has not interacted so far, but inferring from the recommendations also helps in deciding who he should delegate his application. But taking inference from recommendations is not easy because it is based on the subjective judgment of each acquaintance about the attributes of service providers. A practical way is to use a computationally intelligent technique to remove fuzziness and subjectivity from those recommendations. This design goal can be satisfied using FIS which is explained in detail in the following section.

Fuzzy Inference System

The FIS is a computing framework that is based on the concepts of fuzzy set theory (see Figure 15.1). It has been found to be effective in a wide variety of fields, such as decision analysis, expert systems, time series prediction,

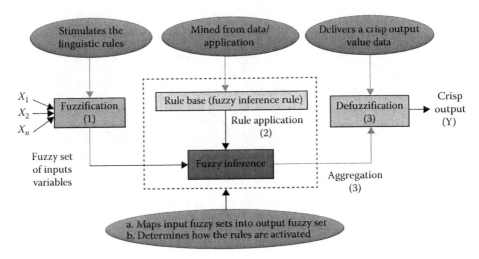

FIGURE 15.1
A fuzzy inference system.

robotics, and pattern recognition, among others. Fuzzy rule-based system [13], fuzzy expert system [13], fuzzy associative memory [14], and fuzzy logic controller are some of the other names of FIS.

Fuzzy Set Theory

The notation of a *fuzzy set* was introduced by Lofti A. Zadeh in 1965 [15] after giving fuzzy logic theory as a generalization of the concept of a classic set. The basic idea of the fuzzy logic theory is to allow not only the values 1 and 0, corresponding to *true* and *false*, but the whole interval (0, 1) as degrees of truth. This leads to a radical extension to classical theory.

Definition 1: If X is the collection of objects denoted by *x*, then a fuzzy set A in X is defined as a set of ordered pairs using the following equation:

$$A = \{(x, \mu_A(x)) \mid x \in X\}, \tag{15.1}$$

where:
   $\mu_A(x)$ is called the *membership function* (MF) of the fuzzy set A
   X is called the universe of discourse, or simply, the universe that consists
      of discrete objects or continuous space

The construction of a fuzzy set depends majorly on two things: (1) the identification of an appropriate universe of discourse and (2) the specification of a suitable MF. The specification of the MF is subjective, which means that MFs defined for the same concept by diverse persons will vary considerably.

This subjectivity is due to individual differences in perceiving the abstract concepts and has little to do with randomness. It also signifies that subjectivity and nonrandomness of the fuzzy set are its primary differences from the probability theory.

Membership Functions

Because universe of discourse X of most fuzzy set consists of real line R, it would be impractical to list all the pairs defining MFs. A more feasible way to define the MF is using mathematical formula. Two of the popular MFs (triangular and trapezoidal) used in the literature [16,17] are presented mathematically using Equations 15.2 and 15.3 and in Figure 15.2

Definition 2: Triangular MF A triangular MF is defined by three parameters (a, b, and c) shown in the following equation:

$$\text{Triangle}(x;a,b,c) = \begin{cases} 0, & x < a \\ \dfrac{x-a}{b-a}, & a \le x < b \\ \dfrac{c-x}{c-b}, & b \le x < c. \\ 0, & c \le x \end{cases} \qquad (15.2)$$

The parameters {a, b, and c} (with $a < b < c$) determine the x coordinates of the three corners of the underlying triangular MF.

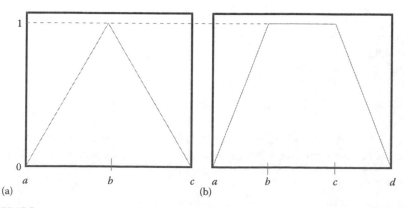

FIGURE 15.2
Membership functions (a) triangular and (b) trapezoidal.

Definition 3: Trapezoidal MF A triangular MF is defined by four parameters (a, b, c, and d) that are written in equation form as follows:

$$
\text{Trapezoid}(x;a,b,c,d) =
\begin{cases}
0, & x < a \\
\dfrac{x-a}{b-a}, & a \le x < b \\
1, & b \le x < c \\
\dfrac{c-x}{c-b}, & c \le x < d. \\
0, & d < x
\end{cases}
\tag{15.3}
$$

Due to their computational efficiency besides simple formulas, triangular and trapezoidal MFs are widely used in many real-time implementations.

The basic FIS consists of conceptual components that function collectively using following three basic steps:

Step 1: Fuzzification

The process of mapping a fuzzy variable to membership of a fuzzy set is called fuzzification and a fuzzy set is a pair (e, μ), where for each e ∈ E, μ(e) is the degree of membership of e as described earlier.

Each of the fuzzy sets defined for the fuzzy variables used in FIS should follow underlined characteristics defined by [18]:

- Distinguishability: Semantic integrity requires that the MFs represent a linguistic concept and be different from each other.
- A justifiable number of fuzzy sets.
- Coverage: Each input data point, x, belongs to at least one fuzzy set and μ(x) > €, where € is the coverage level.
- Normalization: All the fuzzy sets should be normal.
- Overlapping: All the fuzzy sets should significantly overlap.

Once the fuzzy sets are defined for fuzzy variables, a crisp input can be fuzzified to obtain its degrees of membership on which fuzzy rules can be applied.

Step 2: Rule Application

Different kinds of fuzzy rules exist [19]. The most popular ones are conjunctive rules and are defined in the IF–AND–THEN form as follows:

$$R_i : \text{IF } x_1 = A_{i1} \text{ AND } x_2 = A_{i2} \text{ AND} \dots \text{ AND } x_{in} = A_{in} \quad \text{THEN } y = B_i.$$

The format of rules and number of rules to be defined depend on the application in hand. For example, if two input attributes and an output attribute are defined by seven fuzzy sets each, then the system needs at most $3^7 = 2187$ rules in order to enumerate all the possible combinations. However, fuzzy logic systems do not require all possible rules to be explicitly defined for a given system to capture the fuzziness of the input variables [20].

Step 3: Aggregation and Defuzzification

Aggregation is a process whereby the outputs of different fuzzy rules are unified. Aggregation occurs only once for an output variable. The input to the aggregation process is the truncated output fuzzy sets obtained based on the firing strength of each rule, and output is an aggregated fuzzy set in the output space which is later defuzzified to get a crisp numerical value.

As in general form, each of the fuzzy rules has an antecedent and a consequent, separated by "THEN" statement. The antecedent is a conjunction of several fuzzy clauses with operators (AND, OR) among them. The consequent represents the action the system takes, if the antecedent is true (in any degree of membership). The computation of the membership of the antecedent is dependent on fuzzy operators AND, OR, and NOT between fuzzy clauses, which is usually called firing strength of the rule. Figure 15.3

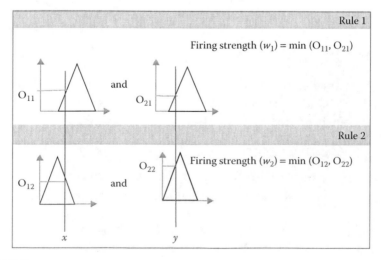

FIGURE 15.3
Computation of firing strength of fuzzy rules.

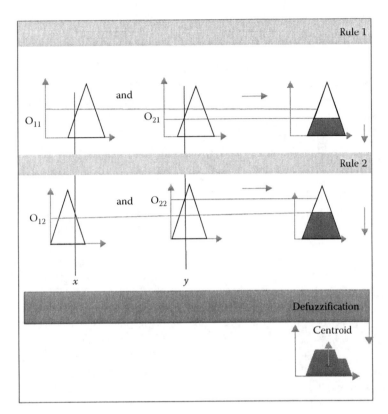

FIGURE 15.4
Computation of the crisp value using defuzzification.

shows the firing strength of two such fuzzy rules where the min. operator is used because fuzzy clauses present in the antecedent of the fuzzy rules are joined using the AND operator.

Each rule firing strength or degree of antecedent is then propagated to get the membership of the consequent part of the fuzzy rules to get the fuzzy output (f_i), which is then aggregated and defuzzified to obtain a crisp output using the following equation, and shown in Figure 15.4.

$$\text{Output} = \frac{\sum_{i=1}^{n} C_i P_i}{\sum_{i=1}^{n} P_i}, \tag{15.4}$$

where:

P_i represents membership of the consequent part of the *i*th fuzzy rules corresponding to f_i fuzzy output set

n is the number of rules that yield output response

C_i is the centroid of the area as obtained after the firing of the *i*th rule

Literature Study

Computational trust models have recently taken the attention of many researchers [21–27], resulting in many of the models getting placed in the literature based on techniques such as statistical analysis, prediction or constraint analysis, game theory, fuzzy theory, and machine learning, and so on. Broadly, these models can be classified as one of the two types: (1) *enforcement models* and (2) *prediction models*. Enforcement models [28–30] try to enforce trustworthy behavior to entities by giving reward or punishment, whereas prediction models try to predict the behavior of entities in order to select trustworthy behavioral entities. The major identified sources of information are an entity's past behavior interaction records collected as direct interaction (firsthand information) and as witness information (secondhand information). The model proposed in Ref. [31] uses only firsthand information for trust prediction with the argument that referral information makes the model vulnerable due to the malicious nature of recommenders.

But, in service-oriented systems where the central control unit is absent and where dynamism (entry and exit at any time) of entities makes it difficult for an entity to have adequate information about interacting entity, referral information plays a significant role. Some of the models based on witness information are SPORAS [32], REGRET [33], Beta reputation system [24], Referral [34], FIRE [35], TRAVOS [36], Repage [37], and CRM [38], which allow the interacting entities to evaluate each other's performance based on witness information as well. Specifically, each time a requesting entity interacts with the provider, it assigns a rating to each attribute on which the provider is assessed. Later, the trustworthiness of the provider is computed as a weighted average of individual interaction evaluations based on direct and indirect information. For each of the specified models the weight assigned to each evaluation is different. For instance, it may be based on time [37] or the trust relationship existing between entities [8], and many more. In Ref. [34] recommendations aggregating algorithm does not address the time efficiency of direct interaction evidences but uses trust relationship as an important weight.

When gathering recommendations from referrals, uncertainty factor exists and different researchers have tried to address this uncertainty differently depending on the system at hand. Focusing only on fuzzy logic-based models used to address this issue, a few models are found in the literature; Ramchurn et al. [29] developed a reputation- and confidence-based model and use fuzzy logic to assess interactions, whereas the model developed by Shuqin et al. [39] uses fuzzy logic for judgment of recommendations. Castelfranchi et al. [40] developed a sociocognitive trust model using fuzzy cognitive maps. Differentiating internal and external attributes, the model also captures the changing behaviors of agents. Song et al. [41] proposed a trust model based on fuzzy logic for securing grid resources by updating and propagating trust values across different sites. Later, in Ref. [42], the model is simulated on eBay data and demonstrated that it is more effective than eigen trust. A fuzzy logic-based framework to decide on the selection of business partner is proposed by Schmidt et al. [43]. Specifically, a fuzzy logic-based customizable trust model integration of postinteraction processes such as interaction reviews and adjustment of credibility is done. In Ref. [44], PATROL-F model is proposed for reputation-based trust incorporating fuzzy subsystems that can be used for any distributed system. This chapter discusses trust-based multiagent framework designed and implemented to address the selection of the service providers' challenge in service-oriented systems.

Method to Find Trustworthy Service Providers

The challenges faced in service-oriented environment are as follows:

- The shared resources may contain a malicious code placed by other users to harm the user application, or the service providers may act deceitfully.
- User applications may contain malicious code that can harm resources of the service providers [45–47].

The above problems are encountered in service-oriented environments due to dynamism (entry and exit at any time) of entities, and as such they do not have knowledge about each other. Focusing only on the first challenge, here, as it happens in real life, people prefer to interact with those who have trustworthy reputation, so that if somehow the reputation of the service providers can be computed, the interactions will be more reliable. The definitions of *reputation* and *trust* used in this chapter are as follows:

> Reputation refers to the value attribute of a specific entity, including agents, services, based on the trust exhibited by it in the past. [8]
> Trust is the subjective probability by which an individual A expects that another individual B performs a given action on which depends its welfare. [48]

Multiagent environment for the proposed framework is best suited because in service-oriented environment, services are geographically distributed and owned by different individuals and multiagent infrastructure can be organized in such a way that every entity in the environment is accompanied by its agent.

In this proposed framework, there are two types of agents:

1. User agent: Users submit the task to their corresponding user agent specifying the requirements, for example, workload, execution deadline, budget limit, minimum reputation of the service

provider required, etc. It is now the responsibility of the user agent to select a suitable service provider for the user satisfying all the requirements specified by her/him. Here in proposed system, the user agents act as a community and help each other at the time of query request from any user. Every agent maintains a database in which trust on various trustworthy agents is stored besides their personal experience about resource providers.

2. Service provider agent: Every service provider communicates with the environment through his agent called service provider agent. The data corresponding to every resource/service such as its availability, its workload, its price are maintained by this agent.

The basic methodology of the system consists of the following steps, which is also shown in Figure 15.5:

1. Generation of a query from the requirements of the user
2. Generating recommendations for a user
3. Handling uncertainty from recommendations

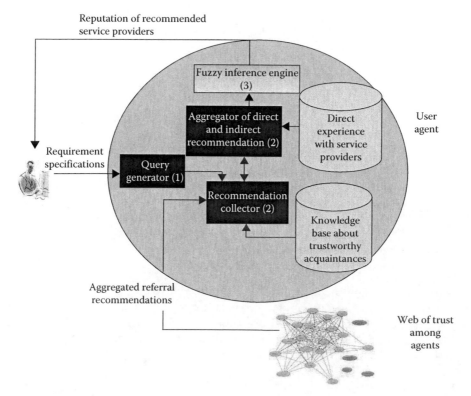

FIGURE 15.5
Selection process of trustworthy service providers by the users.

Generation of a Query from the Requirements of the User

As is the case in the development of any project, requirement specification phase is very crucial, and so is this step. User gives the requirements to his agent, known as the user agent. The requirements can be given as execution deadline, budget limit, minimum reputation of the resource provider, etc. Prior job execution success rates, job turnaround time are many other factors which a user can specify, if required, in his service demand. Once the specification of the requirements is given to the agent, the user agent will prepare a query in the form of a request vector from those specifications generating an identity number to the query and sends this vector with the identification number to his trustworthy agents to get their recommendation.

Generating Recommendations for the User

Here, in our proposed framework, the user agent uses two types of recommendations to decide among service providers, which are explained as follows [49,50]:

- Direct interaction recommendation
- Referral interaction recommendation

Direct Interaction Recommendations

Direct interaction recommendations, being a valuable source of information, have been used in various studies earlier [6,14] for computation of trust, based on direct experience of the agent with the service provider. But in our work we concretely discuss effect various factors on this information before using it. One such factor is temporal-effect factor. We argue that as trust decays with time, therefore, old interaction records become less important than current records, and so we give a time weight to each interaction record. Then the *temporal-effect factor* (f_t) defined using the following equation will assign more weight to the recent experience of the agent.

$$f_t = \exp^{-\alpha \ln(t - t_i)} \alpha \geq 0, \tag{15.5}$$

where:

    variable α is application dependent and can be adjusted according to application requirement, that is, if past interaction is still a valuable source of information then it assigns a smaller value to α, otherwise more emphasis on the recent information can be given using large value of α

    $t - t_i$ shows the time difference between the time of interaction (t) and the time of recommendation (t_i)

Another factor that we used is *cost-service duration effect factor* (f_{cs}). It is used to assign more weight to expansive transactions, that is, the service experience that is acquired after longer duration of service is more valuable than short duration and can be written, mathematically, as follows:

$$f_{cs} = 1 - \exp^{-\beta T_t} \beta \geq 0, \tag{15.6}$$

where:
  β is a constant selected to control the slope of exponential curve
  T_t is the total time for which service was acquired

The importance of cost-service duration effect factor is that it will deter service providers from gaining reputation by providing good-quality services for short duration, but then exploiting this reputation by defrauding on long-term or expensive services.

Applying these two factors on service attribute experience gives a weighted direct experience component (DEC), as written in the following equation, to be used in computation of trust.

$$\text{DEC}\left[s_{imx}(t)\right] = \frac{1}{2}\left[\frac{\sum_{l=0}^{t} f_t(l) * s_{iml}}{\sum_{l=0}^{t} f_t(l)} + \frac{\sum_{l=0}^{t} f_{cs}(l) * s_{iml}}{\sum_{l=0}^{t} f_{cs}(t)}\right], \tag{15.7}$$

where:
  $s_{imx}(t)$ shows average weighted aggregation of s_i attribute of mth service provider at time of request t as done by agent x

The same procedure is repeated for aggregating each of the n attributes of mth service provider and provides it to aggregator module as direct experience of the agent.

Referral Interaction Recommendations

Due to the highly distributed nature of service-oriented systems in hand, direct interaction information is not sufficient to evaluate the trustworthiness of the service provider, therefore, the agent needs help from its trustworthy acquaintances to get recommendations. An added advantage of these referral recommendations is that the agent is also able to get the recommendations from those agents who are not known to it because at the time of the request, the agent takes recommendation from those known to it, and they may further take recommendations from those known to them, and so on. These trustworthy relationships that exist between agents form a web of trust, as shown in Figure 15.6, which depicts that an agent a9 can get recommendations from agent a3.

Justifying the need for referral recommendations in these systems, when these recommendations reach from the recommenders to the requested

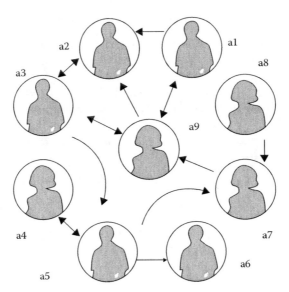

FIGURE 15.6
Web of trust between agents.

agent they get aggregated based on the trust values that requested the agent has on recommenders. The aggregation of the indirect recommendation as referral experience component (REC) is done using the following equation:

$$\text{REC}\left[s_{\text{im}x}(t)\right] = \sum_{k=1}^{n} tl_k s_{\text{im}k},$$

(15.8)

where:

tl_k represents trust level agent x has on its kth trustworthy acquaintances
n represents the total number of agent x's trustworthy acquaintances
$s_{\text{im}k}$ shows the ith service attribute of the mth service provider as recommended by the kth trustworthy acquaintance

Aggregation of Recommendation

The two components DEC [$s_{\text{im}x}(t)$] and REC [$s_{\text{im}x}(t)$] as specified in Equations 15.7 and 15.8 are finally aggregated by the aggregator module using the following equation:

$$S_{\text{im}x}(t) = w_1 \text{DEC}[s_{\text{im}x}(t)] + w_2 \text{REC}[s_{\text{im}x}(t)].$$

(15.9)

Similarly, using the above equation, each of the required n service attributes of different service providers are aggregated and given to fuzzy trust evaluator module. The DEC is always considered more reliable than the REC because it is based on the agent's direct and personal experience, therefore, $w_1 > w_2$. Also $w_1, w_2 \in (0, 1)$ and $w_1 + w_2 = 1$.

Handling Uncertainty in Recommendations Using FIS

The aggregated service parameters representing the experience of various agents with providers contain fuzziness and vagueness due to the observation based on human society in which, when interacting or taking recommendations from someone, recommendations are generally based on his/her mental and social *subjective* characteristics that are different for different persons. Therefore, a mechanism is needed which could quantify subjectivity and vagueness from the attributes given by recommenders. FIS can be employed to do this, whose details were discussed earlier, because of its ability to work with different rules capturing the uncertainty present in the data.

The values of aggregated service attributes are first mapped to their defined MFs and fuzzified; then they are processed by individual configured fuzzy (event)—condition–action rules. The rules define the conditions to be met by fuzzified aggregated attributes for interpreting trustworthy behavior. For example the rule *if job_turnaround_time is low and job_success_ratio is high, then reputation is high* states that reputed service providers should have job_turnaround_time and job_success_ratio as low and high, respectively. Later on, defuzzifying the aggregated results of applied rules yields a crisp value of the reputation of service providers based on the input-recommended, aggregated service parameters.

We have implemented FIS in MATLAB 7.0.1., using Mamdani inference engine, taking two service attributes of service provider (job_ turnaround_ time, job_ success_ ratio) as inputs and reputation as output. The screen snapshots of computation of reputation using FIS are shown in Figure 15.7.

FIS captures and ranks the potential of different resource providers by using a unified metric called reputation, which helps user agents to have personalized and adaptive selection of service providers by writing the fuzzy rules accordingly, for example, optimist agent demand for trustworthy behavior of service provider may not be stricter than pessimist agent.

Periodic Trust Update of Recommenders

This procedure is invoked periodically, after the completion of each transaction, and it helps in sustaining only valuable recommenders [51,52]. For updation, the difference (d) in reputation of a service provider, as obtained from the recommended service attributes by a recommender and as computed from the actual experienced service attribute of the resource provider by the agent itself, is taken. Depending upon whether the difference is below a threshold (ϕ) or not, the user agent updates the trust value on the recommender using the following equation:

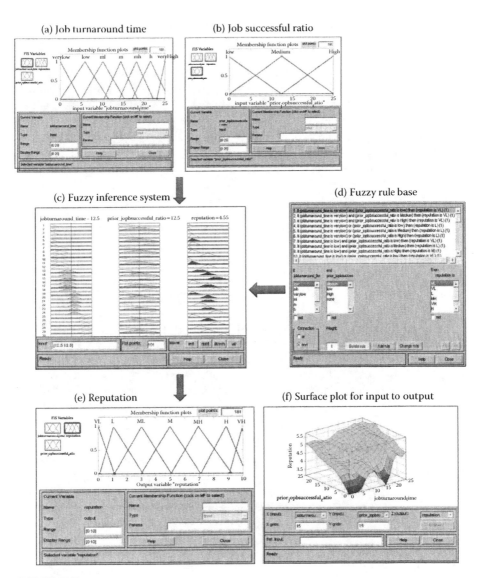

FIGURE 15.7
A fuzzy trust evaluation model. (a), (b) Membership functions of input variables, job turnaround time and job successful ratio; (c) fuzzy inference system; (d) fuzzy rule base; (e) membership function of output variable, reputation; (f) surface plot showing dependence of output variable on input variables.

$$\text{Trust}_{\text{new}}(R) = \text{Trust}_{\text{old}}(R) + \left(\phi^{-d}\right). \tag{15.10}$$

$\text{Trust}_{\text{new}}(R)$ will be the updated value of trust on the recommender (R), whereas $\text{Trust}_{\text{old}}(R)$ is the previous value of trust on the recommender (R). It may also happen that if an actual transaction did not occur with the providers recommended by some of recommenders, then to update trust on them, the difference between the reputation as obtained using aggregated recommended parameters of all other acquaintances and the reputation obtained using parameters recommended by the recommender whose trust is to be updated is taken. So, if there is an agreement in recommendations, that is, if $(\phi - d)$ is positive, then the trust level on the recommender will increase; but if there is difference in opinion, then the trust level will decrease.

Moreover, this update process of the trust level is a continuous process and helps in getting rid of malicious recommenders because the trust level of malicious recommenders will gradually and slowly go below the threshold, and recommendations will not be taken from them in future.

Applications of the Framework

Consider the following two cases where the proposed framework can be used:

1. Business-to-consumer (B2C) search service: A buyer (e.g., user) is trying to select a maintenance service for his laptop. There is no way for the buyer to know whether or not the chosen maintenance company has the buyer's best interest in mind. The uncertainty and lack of complete information makes the decision about the selection of service challenging. The buyer can employ an agent who prioritizes the reputed providers providing concerned service based on query input and the user's preferences using the above-discussed framework.
2. Business-to-business (B2B) loan service: A home dealer also provides financing to its potential buyers. As the dealer does not have any fixed association with any bank, it uses its agent to select loan services from reputed banks based on previous users' experiences.

Summary

In the past few years, service-oriented environments have been widely used by diverse users for their applications. Due to the scale and dynamism of these systems, users need a mechanism by which they are capable of assessing and identifying trustworthy service providers. This chapter proposes a reputation-based trust incorporating fuzzy system that prioritizes the service providers based on their reputation, and hence helps the user in selection of trustworthy service providers. While computing the reputation, direct experience recommendations, referral experience recommendations, and information decay based on time are introduced. The proposed system also takes into account humanistic and subjective concepts such as trust and reputation which are modeled qualitatively using FISs.

References

1. A. Andrade and D. Melo, "A WS-Agreement-Based QoS Auditor Negotiation Mechanism for Grids," in *12th IEEE/ACM International Conference on Grid Computing (GRID)*, Lyon, France, September 21–23, 2011.
2. T. Dillon, C. Wu, and E. Chang, "Cloud Computing: Issues and Challenges," in *24th IEEE International Conference on Advanced Information Networking and Applications*, Perth, WA, April 20–23, 2010.
3. R. Chow, P. Golle, M. Jakobsson, E. Shi, J. Staddon, R. Masuoka, and J. Molina, "Controlling Data in the Cloud: Outsourcing Computation without Outsourcing Control," in *ACM Workshop on Cloud Computing Security (CCSW)*, New York, November 9–13, 2009.
4. M. Alhamad, T. Dillon, and E. Chang, "SLA-Based Trust Model for Cloud Computing," in *13th International Conference on Network-Based Information Systems*, Takayama, Japan, September 14–16, 2010.
5. L. Bing, C. Bu-Qing, W. Kun-Mei, and L. Rui-Xuan, "Trustworthy Assurance of Service Interoperation in Cloud Environment," *International Journal of Automation and Computing*, 8(3): 297–308, 2011.
6. A. Jøsang, R. Ismail, and C. Boyd, "A Survey of Trust and Reputation Systems for Online Service," *Decision Support Systems*, 43(2): 618–6442007.
7. F. Azzedin and M. Maheswaran, "A Trust Brokering System and Its Application to Resource Management in Public Resource Grid," in *18th International Parallel and Distributed Computing Symposium (IPDPS)*, Phoenix, AZ, April 26–30, 2004.
8. B. Alunkal, "Grid Eigen Trust a Framework for Computing Reputation in Grids," MS Thesis, Department of Computer Science, Illinois Institute of Technology, Chicago, IL, 2003.
9. K. M. Khan and Q. Malluhi, "Establishing Trust in Cloud Computing," *IT Professional*, 12(5): 20–27, 2010.
10. H. Simon, *The Sciences of the Artificial*, 3rd ed. Cambridge, MA: The MIT Press, 1996.
11. S. Fiske and S. Taylor, *Social Cognition*, 2nd ed. New York: McGraw-Hill, 1991.
12. J. Lewis and A. Weigert, "Trust as a Social Reality," *Social Forces*, 63(4): 967–985, 1985.
13. A. Kandel and G. Langholz, *Fuzzy Control Systems*. Boca Raton, FL: CRC Press, 1994.
14. B. Kosko, *Neural Networks and Fuzzy Systems*. Englewood Cliffs, NJ: Prentice Hall, 1992.
15. L. Zadeh, "Fuzzy Sets," *Information and Control*, 8: 338–353, 1965.
16. D. DuBois and H. M. Prade, *Fuzzy Sets and Systems: Theory and Applications*. New York: Academic Press, 1980.
17. H.-J. Zimmermann, *Fuzzy Set Theory and Its Applications*, 3rd ed. Norwell, MA: Kluwer Academic Publishers, 1996.

18. S. Guillaume and B. Charnomordic, "Fuzzy Inference Systems: An Integrated Modeling Enviornment for Collabortion between Expert Knowledge and Data Using FisPro," *Expert Systems with Applications*, 39: 8744–8755, 2012.

19. D. Dubois, H. Prade, and L. Ughetto, "A New Prespective on Reasoning with Fuzzy Rules," in *Advances in Soft Computing (AFSS), International Coference on Fuzzy Systems*, February 3–6, Calcutta, India, 2002.

20. T. Munakata and Y. Jani, "Fuzzy Systems: An Overview," *Communications of the ACM*, 37(3): 69–76, 1994.

21. A. Archer and E. Tardos, "Truthful Mechanisms for One-Parameter Agents," in *42nd IEEE Symposium on Foundations of Computer Science (FOCS)*, October 8–11, Las Vegas, NV, 2001.

22. R. Dash, S. Ramchurn, and N. Jennings, "Trust-Based Mechanism Design," in *Third International Joint Conference on Autonomous Agents and Multi-Agent Systems*, New York, July 19–23, 2004.

23. A. Moukas, G. Zacharia, and P. Maes, "Collaborative Reputation Mechanisms in Electronic Marketplaces," *Decision Support Systems*, 29(4): 371–388, 2000.

24. A. Josang and R. Ismail, "The Beta Reputation System," in *15th Bled Conference on Electronic Commerce*, Bled, Slovenia, June 17–19, 2002.

25. Y. Kim and H. Song, "Strategies for Predicting Local Trust Based on Trust Propagation in Social Networks," *Knowledge-Based Systems*, 24(8): 1360–1371, 2011.

26. S. Marti and H. Garcia-Molina, "Limited Reputation Sharing in P2P Systems," in *EC'04, 5th ACM Conference on Electronic Commerce*, New York, May 17–20, 2004.

27. W. Yuan, D. Guan, Y. Lee, S. Lee, and S. Hur, "Improved Trust-Aware Recommender System Using Small-Worldness of Trust Networks," *Knowledge-Based Systems*, 23(3): 232–238, 2010.

28. C. Burnett, T. J. Norman, and K. Sycara, "Trust Decision-Making in Multiagent Systems," in *22nd International Joint Conference on Artificial Intelligence*, Barcelona, Spain, July 16–22, 2011.

29. S. Ramchurn, N. Jennings, C. Sierra, and L. Godo, "Devising a Trust Model for Multi-Agent Using Confidence and Reputation," *Applied Artificial Intelligence*, 18: 833–852, 2004.

30. R. Jurca and B. Faltings, "Obtaining Reliable Feedback for Sanctioning Reputation Mechanisms," *Journal of Artificial Intelligence Research*, 29: 391–419, 2007.

31. H. Yahyaoui, "A Trust-Based Game Theoretical Model for Web Services Collaboration," *Knowledge-Based Systems*, 27: 162–169, 2012.

32. G. Zacharia and P. Maes, "Trust Management through Reputation Mechanisms," *Applied Artificial Intelligence*, 14(9): 881–907, 2000.

33. J. Sabater and C. Sierra, "REGRET: Reputation in Gregarious Societies," in *Fifth International Conference on Autonomous Agents*, Montreal, QC, May 28–June 1, 2001.

34. B. Yuan and M. Singh, "An Evidential Model of Distributed Reputation Management," in *International Joint Conference on Autonomous Agents and Multiagent Systems*, Estoril, Portugal, May 12–16, 2008.

35. T. Huynh, N. Jennings, and N. Shadbolt, "An Integrated Trust and Reputation Model for Open Multi-Agent Systems," *Journal of Autonomous Agents and Multi-Agent Systems*, 13(2): 119–154, 2006.

36. W. Teacy, J. Patel, N. Jennings, and M. Luck, "Travos: Trust and Reputation in the Context of Inaccurate Information Sources," *Journal of Autonomous Agents and Multi-Agent Systems*, 12(2): 183–198, 2006.

37. J. Sabater, M. Paolucci, and R. Conte, "Repage: REPutation and ImAGE among Limited Autonomous Partners," *Journal of Artificial Societies and Social Simulation*, 9(2): 3, 2006.
38. B. Khosravifar, J. Bentahar, M. Gomrokchi, and R. Alam, "CRM: An Efficient Trust and Reputation Model for Agent Computing," *Knowledge-Based Systems*, 30: 1–16, 2012.
39. Z. Shuqin, L. Dongxin, and Y. Yongtian, "A Fuzzy Set Based Trust and Reputation Model in P2P Networks," *Lecture Notes in Computer Science*, 3177: 211–217, 2004.
40. C. Castelfranchi, R. Falcone, and G. Pezzulo, "Trust in Information Sources as a Source for Trust: A Fuzzy Approach," in *Second International Joint Conference on Autonomous Agents and Multiagent Systems*, Melbourne, VIC, July 14–18, 2003.
41. S. Song, K. Hwang, and Y.-K. Kwok, "Trusted Grid Computing with Security Binding and Trust Integration," *Journal of Grid Computing*, 3: 53–73, 2005.
42. S. Song, K. Hwang, R. Zhou, and Y. Kwok, "Trusted P2P Transactions with Fuzzy Reputation Aggregation," *IEEE Internet Computing*, 9: 24–34, 2005.
43. S. Schmidt, E. Chang, T. Dillon, and R. Steele, "Fuzzy Decision Support for Service Selection in E-Business Environments," in *IEEE Symposium on Computational Intelligence in Multicriteria Decision Making*, Honolulu, HI, April 1–5, 2007.
44. A. Tajeddine, A. Kayssi, A. Cheha, and H. Artail, "Fuzzy Reputation-Based Trust Model," *Applied Soft Computing*, 11: 345–355, 2011.
45. P. Bedi, H. Kaur, and B. Gupta, "Access Control for Cloud Resources Using Radial Basis Neural Network," in *Computer, Communication, Control and Information Technology (c3it)*, West Bengal, India, February 25–26, 2012.
46. B. Gupta, H. Kaur, and P. Bedi, "Predicting Grid User Trustworthiness Using Neural Networks," in *Int'l World conference on Information and Communication Technologies (WICT)*, Bombay, India, December 11–14, 2011.
47. B. Gupta, H. Kaur, and P. Bedi, "Trust Based Access Control for Grid Resources," in *Int'l Conference Communications Systems and Network Technologies (CSNT)*, Jammu, India, June 3–5, 2011.
48. S. M. Habib, S. Ries, and M. Muhlhauser, "Cloud Computing Landscape and Research Challenges Regarding Trust and Reputation," in *7th International Conference on Ubiquitous Intelligence Computing and 7th International Conference on Autonomic Trusted Computing*, Shaanxi, China, October 26–29, 2010.
49. B. Gupta, H. Kaur, and P. Bedi, "A Balanced Reputation Based Service Provider Selection for Farmers," *Journal of Indian Society of Agricultural Statistics*, 67(1): 1–9, 2011.
50. B. Gupta, H. Kaur, and P. Bedi, "Trust Based Personalized Ecommerce System for Farmer," in *4th Indian International Conference on Artifical Conference (IICAI)*, Pune, December 14–16, 2011.
51. P. Bedi, H. Kaur, and B. Gupta, "An Agent Based Reputation System for Unreliable Grid Environment," *Journal of Information Modeling and Optimization*, 3(1): 74–79, 2012.
52. B. Gupta, H. Kaur, and P. Bedi, "A Reputation Based Service Provider Selection for Farmers," *Journal of Information Assurance and Security*, 5: 515, 2010.
53. K. A. Hribernik, K.-D. Thoben, and Michael Nilsson, *Encyclopedia of E-Collaboration*. Hershay, PA: IGI Publishing, pp. 308–313, 2008.
54. D. M. Osher, "Creating Comprehensive and Collaborative Systems," *Journal of Child & Family Studies*, 11(1): 91–99, 2002.

16

Trust Issues in Modern Embedded Computing

Alex Pappachen James and Sugathan Sherin

CONTENTS

Introduction

The growth of the Internet since 2000, which was expected to happen in the next decade, brings several new challenges in embedded computing. With most embedded devices getting connected to the Internet, the information risk due to its value, and the risk associated with information storage, transfer, and duplication is high. The critical aspect of this information in the embedded world is that the majority of this information has a direct impact on the physical world. For example, consider an Internet-enabled control system that has access to the information acquired from the sensors; this information can be used to manipulate, undo, or modify the operating conditions of the system, and would have a significant impact on the physical world. In the past, the embedded systems, due to its isolated existence, had less worries on the need for trust and information security. The access to the embedded devices was usually restricted and different from the microprocessor pathway. However, beginning in 2009, this scenario started changing mainly because embedded devices are becoming Internet enabled, resulting in evolution of different communication pathways to access the information and the devices. The trust in the modern embedded solutions faces the challenges of multicore computing capabilities, increasing overlap with general computing architectures, and data analytic-aware computing hardware. In this chapter, we provide a modern perspective on the trust issues in embedded computing and devices. In addition, it should be noted that there is a high level of trust placed in many mission critical applications that use embedded systems. This makes this topic highly relevant and of prime importance for ensuring the robustness of embedded computing solutions in the future.

Principles of Embedded Trust

Trust can be seen as a confidence that one has in the integrity, availability, confidentiality, strength, etc., of an embedded system. The concept of trust is often confused with security and sometimes both the terms "trust" and "security" are equated. It is highly required that the developers should also consider maximizing the likelihood of the embedded system accomplishing its goal. Developing a trust is very essential for a system and trustworthiness can never be guaranteed by applying high amount of care, concern, or validation [1]. Also, one can never be confident about the trustworthiness of a system without proper evidence. The evidences that support the trustworthiness of a system can come from different sources such as past performance or quality of outcomes for designers, implementers, etc. However, no amount of evidence can guarantee trust in an absolute manner. Because we live in a dynamic world, trust should be dynamically confirmed. Adding more constraints and requirements to a system can lead to weakening of the trustworthiness as more constraints imply less number of feasible solutions.

Trusted platform module (TPM) is an example of hardware-based trusted computing platforms. TPMs provide various security features such as hashing, generating random numbers and asymmetric cryptographic keys, and storing keys with confidentiality and integrity without causing the vulnerabilities that are inherent in software implementations. TPMs provide security isolation functions that cannot be compromised by software. Platforms such as TPM are especially good in embedded systems as they are efficient in terms of weight and power requirements. Trusted computing solutions can be best deployed by improving CPU architectures and by the use of these trusted computing platforms.

The trust in the embedded systems is a combination of people, process, and technology (Figure 16.1). It may be seen far from being a technical concern; however, in most cases the culprit is often engineers going down to the technical details. The users of the embedded systems often have some means to access the information and this has to be carefully dealt with, as irrespective of the strength of the authentication technique, it is

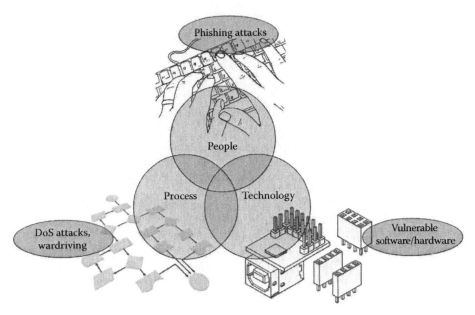

FIGURE 16.1
Examples of relationships of people, process, technology, and security.

largely evident that the user has a major role in knowledge building on the use and reuse of the authentication method. The process in an embedded trust would indicate the set of rules and steps that must be followed by the concerned people and system, to ensure the proper use of the embedded system. For example, the way in which the authentication is done by a user is largely limited to the extent of the process that goes in designing the rules. Technology is another major component of trust, where it can play a significant role in securing the processes and information through automated tools, and automated machine-learning methods that can track and object to anomaly.

Security Requirements of an Embedded System

The migration of heavy applications to handheld devices is making security an important concern in embedded systems. Embedded software are also prone to attacks for which security measures need to be hardwired into the architecture in order to prevent the attacks. The security demands that can be met by improving the hardware architecture are again constrained by battery-driven embedded systems. Limited storage capacity is another factor that limits an embedded system from improving its security

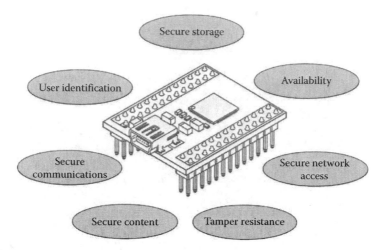

FIGURE 16.2
Security requirements of an embedded system. (From Kocher, P. et al., *Proceedings of the 41st Design Automation Conference*, 2004.)

features. The security requirements of an embedded system are depicted in Figure 16.2.

Cryptographic algorithms can also be brought into place to increase the security of an embedded system. The cryptographic algorithms that have been classified broadly as symmetric ciphers, hashing algorithms, and asymmetric algorithms can be deployed for checking the integrity of the data. In addition to these, one can also consider using secure communication protocols such as IPSec, SSL, and VPN. For an introduction to cryptographic methods, see Refs. [2,3].

Attacks and Countermeasures

Managing vulnerabilities in software is a challenging task as it grows large. [4] Risk management in a software is a challenging one primarily because of three factors—complexity, extensibility, and connectivity. Embedded systems usually advocate the use of programming languages such as C or C++ because the efficiency and the use of such unmanaged programming languages would lead to several vulnerabilities such as buffer overflows. When these vulnerable applications get connected with the Internet, they become more complex such that the vulnerabilities become hard to predict. Extensibility is another factor that makes a system vulnerable to attack. Modern software built using java and .NET can be easily extended by adding external modules/functionalities which may be malicious. As more embedded systems get connected to the Internet, the problems/failures start spreading to the entire network and cause massive damage. There are so many other software defects that make an embedded system vulnerable to attacks [4].

The infection caused by a hardware virus in an embedded system persists even after system reboot or system reinstallation. Hackers can inject their code in the Flash ROM of the system and thereafter take control of the system. Sometimes the infection may not alter the normal working of a system, but it can impart false data to the system. Most of the flash-ROM chips are not fully utilized and which gives enough space for hackers to store backdoor information and viruses.

Countermeasures should begin from the software life cycle model itself. Most people lose their concern about security and concentrate on the functionality of the system. The security measures should be applied at the requirements level, the design level [5], and the code level. Static analysis tools are available to discover implementation bugs in the code. Realizing the importance of security, now practitioners have started using the best practices in embedded security [5,6]. Figure 16.3 depicts the phases in the software development life cycle and the security measures to be taken at each stage.

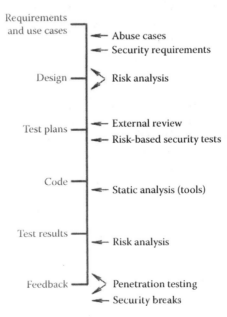

FIGURE 16.3
Software security in the SDLC. (From Kocher, P. et al., *Proceedings of the 41st Design Automation Conference*, 2004.)

Physical Attacks

Physical attacks [7–14] are generally classified into invasive and noninvasive attacks. Invasive attack includes getting a direct access to the system and a person can alter the internals of the system. However, noninvasive attack does not include a direct contact with the system. Invasive physical attacks are usually done by using probes for eavesdropping on the data flowing between components. However, this can be discouraged to a certain level by the use of a system-on-chip strategy. For a system-on-chip, sophisticated microprobing [8,9] techniques are required which include complete de-packaging of the chip for internal inspection. The attackers are reluctant to that kind of breach because expensive infrastructure is required. Another interesting attack strategy is based on timing analysis. Timing analysis involves analyzing variations in the time required for cryptographic computations. Timing analysis includes statistical techniques to predict the keys or individual key bits. In timing analysis, the attacker finds every bit in the key by comparing the execution time of a known bit with the test bits and selects a bit or group of bits showing strong correlation. As a countermeasure, one can make all computations take equal amount of time for any input by introducing delays in some paths. Unfortunately, few modern microprocessors operate exactly in constant time which makes it tricky and cumbersome to write constant-time code particularly in high-level languages. This method has significance because the countermeasures often do not work.

In addition to timing analysis, power analysis can also be done which is another interesting attack strategy. The power drawn by the hardware while performing computations is analyzed to find the correct combination of bits in a key. The two main power analysis techniques include single power analysis (SPA) and differential power analysis (DPA). There are several cases in which SPA attacks are used to simplify brute-force attacks [15]. The adversary is required to perform complex statistical analysis based on the power consumption measurements. This method of attack also relies upon the analysis of differences in signals such as in timing attack. Countermeasures for this type of attack are less effective as compared to the timing attack. As a countermeasure the quality of power measurements can be reduced by running other circuits simultaneously, but this will not help much because this will only make the adversary to collect more samples to achieve the goal and will not help in preventing the attack.

Electromagnetic analysis attacks [16,17] try to measure the electromagnetic radiations emitted by a device to reveal sensitive information. Embedded security has a pervasive requirement in so many areas which include automobiles. The vendors and the customers are not very interested in a full integration of embedded solutions in automobiles because of security considerations. Security concerns are more when devices in automobiles connect to a public network. Modern vehicles usually have an in-vehicle network consisting of several electronic control units which control other systems such as collision warning system and brake system. The potential vulnerabilities in such vehicles and the possible solutions have been discussed in Ref. [18].

Embedded Security for Internet of Things

Internet of Things (IoTs) is going to be the one of the dominating technologies in the future by enabling communication and networking accessible anytime, anywhere. Security for IoTs [19] will be given more focus as the principle needs to unite different technologies and has to communicate with diverse kind of networks and protocols. One of the main challenges in IoT will be associated with low-power devices in which the available computational power is very limited. There is high requirement for optimized lightweight cryptographic algorithms in such devices. Another interesting fact is that there is no "correct solution" for ensuring security, and for IoT things become more complex as it is designed to seamlessly connect, interact, and exchange information with others in the environment. In IoT, security is based on applications and it varies from application to application. The heterogeneity in IoT will worsen the need of sophisticated security measures. The IoTs are prone to different types of security issues, which are depicted in Figure 16.4.

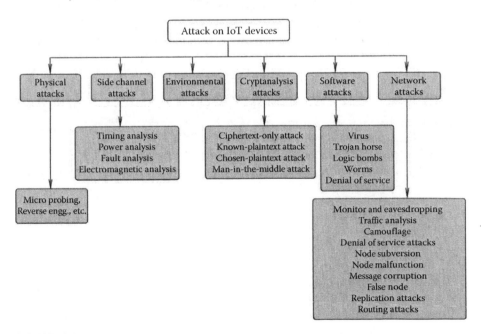

FIGURE 16.4
Different attacks on IoT devices. (From Stango, A. et al., *Vehicular Technology, Information Theory and Aerospace & Electronic Systems Technology*, 2011.)

Existing solutions to implement security in IoT are classified into three, which are as follows:

1. Software approach
2. Hardware approach
3. Hybrid approach

In software-based approach, programmability of embedded general purpose processors (GPPs) is used to implement security. In a hardware-based approach, application-specific integrated circuits (ASICs) are used to implement a given cryptography algorithm in hardware.

Conclusion

The highly dynamic environment and the continuously emerging security needs are posing huge challenges toward achieving the desired level of embedded system security. The limited hardware resources in an embedded system naturally demand the use of optimized security implementations. The high degree of optimization sometimes affects the flexibility of a system to change. A combination of advances in architectures and design methodologies would help in improving the future embedded systems especially in IoTs.

References

1. D. A. Fisher, *Principles of Trust for Embedded Systems*. Pittsburgh, PA: Software Engineering Institute, 2012.
2. W. Stallings, *Cryptography and Network Security: Principles and Practice*. Upper Saddle River, NJ: Prentice Hall, 1998.
3. B. Schneier, *Applied Cryptography: Protocols, Algorithms, and Source Code in C*. New York: John Wiley & Sons, 1996.
4. G. Hoglund and G. McGraw, *Exploiting Software: How to Break Code*. Boston, MA: Addison-Wesley, 2004. http://www.exploitingsoftware.com
5. J. Viega and G. McGraw, *Building Secure Software*. Boston, MA: Addison-Wesley, 2001. http://www.buildingsecuresoftware.com
6. G. McGraw, "Software Security," *IEEE Security & Privacy*, 2: 80–83, 2004.
7. R. Anderson and M. Kuhn, "Low cost attacks on tamper resistant devices," in *International Workshop on Security Protocols (IWSP), Lecture Notes on Computer Science*, Cambridge, UK. Berlin/Heidelberg: Springer, pp. 125–136, 1997.

8. O. Kommerling and M. G. Kuhn, "Design principles for tamper-resistant smartcard processors," in *Proceedings of the USENIX Workshop on Smartcard Technology (Smartcard)*, USENIX Association, Chicago, IL, May, pp. 9–20, 1999.

9. W. Rankl and W. Effing, *Smart Card Handbook*. 1st ed. New York: John Wiley & Sons , Inc., 1997.

10. E. Hess, N. Janssen, B. Meyer, and T. Schutze, "Information leakage attacks against smart card implementations of cryptographic algorithms and counter-measures," in *Proceedings of the EUROSMART Security Conference*, Marseilles, France, June, Vol. 130, pp. 55–64, 2000.

11. J. J. Quisquater and D. Samyde, "Side channel cryptanalysis," in *Proceedings of the SECI*, Tunis, Tunisia, September, pp. 179–184, 2002.

12. J. Kelsey, B. Schneier, D. Wagner, and C. Hall, "Side channel cryptanalysis of product ciphers," in *Proceedings of the European Symposium on Research in Computer Security (ESORICS)*, Berlin/Heidelberg: Springer, September, pp. 97–110, 1998.

13. S. Ravi, A. Raghunathan, and S. Chakradhar, "Tamper resistance mechanisms for secure embedded systems," in *Proceedings of the International Conference on VLSI Design*, Mumbai, India, January 2004.

14. R. Anderson and M. Kuhn, "Tamper resistance—A cautionary note," in *Proceedings of the 2nd Workshop on Electronic Commerce*, Oakland, CA, November 18–20, pp. 1–11, 1996.

15. T. S. Messerges, E. A. Dabbish, and R. H. Sloan, "Examining smart-card security under the threat of power analysis attacks," *IEEE Transactions on Computers*, 51: 541–552, 2002.

16. W. van Eck, "Electromagnetic radiation from video display units: An eaves-dropping risk?" *Computers and Security*, 4(4): 269–286, 1985.

17. M. G. Kuhn and R. Anderson, "Soft tempest: Hidden data transmission using elec-tromagnetic emanations," in *Proceedings of the International Workshop on Information Hiding (IH)*, Portland, OR, April, Berlin/Heidelberg: Springer, pp. 124–142, 1998.

18. H. Chaudhry and T. Bohn, "Security concerns of a plug-in vehicle," in *Innovative Smart Grid Technologies (ISGT), IEEE PES*, IEEE, Washington, DC, pp. 1–6, 2012.

19. A. Stango, N. Prasad, J. Sen, and R. Prasad, "Proposed embedded security framework for internet of things (IoT)", *Vehicular Technology, Information Theory and Aerospace & Electronic Systems Technology (Wireless VITAE)*, 2011.

20. P. Kocher, R. Lee, G. Mcgraw, and S. Ravi, "Security as a new dimension in embedded system design," in *Proceedings of the 41st Design Automation Conference (DAC)*, San Diego, CA: ACM, 2004.

17

A Framework of Content- and Context-Based Network Monitoring

Eliot Foye, Shivani Prasad, Steven Sivan, Stephen Faatamai,
Wei Q. Yan, and William Liu

CONTENTS

Introduction

Network security is an increasingly growing and crucial aspect to any computer network. Security is a growing problem not just within the networking area but also within the computing industry as a whole. Many businesses and organizations base their entire existence around large amounts of rich data [1], which can contain vital business and confidential information. Enabling this data to remain confidential and secure, businesses and organizations alike emphasize on the importance of securing the data from the outside world [2,3]. To achieve this, the communication medium from within the core of the business to the outside world must be secured [4]. In computing terms, this means securing the computer network.

To achieve this, networking professionals spend hours, implementing various monitoring tools and strategies to provide and maintain a robust and secure network environment [5]. However, in reality, the vulnerabilities and threats to these implementations always seem to be one step ahead. It was reported that within the first 6 months in 2005 more than 1000 variants of worms and viruses were discovered [6]. An important statistic, emphasizing the strong need for change, is discussed and debated within the network security research and industrial communities. The basis of this chapter is to expand from the idea of needed change and propose a new strategy for defending computer networks against possible threats and vulnerabilities. The strategy that we intend to propose is that of content- and context-based network monitoring (CCBNM).

The rest of this chapter is structured as follows. In the section "The Fundamentals of Network Monitoring," we introduce network monitoring, then we present content-based network monitoring in the section "Content-Based Network Monitoring." In the section "Context-Based Network Monitoring," we present context-based network monitoring. In the section "Techniques, Tools, and Systems of the CCBNM," we introduce techniques, tools, and systems of CCBNM. We summarize the trends of CCBNM in the section "The Trends of CCBNM." Finally, we draw conclusion and future work in the section "Conclusions." Within the rest of this chapter, we will explore in greater detail the means behind what CCBNM is and how it can be successfully implemented to achieve successful results. This chapter will then act as the fundamental building block of CCBNM and will allow for future work to be conducted in this area. It then allows materials to be added to the suggested strategy and also suggests implementation of the strategy in existing or new tools that perform network security and intrusion detection tasks.

The Fundamentals of Network Monitoring

The management and control of the communication medium among the widespread computer networks can be devious, time consuming, and a very challenging task. Many researchers have implemented strategies and tools to help assist with securing the flow of information from the unlawful interception of network packet data [7]. As securing the network from outside attacks is such a vital aspect of network security, understanding flow and content of the traffic is just as important [8,9]. To achieve this, the researchers take full advantage of tools that allow them to manage and visualize the communication packets. The packet sniffers or network protocol analyzers are software programs that run on a networked device and can monitor the traffic through it. The program sets the network interface card (NIC) of the attached device to operate in promiscuous mode, allowing that device to passively receive all data link layer frames passing through its network adapter. The software program then captures the data that are addressed to other machines present on the network and then saves them for later traffic analysis [10–13]. Figure 17.1 shows how the information flows from the NIC to the packet sniffing program and vice versa.

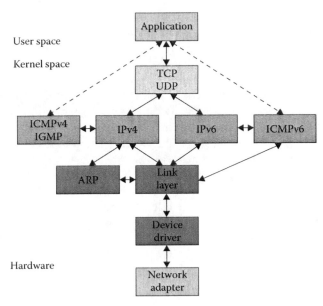

FIGURE 17.1
Flow of information from NIC to application.

Figure 17.1 shows the process involved in the packet sniffing cycle, and we can see that inside the kernel space, the packets can be filtered on a range of contextual information. Therefore, they enable users to navigate easily through captured packets, potentially allowing quicker and more efficient discovery of anomalies and possible threats [14]. Some examples of packet-sniffing programs which are easy to use and easily downloadable are Wireshark [15], Netpy [16], and Nagious [17]. The comparison of these systems will be given later.

Although packet sniffers are seen as a great approach to use by researchers to understand the content and context of information flowing over the network, the process required to find any possible anomalies or threats can be a very complicated and time-consuming task [18]. Packet sniffers operate in real time; therefore, the program itself operates its functions simultaneously as the packets flow into it. It means that all packets are saved as soon as they are picked up from the NIC [19]. We would then have to manually go back through and review all this saved information. A possible solution is to carry out data mining techniques to correlate information to try and find any possible threats or anomalies [20,21]. This whole process is known as network forensics [3] and requires a large amount of time, a sound knowledge of the normal network behaviors and also the protocols' standards that the network uses [22–25].

To overcome this complicated and time-consuming problem, other devices and programs can be added to the network as another layer of security.

Intrusion detection systems (IDSs) and intrusion prevention systems (IPSs) are both dedicated devices or software applications which are fully automated and run on a network [26,27]. How we distinguish one from the other is that IDS recognizes any anomalies or threats present in the network traffic and prompts a message to the control center, and the network administrator who can then either allow the traffic or block it as he/she wishes. If blocked, the packet information of that potentially hazardous threat will then be added to the blacklist and will be blocked again if trying to enter the network [10]. IPSs operate a little differently; they already have a predefined blacklist that has a fully automated update procedure, allowing the blacklist to stay up to date with information on the latest threats. Although IPSs are fully automated and seem like the less stressful option to use, vendors of the devices can take hours or maybe even days to release patches for these devices, therefore potentially allowing malware and threats alike straight into one network [24,28]. IDS, on the other hand, can operate in many different ways; they can be set up to use either signature-based detection or behavior-based detection or even both if needed. Signature-based detection is the technique of detecting known threats or attacks by scanning the content of the packet information for a certain string; strings have already been predefined and stored in a database, which is then used as a predefined blacklist and will block the traffic [29]. Behavior-based detection, on the other hand, is a technique that uses a predefined algorithm which is set by the network administrators and defines the normal operating behavior of their network, as soon as there is abnormal behavior in the network, for example, a spike in traffic or traffic destined for a large amount of hosts an alert is raised, allowing the network administrator to take action [30,31].

Although both IPS and IDS seem like a great extra layer of security to add to one network, both can potentially be a huge problem. In today's large volume of data traffic transversing is assessed through networks as it is seen as the modern-day communication medium. With every single packet having to pass through either an IDS or IPS, it could potentially end up being a bottleneck in one's network [32]. Not only is this a problem, but IDSs and IPSs both generate a very high volume of false positives, making the task of manually analyzing these alerts extremely difficult and inefficient [33].

In history, the above-discussed network monitoring techniques have significant success in preventing threats and attacks from entering a network in the past. While in today's modern communication era, the new emerging attacks and threats are becoming smarter and avoid the countermeasures to escaping from the traditional ways of monitoring, causing a huge alert for concern. The new attacking techniques such as port scans, denial of service (DoS) [34], and malware attacks are some of the most common threats in current network environment. All of these can be easily detected and potentially stopped by implementing network monitoring techniques such as packet sniffing, IDS, or IPS. Network attackers mainly start their attacks on a network by implementing a port scan to attempt a way for the network or a shortcut way into host/device present in the network. The attackers use a range of port scanning techniques to achieve this such as SYN scan, known as a half-open scan where the attacker does not make a complete transmission control protocol (TCP) connection but checks for open ports by sending SYN packets and checking the responses (SYN-ACK = open ports; RST = closed ports). The Connect Scan, this is where the attacker attempts to make a full connection to each port by issuing the connect call system, potentially again to find open ports; FIN scan, this technique sets the FIN flag inside the TCP header of a packet to be able to bypass some firewalls that do not block FIN packets to see whether a port is open. (If closed, an RST will be returned. Otherwise, it will ignore the FIN.) ACK scan is a basic technique used to differentiate whether a port is closed or filtered by a firewall [35]. If used correctly, port scans can potentially be the building blocks to bigger attacks, which can cause serious harm to the network environment. In addition, DoS attack [34] is a popular attack among the most common network attacks preceding a port scan. They are among one of the oldest types of attacks used by hackers, but yet still very harmful if successfully completed [36] nowadays. It works by penetrating a host within the network with large amounts of traffic, causing the host to overload and crash. It also causes an outage and potentially makes vital resources unavailable to its intended users. As all network administrators know uptime is vital to running a successful and robust network environment, to have a host or potentially a server to go down is a frightening thought.

Content-Based Network Monitoring

There are a prominent class of applications that are emerging daily, such as information sharing, auctioning online, and multimedia games. These applications are characterized by large and dynamically varying numbers of autonomous clients engaged in multiparty communication, wide geographical distribution, dynamic unpredictable interaction patterns, and high rates of message traffic. This has led to the development of a new kind of advanced communication infrastructure to support these class applications called content-based network monitoring. A content-based network is a communication network based on a novel connectionless service model [12] (Figure 17.2).

The figure above shows that content-based network monitoring involves monitoring all the packets in and out of the network. It also tracks and restores protocols. In order to monitor the internal users and track the network status, the system's function mainly includes six modules. IP packet transmitting occurs when internal hosts attempt to access external network resources; they must send the data request to the content filter gateway first. The gateway will then decide whether to transmit these data or not. Real-time

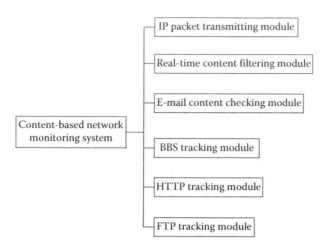

FIGURE 17.2
The function modules of the system.

content filtering is a core module, and it can check all data content in and out the gateway. E-mail content checking is another module that focuses on e-mails sent to the internal network users through SMTP protocol. The e-mails will be redirected to the e-mail filtering server by the gateway and the e-mail filter server will check the title, texts, and attachments separately. BBS stands for bulletin board system, and is the module that can record all the operations when the user logs in, browses data, and submits data. HTTP tracking module can store all Web pages browsed by users in the hard disk. This allows an administrator to track which information has been entered by a user and which Web sites have been visited. It can also extract the secret content in the user submission information. FTP tracking module is a content filtering gateway that can record all of the user's FTP operations, including the login username, password, file name, and file content [24].

Clients declare their characteristics to the network by means of predicates; this is simply a profile that describes the interests of a client. A particular message is delivered to a particular client if the client's predicate applied to the information content of the message is true. It is the predicate that determines which messages will flow to the client through the network, and forms the basis of routing and forwarding functions. Therefore, the content-based network is a new network paradigm that has transited from the traditional destination address configuration (i.e., address-centric) to being more identifiable of the interests of the intended recipients, which has more informative meanings for applications. An example of content-based routing would be the mechanism used in clusters of the Web servers for mapping HTTP requests to individual servers within a cluster. We call this mechanism content-based routing because each HTTP request is directed by a gateway of some kind (e.g., a master server, a scheduler, or a switch) to a particular server based on the content of the request. Content analysis is very important within network monitoring, as this allows operations of network to take place successfully by receiving and sending information. All IP packets are contents within them such as version, type of service, total length, identification, flags, fragment offset, time to live, protocol, header checksum, source IP address, destination IP address, options, and padding information. This information is necessary so that communication is monitored informatively and carefully, according to the content of the packets it is sent and received correctly. IDSs are signature based; this is the rule that examines a packet or series of packets for certain "contents" such as matches on packet header or data payload information. The signatures are developed with content to examine packet contents for a match; this involves looking inside the payload of a packet as well as the packet headers. There are several benefits of content-based network monitoring; content-aware congestion control can help packets minimize distortion impacts and delay deadlines of packets for better throughput in a network [29]. Because of similarities among predicates, if a new client is added to the network whose predicate is similar to another predicate, then there is no need to advertise the new predicate as it already exists. In terms of routing benefits in content-based network monitoring, existing protocols are leveraged for discovery and maintenance of basic network topology information (e.g., distance vector and link-state protocols). Routing decisions are distilled into forwarding tables located in each content-based router. In traditional networks, the basic content-based network is a best-effort service.

Context-Based Network Monitoring

Context is any information that can be used to characterize the situation of an entity [2]. For example, in terms of data prioritization within a network, data from the first responder may be more critical than from a volunteer. The sender's activity location, network condition, proximity of other team members, and battery level are examples of context parameters. Contextual data generally come from collecting packet-level details of the event-related network traffic [1]. Network analysis tools such as tcpdump focus on extracting this vital, detailed information from individual packets, but such tools lack the mechanism of providing simultaneous "big picture" view of data. If the analysts want to understand the details of packets within the larger context surrounding network monitoring, they must continually shift their attention to new and innovative ideas. Such tools excel at filtering and searching for details, provided the analysts know exactly what they are looking for in the data. In IDS, signatures are used to detect attacks; signatures can be implemented using context. Context signatures examine only the packet header for information when looking for a match. This information can include the IP address fields, the IP protocol field, IP options, IP fragment parameters, IP/TCP/UDP checksums, IP and TCP port numbers, TCP flags, ICMP message types, and others. Context can be defined in many ways, one of which could be related to location, nearby person, hosts of objects as well as changes of them over time. The people, orientation and objects, date, and time of the user's environment (or can say the time and space and their relationship, etc.) can be viewed as the aspects of context. Study shows that we could divide context into three parts. The first part is the computing context, which focuses on available processors, devices accessible for user input and display, nearby resources such as printers, workstations, network capacity, connectivity, costs of computing and communication, and bandwidth. The second part is user context, which focuses on location, collection of nearby people, user profiles, and social situation. The third part is physical context, which is lighting, temperature, noise, humidity level, and traffic conditions [19]. Context-based network monitoring provides many benefits such as context parameters are communicated at all layers and between nodes. The load is distributed over the entire network to achieve load balancing and lengthened network lifetime. Context-aware models can be used to exploit wireless sensor network's full potential by getting the running applications involved in lower layer decisions and giving them the ability to control network behavior [37,38]. In context-aware networks, new applications may be composed from existing network applications.

Figure 17.3 shows how context information flows within the system. First, the sender sends data packets and context information to the receiver. After this, the intermediate nodes detect congestion. Hop-by-hop and centralized congestion avoidance is applied to mitigate incipient congestion. The receiver determines the critical level of the packet using sender context information.

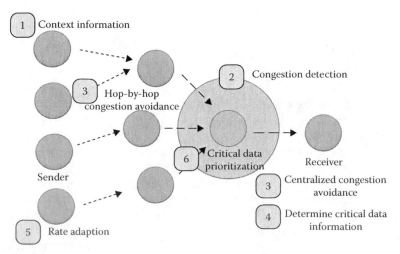

FIGURE 17.3
Context information flow.

The critical level and network rate feedback is relayed to the sender. After this, the sender determines the sending rate and attaches the critical level to outgoing packets. Once all are completed, then intermediate nodes perform packet prioritization [39].

Techniques, Tools, and Systems of the CCBNM

Data mining is one of the techniques used in CCBNM framework. Data mining is a process of analyzing data from different perspectives and summarizing it into useful information [40]. There are data mining software tools that are known as analytical tools that help to analyze data. These tools help the user to analyze data from different dimensions or angles, categorize it, and summarize the relationships identified. It is the process of finding correlations or patterns among dozens of fields in large relational databases. Data mining algorithm applied to intrusion detection mainly has four basic patterns: (1) association, (2) sequence, (3) classification, and (4) clustering. Data mining technology can process large amount of data and does not need the user's subjective evaluation and is more likely to discover the ignored and hidden information. The patterns in anomaly detection are pattern comparison and clustering algorithms [41]. Pattern comparison first establishes normal behavior pattern under association rules and sequence rules then distinguish from normal behavior and association rules. Sequence analysis is designed for the purpose of mining the links among data. Clustering algorithm involves detection for anomaly with no supervision and it detects intrusion by training the unmarked data; it needs no training process, so it can discover new and unknown intrusion types [3,42].

Database auditing is another important technique involved with CCBNM. Database auditing involves observing a database to be aware of the actions of database users. This is often for security purposes, for example, to ensure that information is not accessed by those without permission. Database auditing can be done by many methods. The first method consists of passive server or device, which looks into the network and monitors traffic flowing into and out of the database system. The second method for logging database activities based on network is to put devices directly in line with the network, so that all packets going in and out are intercepted and passed by the device. The third method is agent-based monitoring where the extraction of logs is done directly from the database. The fourth method of logging and monitoring is a system integrated method, which uses monitoring methods provided by Oracle, SQL, and other databases. For the purpose of our simulation task, we will be using a built-in database, within our network monitoring tool to capture packets; information in the database will be audited. Protocol reverse engineering is the process of extracting application-level protocol specifications. The detailed

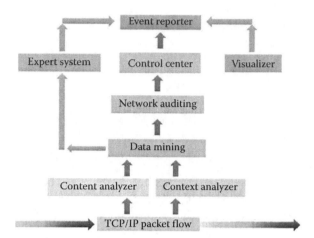

FIGURE 17.4
The framework of CCBNM.

knowledge of such protocol specifications is useful for addressing a number of security problems. This is essential for protocol analysis [43,44]. Our framework of CCBNM is summarized in Figure 17.4.

The Trends of CCBNM

As network security has only recently become a major concern of network technologies, this novel CCBNM is proposed as a new and promising attack monitoring and detection framework, which can effectively prevent various attacks in the error-prone network environment. Many traditional techniques such as the ones discussed earlier in this chapter are outdated and has increasing vulnerabilities and risks, in term of network security. To confront this growing concern, many new network monitoring and security tools are being developed to incorporate all of the existing techniques such as packet sniffing, IDS, and IPS to create and support a more secure and unified network environment. In addition, a new technique of data security visualization has also been added to ensure that it stays one step ahead. Data security visualization is seen as the turning point in CCBNM as it is assisting network administrators to cover threats and anomalies that might have otherwise gone a miss if they were using traditional techniques [45]. Recent research has focused on using visualization techniques to help analysts gain a mental image of network behavior. Visualization is powerful because it allows us to see significant amount of data at once and utilize our cognitive and preattentive processing abilities to find patterns more quickly than sifting through packets or flow records of data [46–48].

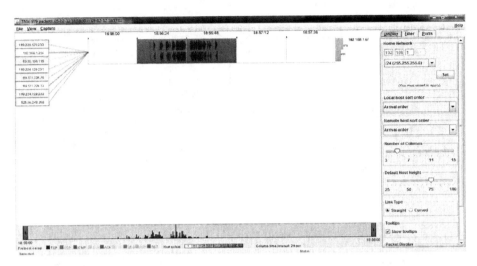

FIGURE 17.5
TNV: network security monitoring tool.

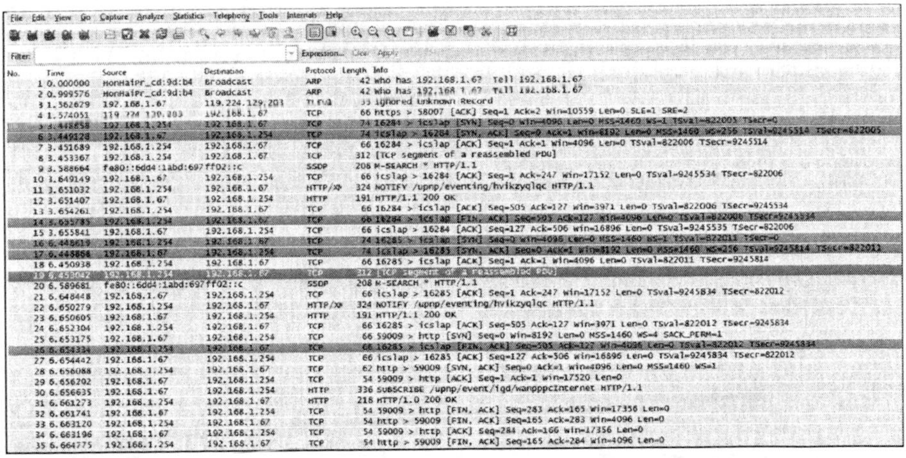

FIGURE 17.6
Wireshark: packet sniffing tool.

The task of network monitoring has also become a demanding and time-consuming task because of having to interoperate movements, and detecting changes of surroundings. It can get complicated if we deal with highly dynamic network traffic; however, reduction of the complicated network traffic data into simple information for visualizing is a suitable platform for a network administrator [49]. This idea can be backed up below as we can see in Figures 17.5 and 17.6 [50].

From Figures 17.5 and 17.6, we are able to see how data security visualization can play a huge role in helping network administrators understand and interoperate network traffic. Both figures show the identical network flow in which we are able to use our cognitive and preattentive processing abilities to spot any anomalies in the data set easier, therefore, allowing us to uncover possible threats.

Even though this is a very new field, many tools already incorporate the idea of visualization; tools such as TNV, InetVis, and RUMINT [17,51] can be successfully used to model this technique. In the near future, we hope to see that visualization can be used fully by going one step further and providing tools that allow 3D visualization capabilities as well as support for IPV6 which is slowly being rolled out worldwide and will one day be the defector standard [52].

Conclusions

Network security is a rapidly growing area in the field of ICT area. Thousands of new threats and compromisation strategies have been found over the last few months alone, leaving us to believe this number will just continue to grow at an alarming rate. In this chapter, after a broad survey of existing network monitoring and detections mechanism, we propose our novel CCBNM, a new monitoring framework with which a researcher can protect against various threats and compromisation risks by using more informative content and context components and also data mining techniques; CCBNM is seen as the future trend of network security monitoring and detection framework, and incorporates traditional effective techniques such as packet sniffing, IDSs, and IPSs. These techniques can be advanced by integrating novel components such as content- and context-aware packet, data mining, and visualization to provide a promising security framework for future vital network.

References

1. Rui Lu, Jia Mi, Bo Huang, "Design and Implementation of Instant Messenger Security Monitoring System Based on Protocol Analysis," *Chinese Control and Decision Conference (CCDC)*, Jiangsu, China, May 26–28, pp. 4290–4293, 2010.
2. Lei Xu, Zhihong Tian, Jianwei Ye, Hongli Zhang, "IR4CF: An Intrusion Replay System for Computer Forensics," *IEEE 2nd International Conference on Computing, Control and Industrial Engineering (CCIE)*, Hubei, China, August 20–21, Vol. 1, pp. 66–69, 2011.
3. Natarajan Meghanathan, Sumanth Reddy Allam, Loretta A. Moore, "Tools and Techniques for Network Forensics," *International Journal of Network Security and Its Applications*, 1(1): 14–25, 2009.

4. Di Wu, Ying Yin, Xiao-hua Chen, Ning Bu, "An Access-Context Based Method to Detect Network Scanning Event in LAN," *International Conference on Machine Learning and Cybernetics*, Hebei, China, July 12–15, Vol. 5, pp. 2781–2786, 2009.

5. Mohammed A. Qadeer, Nadeem Akhtar, Faraz Khan, Faridul Haque, "Monitoring and Analysis of Data Packets Using Data Stream Management System," *International Conference on Computer and Electrical Engineering (ICCEE)*, Phuket, Thailand, December 20–22, pp. 214–218, 2008.

6. Kulsoom Abdullah, Chris Lee, Gregory Conti, John Copeland, "Processing Data to Construct Practical Visualizations for Network Security," *Information Assurance Newsletter*, 9: 3–7, 2006.

7. Joshua Broadway, Benjamin Turnbull, Jill Slay, "Improving the Analysis of Lawfully Intercepted Network Packet Data Captured for Forensic Analysis," *The Third International Conference on Availability, Reliability and Security (ARES)*, Barcelona, Spain, March 4–7, pp. 1361–1368, 2008.

8. John R. Goodall, Wayne G. Lutters, Penny Rheingans, Anita Komlodi, "Preserving the Big Picture: Visual Network Traffic Analysis with TNV," *IEEE Workshop on Visualization for Computer Security (VizSEC)*, October 26, Minneapolis, MN, IEEE, pp. 47–54, 2005.

9. Fuxiang Gao, Shauman Liu, "Design and Implementation of an IPV6-Supported Network Behavior Analysis System," *International Conference on Internet Technology and Applications (iTAP)*, Hubei, China, August 16–18, pp. 1–3, 2011.

10. Mohammed A. Qadeer, Mohamad Zahid, Arshad Iqbal, Misbahur Rahman Siddiqui, "Network Traffic Analysis and Intrusion Detection Using Packet Sniffer," *IEEE Second International Conference on Communication Software and Networks*, Singapore, February 26–28, pp. 313–317, 2010.

11. Wentao Liu, "Design and Implement of Common Network Security Scanning System," *International Symposium on Intelligent Ubiquitous Computing and Education (IUCE)*, Sichuan, China, May 15–16, pp. 148–151, 2009.

12. Mohsen Beheshti, Jianchao Han, Kazimierz, Joel Ortiz, Johnly Tomelden, Damian Alvillar, "Packet Information Collection and Transformation for Network and Intrusion Detection and Prevention," *International Symposium on Telecommunications (IST)*, Tehran, Iran, August 27–28, pp. 42–48, 2008.

13. Naveed Ahmed, Noman Ahmed, Abdul Qadir Khan Rajput, "TCP/IP Protocol Stack Analysis Using MENet," *International Conference on Convergent Technologies for Asia-Pacific Region (TENCON)*, October 15–17, Vol. 4, pp. 1329–1333, 2003.

14. Tiago H. Kobayashi, Aguinaldo B. Barista, Agostinho M. Brito, Paulo S. Motta Pires, "Using a Packet Manipulation Tool for Security Analysis of Industrial Network Protocols," *IEEE Conference on Emerging Technologies and Factory Automation (ETFA)*, Patras, Greece, September 25–28, pp. 744–747, 2007.

15. Sindhu Kakuru, "Behavior Based Network Traffic Analysis Tool," *IEEE IIIrd International Conference on Communication Software and Networks (ICCSN)*, Shaanxi, China, May 27–29, pp. 649–652, 2011.

16. Andrcca Cirncci, Stefan Bobe, Cristian Estan, "Netpy: Advanced Network Traffic Monitoring," *International Conference on Intelligent Networking and Collaborative Systems (INCOS)*, Barcelona, Spain, November 4–6, pp. 253–254, 2009.

17. Sam Abbott, A. J. Newtson, Robert Ross, Ralph Ware, Gregory Conti, "Free Visualization Tools for Security Analysis and Network Monitoring," InSecure, Issue 15, February 2008, pp. 1–8, 2008.

18. Guohui Yin, Wei Gong, "Application Design of Data Packet Capturing Based on Sharpcap," *The Fourth International Joint Conference on Computational Sciences and Optimization (CSO)*, Yunnan, China, April 15–19, pp. 861–864, 2011.
19. Fenggen Jia, Weiming Wang, Ming Gao, Chaqi Lv, "A Real-Time Rule-Matching Algorithm for the Network Security Audit System," *7th International Conference on Information, Communications and Signal Processing (ICICS)*, Piscataway, NJ: IEEE Press, pp. 1–4, 2009.
20. Anita D. D'Amico, John R. Goodall, Daniel R. Tesone, Jason K. Kopylec, "Visual Discovery in Computer Network Defense," *IEEE Computer Graphics and Applications*, 27(5): 20–27, 2007.
21. Abdallah Ghourabi, Tarek Abbes, Adel Bouhoula, "Data Analyzer Based on Data Mining for Honeypot Router," *IEEE/ACS International Conference on Computer Systems and Applications (AICCSA)*, Hammamet, Tunisia, May 16–19, pp. 1–6, 2010.
22. Vicka Corey, Charles Peterman, Sybil Shearin, Micheal S. Greenberg, James Van Bokkelen, "Network Forensic Analysis," *IEEE Internet Computing*, 6(6): 60–66, 2002.
23. Nuttachot Promrit, Anirach Mingkhwan, Supporn Simcharoen, Nati Namvong, "Multi-Dimensional Visualization for Network Forensic Analysis," *The 7th International Conference on Networked Computing (INC)*, Gyeongsangbuk-do, South Korea, September 26–28, pp. 68–73, 2011.
24. Zhenqi Wang, Xinyu Wang, "The Research and Design of Content-Based Network Monitoring System," *2nd International Conference on Power Electronics and Intelligent Transportation System (PEITS)*, Guangdong, China, December 19–20, Vol. 1, pp. 68–71, 2009.
25. Kevin Butler, Toni Farley, Patrick McDaniel, Jennifer Rexford, "A Survey of BGP Security Issues and Solutions," *Proceedings of the IEEE*, 98(1): 100–122, 2010.
26. Lothar Bruan, Gerhard Munz, Georg Carle, "Packet Sampling for Worm and Botnet Detection in TCP Connections," *IEEE Network Operations and Management Symposium (NOMS)*, Osaka, Japan, April 19–23, pp. 264–271, 2010.
27. Alexander S. Seewald, Wilfried N. Gansterer, "On the Detection and Identification of Botnets," *Computer and Security*, 29: 45–58, 2010.
28. Yuxin Meng, Lam-for Kwok, "Adaptive Context-Aware Packet Filter Scheme Using Statistic-Based blacklist Generation in Network Intrusion Detection," *The 7th International Conference on Information Assurance and Security (IAS)*, Melaka, Malay Peninsula, December 5–8, pp. 74–79, 2011.
29. Farxaneh Izak Shiri, Bharandiharan Shanmugam, Norbik Bashah Idris, "A Parallel Technique for Improving the Performance of Signature-Based Network Intrusion Detection System," *IEEE 3rd International Conference on Communication Software and Networks (ICCSN)*, Shaanxi, China, May 27–29, pp. 692–696, 2011.
30. S. H. C. Haris, R. B. Ahmad, M. A. H. A. Ghani, Ghossoon M. Waleed, "Packet Analysis Using Packet Filtering and Traffic Monitoring Techniques," *International Conference on Computer Applications and Industrial Electronics (ICCAIE)*, Kuala Lumpur, Malaysia, December 5–8, pp. 271–275, 2010.
31. Hong Wang, Zhenghu Gong, Qing Guan, Baosheng Wang, "Detection Network Anomalies Based on Packet and Flow Analysis," *Seventh International Conference on Networking*, Cancun, Mexico, April 13–18, pp. 497–502, 2009.

32. Hamed Salehi, Hossein Shirazi, Reza Askari Moghadam, "Increasing Overall Network Security by Integrating Signature-Based NIDS with Packet Filtering Firewall," *International Joint Conference on Artificial Intelligence (IJCAI)*, Hainan, China, April 25–26, pp. 357–362, 2009.

33. Emmanuel Hooper, "An Intelligent Detection and Response Strategy to False Positives and Network Attacks," *Fourth IEEE International Workshop on Information Assurance (IWIA)*, London, April 13–14, pp. 1–20, 2006.

34. Glenn Carl, George Kesidis, Richard Brooks, Rai Suresh, "Denial-of-Service Attack-Detection Techniques," *IEEE Internet Computing*, 10(1): 82–89, 2006.

35. Atul Kant Kaushik, Emmanuel S. Pilli, R.C. Joshi, "Network Forensic System for Port Scanning Attack," *IEEE 2nd International Advance Computing Conference (IACC)*, Patiala, Punjab, India, February 19–20, pp. 310–315, 2010.

36. Tadashi Kiuchi, Yoshiaki Hon, Kouichi Sakurai, "A Design History Based Traffic Filtering with Probablistic Packet Marking against Dos Attacks," *10th IEEE/IPSJ International Symposium on Applications and the Internet (SAINT)*, Seoul, South Korea, July 19–23, pp. 261–264, 2010.

37. Hu Han, "Performance Improvement of TCP_Reno Based on Monitoring the Wireless Packet Loss Rate," *IEEE 3rd International Conference on Communication Software and Networks (ICCSN)*, Shaanxi, China, May 27–29, pp. 469–472, 2011.

38. Ke Meng, Yang Xiao, Susan V. Vrbsky, "Building a Wireless Capturing Tool for WiFi," *Security and Communication Networks*, 2: 654–668, 2009.

39. Faisal Luqman, "TRIAGE: Applying Context to Improve Timely Delivery of Critical Data in Mobile Ad Hoc Networks for Disaster Response," *IEEE International Conference on Pervasive Computing and Communications Workshops (PERCOM)*, Seattle, WA, March 21–25, pp. 407–408, 2011.

40. Jiawei Han, Micheline Kamber, *Data Mining Concepts and Techniques* (2nd Edition), San Francisco, CA: Morgan Kaufmann Publishers, 2006.

41. Rick Dove, "Self-Organizing Resilient Network Sensing (SornS) with Very Large Scale Anomaly Detection," *IEEE International Conference on Technologies for Homeland Security (HST)*, Waltham, MA, November 15–17, pp. 487–493, 2011.

42. Ming Xue, Changjun Zhu, "Applied Research on Data Mining Algorithm in Network Intrusion Detection," *IEEE International Joint Conference on Artificial Intelligence (JCAI)*, Hainan, China, April 25–26, pp. 275–277, 2010.

43. Hugang Qiang, Lianzhong Liu, "A Logging Scheme for Database Audit," *IEEE International Workshop on Computer Science and Engineering (WCSE)*, Shandong, China, October 28–30, pp. 390–393, 2009.

44. Rafa Marin-Lopez, Fernando Pereniguez-Garcia, Antonio Gomez-Skarmeta, Yoshihiro Ohba, "Network Access Security for the Internet: Protocol for Carrying Authentication for Network Access," *IEEE Communications Magazine*, 50(3): 84–92, 2012.

45. Guizhong Guo, Xinhua Mao, "Research on the Milling Tool Monitoring System Based on Wavelet Neural Network," *International Conference on Electronic and Mechanical Engineering and Information Technology (EMEIT)*, Harbin, Heilongjiang, China, August 12–14, Vol. 3, pp. 1421–1423, 2011.

46. Mai El-Shehaly, Denis Gracanin, Ayman Abdel-Hamid, Kresimir Matkovic, "A Visualization Framework for Traffic Data Exploration and Scan Detection," *IEEE 3rd International Conference on New Technologies, Mobility and Security (NTMS)*, Cairo, Egypt, December 20–23, pp. 1–6, 2009.

47. Teryl Taylor, Diana Paterson, Joel Glanfield, Carrie Gates, Stephen Brooks, John Mchugh, "FloVis: Flow Visualization System," *IEEE Computing Society, IEEE the 2009 Cybersecurity Applications & Technology Conference for Homeland Security (CATCH)*, Washington, DC, March 3–4, pp. 186–198, 2009.
48. Seung-Hoon Kang, Juho Kim, "Network Forensic Analysis Using Visualization Effect," *International Conference on Convergence and Hybrid Information Technology (ICHIT)*, Daejeon, South Korea, August 28–30, pp. 466–473, 2008.
49. Doris Wong Hooi Ten, Selvakumar Manickam, Sureswaran Ramadass, Hussein Al Bazar, "Study on Advanced Visualization Tools in Network Monitoring Platform," *The 3rd UKSim European Symposium on Computer Modeling and Simulation (EMS)*, Athens, Greece, November 25–27, pp. 445–449, 2009.
50. John R. Goodall, "Visualization Is Better! A Comparative Evaluation, Visualization for Cyber Security," *IEEE 6th International Workshop on Date (VizSec)*, Atlantic City, NJ, October 11, pp. 57–68, 2009.
51. Russ McRee, "Security Visualization: What You Don't See Can Hurt You," *Information Systems Security Association (ISSA) Journal*, 38–41, 2008.
52. Bruce J. Nikkel, "An Introduction to Investigating IPv6 Networks," *Digital Investigation*, 4: 59–67, 2007.

18

A Comparison of Three Sophisticated Cyber Weapons

Makkuva Shyam Vinay and Manoj Balakrishnan

CONTENTS

Introduction

Cyber malwares have been steadily improving in terms of sophistication of design as well as the ability to cause substantial damage to large-scale IT systems. Today's malwares exploit innumerable vulnerabilities to create a back door for bypassing authentication or securing remote access to computers in order to carry out their objectives. Three recent malwares that display high design sophistication are Stuxnet, Flame, and Duqu. Stuxnet malware was created to cause physical damage by reprogramming industrial control systems (ICSs) and programmable logic controllers (PLCs). They are capable of reprogramming ICSs and PLCs in such a way as to make them work in a manner the attacker intends. Flame, on the other hand, is a 20-MB-size malware designed for confidential information theft from the victimized machines. Duqu, a Stuxnet like malware, discovered by CrySyS labs is also an information stealer malware. Flame looks for information that could help in attacking ICSs, though its purpose is not destructive. The aim of this chapter is to study these malwares, their properties, mechanisms, and exploits used and to conduct a qualitative comparison among them.

The rest of this chapter is organized as follows: The second section discusses Stuxnet malware while the third section presents Flame malware. The fourth section explains the operation of Duqu malware and the fifth section provides a qualitative comparison of the three malwares. This chapter ends with the "Conclusion."

Stuxnet Malware

Stuxnet is a computer worm discovered in 2010 by the Internet security company VirusBlokAda [1,2]. Media reports indicated that Stuxnet was created to sabotage the nuclear program of a nation under the code name "Operation Olympic Games" [3]. Stuxnet mainly targets Siemens industrial software and equipment, due to the serious security holes in it. To increase its chances of success Stuxnet uses a number of vulnerabilities such as zero-day exploits, windows rootkit, PLC rootkit, antivirus evasion technique, process injection and hooking code techniques besides a command and control interface. Due to the sophisticated design of Stuxnet, it was able to carry out several stealth attacks including the destruction of the nuclear centrifuges in Iran.

Timeline of Major Stuxnet Activities

Table 18.1 shows the timeline of major activities of Stuxnet malware [4]. Stuxnet caused three main attack waves. First was on June 22, 2009, the second happened on March 1, 2010, and the third on April 14, 2010.

Stuxnet: Installation and Spread

Stuxnet targets only computers with Windows operating system (OS) [3–6]. That is, Stuxnet exits if the computer is of some other OS. Because most process control systems (PCSs) use Windows OS, the impact of Stuxnet can be disastrous

TABLE 18.1

Timeline of Major Stuxnet Activities

Date	Event
June 2009	Earliest Stuxnet sample seen. Sample does not have signed drivers.
January 2010	Stuxnet signed driver with a valid signature belonging to Realtek Semiconductors.
March 2010	First variant of Stuxnet to exploit vulnerability that could do remote code execution.
June 2010	VirusBlokAda first identifies Stuxnet.
July 17, 2010	ESET, an antivirus company, identifies a new Stuxnet driver, this time signed by JMicron Technology.
July 20, 2010	Symantec monitors Stuxnet's command and control traffic.
August 2, 2010	Microsoft issues MS10-046, which patches Windows shell shortcut (lnk) vulnerability.
September 4, 2010	Microsoft releases MS10-061, to patch up printer spooler vulnerability identified by Symantec.

in future attack waves. The reported infections of Stuxnet that resulted in the above-mentioned attack waves are likely to have happened through removable storage devices such as USB drives because many of the affected PLCs were controlled by Windows computers that are not typically connected to the Internet.

Once Stuxnet infects one Windows computer in an organization, it spreads by searching and infecting Field PGs (SIMATIC rugged laptops computers, which are used for industrial purposes). Once infected to the Field PGs, Stuxnet attempts controlling the PCSs. Stuxnet requires the following key access or authorization information to control the PCSs: (1) passwords to access the PCS, (2) digital signatures for drivers, and (3) design documentation of the victim system.

First, Stuxnet uses the default passwords Siemens to gain access to systems that run WinCC and PCS-7 programs after completing installation in a Windows-based Field PG computer. WinCC is a human–machine interface system and PCS-7 is a PCS [7]. Second, the attacker's malicious binaries contained driver files that need to be signed digitally to avoid suspicion, detection, or prevention of installation of the binaries. Finally, in order to reprogram ICS, Stuxnet needs the design schematics of ICS. As of today, it remains a mystery how Stuxnet gathered the design schematics of the victim ICSs. It is assumed that the ICS design documents might have been supplied by a company insider or stolen from the company through espionage. It is unclear whether ICS design details have already been available with an earlier version of Stuxnet. Such availability of ICS design information shows the ability of the Stuxnet designers to prepare and gather very sophisticated and deeply targeted attacks.

Once Stuxnet spreads to a victim computer connected to an ICS, it begins its operation that can be categorized into three phases. In the first phase,

Stuxnet detects a Siemens PCS computer that runs PCS-7 process and gathers all the configuration and design information about the victim. In the second phase, it transfers the gathered configuration and design information to the remote command-and-control server (CCS) maintained by the perpetrators of the attack. In the final phase, Stuxnet interacts with the victim PCS and reprograms it based on the instructions from the CCS. In parallel to the above three phases, Stuxnet searches the local network for additional victims and repeats the above-mentioned spreading and three-phase operational activities.

Stuxnet: Architecture and Organization

Stuxnet's architecture is one of the most sophisticated. Stuxnet consists of a large dynamic linked library (DLL) file, with file type extension ".dll" that contains about 32 exports and 15 resources. Stuxnet dropper is a file, named ~WTR4132.TMP, containing all of the above in a specific section of the file called "stub." Figure 18.1 shows the life cycle of Stuxnet operation which presents the key operations during Stuxnet loading. First, it loads a file named ~WTR4141.TMP followed by another DLL file ~WTR4132.TMP. The ~WTR4132.TMP file loads the main Stuxnet.dll. The important functions invoked by the loading of these files are discussed in what follows.

A USB external drive infected with the Stuxnet worm typically contains the following six files [5] whose functions are explained as follows:

1. ~WTR4141.TMP
2. ~WTR4132.TMP
3. "Copy of shortcut to .lnk"
4. "Copy of copy of shortcut to .lnk"

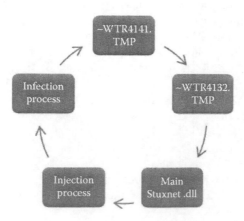

FIGURE 18.1
Life cycle of Stuxnet.

5. "Copy of copy of copy of shortcut to .lnk"
6. "Copy of copy of copy of copy of shortcut to .lnk"

Stuxnet spreads to all removable drives by dropping shortcut files (.lnk) that automatically run when the removable drive is accessed using an application that displays shortcut icons. The external drive also contains copy of shortcut to this ".lnk" file. It also contains copy of copy of this shortcut, copy of copy of copy of this shortcut, and also copy of copy of copy of copy of this shortcut.

~WTR4141.TMP

As soon as this file is loaded, it hooks with a number of functions in both *kernel32.dll* and *ntdll.dll*. In kernel32.dll, the following functions are invoked: (1) FindFirstFileW(.), (2) FindNextFileW(.), and (3) FindFirstFileExW(.). The functions invoked from the file ntdll.dll are NtQueryDirectoryFile(.) and ZwQueryDirectoryFile(.).

~WTR4132.TMP

~WTR4132.TMP is a DLL file loaded into explorer.exe. After loading, it begins execution by searching for a section called ".stub" in it. This .stub section contains main .dll file which has all functions and mechanisms such as rootkits. Then it loads main .dll file by first allocating memory buffer and then patches the six ntdll.dll files with these names:

1. ZwMapViewOfSection(.)
2. ZwCreateSection(.)
3. ZwOpenFile(.)
4. ZwClose(.)
5. ZwQueryAttributesFile(.)
6. ZwQuerySection(.)

These patches make DLL file to be loaded not from hard disk but from memory. The sequence of file loading and the entire life cycle of Stuxnet malware are shown in Figure 18.1 [5].

Main Stuxnet .dll

Once loaded, the Stuxnet .dll checks if it has administrator rights. If necessary, Stuxnet uses one of the following zero-day vulnerabilities to escalate the user privileges to administrator level: (1) Win32k.sys keyboard layout vulnerability and (2) Windows task scheduler vulnerability [4]. After gaining administrator privilege, it injects into a new process in order to install

TABLE 18.2

Injection Targets

Security Product	Injection Target
McAfee	Winlogon.exe
KAV v1 to v7	Lsass.exe
KAV v8 and v9	KAV process
Symantec	Lsass.exe
ETrust v5 and v6	Injection is not possible
BitDefender	Lsass.exe

itself. This injection is done into an antivirus application process running in the machine as given in Table 18.2. For each popular antivirus product, the specific injection target is provided in Table 18.2.

Schematic representation of the injection process is given in Figure 18.2. After creating a process it injects by unloading the program (of chosen antivirus) from memory and loads another file from Stuxnet.dll into the same place. Before unloading, Stuxnet modifies by adding a new section called ".verif." This makes the size of loaded file equal to that of the unloaded one. After this Stuxnet installs itself by writing the following six files in C:\Windows\directory:

- C:\WINDOWS\inf\oem7A.PNF
- C:\WINDOWS\inf\oem6C.PNF
- C:\WINDOWS\inf\mdmcpq3.PNF
- C:\WINDOWS\inf\mdmeric3.PNF
- C:\WINDOWS\system32\Drivers\mrxnet.sys
- C:\WINDOWS\system32\Drivers\mrxcls.sys

Stuxnet: Spreading Mechanisms

Stuxnet is found to be spreading using the following infection mechanisms: (1) USB-driven infection, (2) infection through network, and (3) infection through server message blocks (SMBs).

1. *USB-driven infection.* Stuxnet spreads via USB drives by exploiting the Windows shortcut vulnerability. It is a well-known vulnerability for Windows OS to load icons for .lnk files which results in the vulnerability. This vulnerability allows auto-execution of software without the users' intervention. This vulnerability can be prevented by using Windows OS patches.

2. *Infection through the network.* Stuxnet spreads via network by exploiting one of the following vulnerabilities: (a) Windows Server

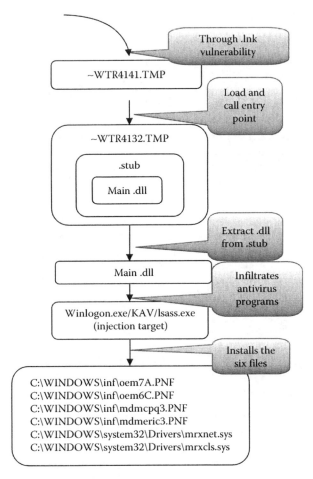

FIGURE 18.2
Injection process of Stuxnet malware.

Service NetPathCanonicalize(.) vulnerability which allows remote code execution if an affected system receives a specially crafted RPC request, or (b) Windows print spooler service vulnerability. The impact of this vulnerability can be minimized by breaking the network into security zones thereby limiting the reach of the printer spool service.

3. *Infection through SMB.* SMB is an application layer protocol used by Windows Remote Procedure Call (RPC) server for providing shared access to files and miscellaneous communication between nodes of a network. Stuxnet also spreads through SMB, by exploiting Microsoft Windows Service Server's remote code execution vulnerability.

Stuxnet: Updating Methods

The two main methods through which Stuxnet updates itself are (1) via the Internet and (2) peer-to-peer connections [6].

1. *Updating via the Internet.* Stuxnet updates itself via the Internet by establishing an HTTP connection through port 80 to a set of CCS. Some known Web sites to which Stuxnet has been found communicating include the following [4]: (a) www.mypremierfutbol.com and (b) www.todaysfutbol.com. Stuxnet also gathers and sends certain basic information such as the IP address and network interface card name of the victim computer to the CC Web sites. The Web sites mentioned above were among the suspected CCSs.

2. *Updating via peer-to-peer connection.* Stuxnet creates an RPC server in the victim's machine and listens to any connections coming from any PC on the network. On finding incoming connection requests from any PC, it sends the Stuxnet code into the PC, injects into a process, and begins the installation process mentioned above [5]. Thus, through the RPC service, Stuxnet spreads into other PCs in the network.

The Flame Malware

Flame is another malware which was recently detected by Kaspersky Labs. Flame is widely considered as a cyber espionage tool that affects only computers that run Microsoft Windows OS. It was named by Kaspersky Labs as "Flame" because the main module that is responsible for the attack and infection of additional machines is known as Flame [8]. Flame is also known by the names *Flamer* and *SkyWiper* as named by various other organizations that claim to have detected it. In comparison to other malwares, Flame is rather large in size; however, it is designed in such a way that detection is nearly impossible. Flame's huge size is attributed to the fact that it contains several modules such as decompressed libraries, SQLite database, and LUA virtual machine that is not seen in any other malware [9]. The most striking aspect of Flame is its elusive nature for years. The fact that this dangerous cyber espionage malware evaded security controls for so long demonstrates how its similar structure to a commercial software program and its use of off-the-shelf techniques, such as SSL, SSH, and an SQL database, helped it blend in with other application traffic.

Flame is mainly designed for stealing information; therefore, it is considered a targeted cyber espionage weapon. It is written in Lua scripting language in order to avoid reverse engineering. Flame can record environmental audio through microphone, capture screen shots, log keyboard activity, monitor network traffic, and record Skype conversations, thereby, making it one of the most potent espionage tool ever detected. Further, it also makes the infected machine discoverable via Bluetooth, thereby, making the device further vulnerable. Flame can also infect a fully patched Windows 7 computer, which indicates the possible presence of an unknown high zero-day attack [10]. A significant difference from other malwares is Flame's support of a kill command. Flame's "kill" command wipes out all traces of malware from the infected computer. Flame uses about five encryption methods thus becoming one of the large-size malwares. Flame's large size made it difficult to propagate. However, even with the large size, the complex encryption process employed by Flame helped evade detection.

Flame communicates to remote servers for both transferring the data gathered from the victim machine as well as to receive instructions. There were about 50 different domains used by the malware to contact command and control server (CCS) and more than 15 distinct IP addresses [11]. CCSs are changed frequently by changing the IP address of a particular host by using the well-known fluxing technique that allows having numerous IP addresses for a single fully qualified domain name [12] and is used by Botnets. Due to these sophistications, it is very difficult to track usage of the deployment of Flame's CCSs.

Flame: Timeline of Activities

Table 18.3 shows the timeline of the events of Flame malware. Flame's first activity was in the year 2007, though it went unnoticed. Subsequently, it reappeared multiple times in 2010. Due to lack of reports from victimized nations, the impact of Flame is clearly not known at the time of writing this.

Flame: Architecture and Organization

The main module of Flame is a DLL file called MSSECMGR.OCX. Two versions, a bigger and a smaller, of this file are typically found. The bigger version

TABLE 18.3

Flame's Timeline of Activities

Time	Activity
December 2007	CrySyS cited appearance of Flame's main component as possible proof of an early development.
February 2010	According to Kaspersky, Flame had been operating in wild since at least February 2010.
August 2010	Microsoft patched Windows shortcut vulnerability which Flame exploited.
September 2010	Microsoft patched print spooler vulnerability which Flame exploited.
May 28, 2010	Flame's discovery was confirmed by CERT, Kaspersky, and CrySyS lab.

contains additional modules and is about 6 MB in size. The smaller, with no additional modules, is found to be of only 900 KB size. The smaller version module downloads and installs other components in%windir%\system32\ directory from CCS servers after its installation, as can be found from Figure 18.3.

The functionality of the modules in Flame is divided into different "units" based on their function. These units are used extensively in the code. Table 18.4 presents some of the different units in Flame and their functions in brief [13].

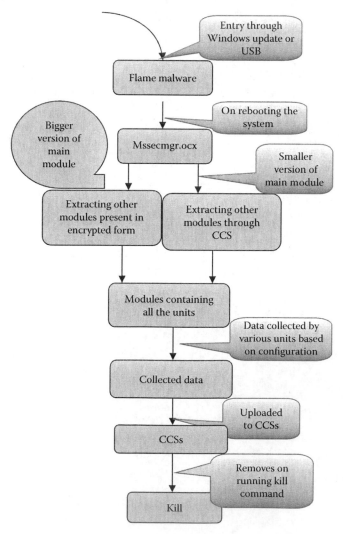

FIGURE 18.3
Lifetime of operation of Flame.

TABLE 18.4

Functionality of Various Units in Flame

Unit	Function
Beetlejuice	Enumerates bluetooth devices around the infected machine and makes the infected machine discoverable to all other devices.
Microbe	Selects suitable recording device from existing audio sources and records audio.
InfectMedia	Selects one of the following two methods for infecting USB drives: (1) Autorun_infector and (2) Euphoria.
Autorun_infector	Creates autorun.inf
Euphoria	Creates junction point directory with "desktop.ini" and "target.lnk." The directory acts as shortcut for launching Flame.
Gator	It downloads new modules and uploads collected data by connecting to CCSs.
Munch	It is the HTTP server module of Flame that responds to "/view.php" and "/wpad.dat" requests.
Limbo	When administrator rights are available, it creates backdoor accounts with login "HelpAssistant" on machines within network domain.
Snack	It creates raw network socket and begins to receive all network packets and save NBNS (NetBios Naming Service) in a log file. It has the option to start only when "Munch" is started.
Boot_dll_loader	This contains a list of additional modules that must be loaded and started.
Gadget	This module is used to spread within a network from a machine that is already infected with the malware.
Transport	Replicates the modules.
Spotter	Scans the modules.

Flame: Spreading Mechanisms

Figure 18.3 shows the lifetime operation of Flame starting from the entry into the system until kill command is employed. It initially enters through a Windows update or USB. On rebooting, it loads main module Mssecmgr.ocx which could be either in the bigger or the smaller version. The smaller version extracts other modules through command and control servers. These modules carry out their respective functions and send the collected data to CCSs. Finally, it removes all its traces from the system by running kill command.

Flame exploits a weakness in an old cryptographic hash function, MD5, to masquerade as a legitimate Microsoft Windows update [9]. Flame also uses printer spooler vulnerability and remote job tasks which have already been exploited by Stuxnet malware. The initial entry point of Flame is unknown, but there is a possibility of it using of MS10-033 vulnerability [14]. This vulnerability could allow remote code execution if a user opens a special media file or receives specially crafted streaming content from a Web site or any application that delivers Web content.

Flame used man-in-the-middle attack against other computers in the network [15]. When a system tries to connect to Microsoft Windows update service, it redirects the connection through the infected system and sends a fake harmful Windows update to the client. Two of the three certificates Microsoft revoked in this update used MD5 hashing scheme which is prone to collisions [9]. Microsoft accidentally issued certificates that can be used to sign code using outdated ciphers. Flame authors are thus successful in making the malware look like it was actually from Microsoft. The malware also spreads through printer spooler vulnerability through LAN, ".lnk" exploit, or by using a ".BAT" file.

On the next system boot, the main module (mssecmgr.ocx) is automatically loaded by the OS. After the Windows registry is updated, mssecmgr.ocx extracts additional modules that are present in encrypted and compressed form and installs them. Once the installation is complete, mssecmgr.ocx loads available modules and starts many threads to implement a channel to CCSs and Lua interpreter host [16].

The Duqu Malware

Duqu is a computer malware which is very much identical to Stuxnet; however, Duqu has been written with a different purpose. Discovered in September 2011, this malware is called Duqu because it has some information stealer components that create files in the infected system with filenames starting with the string "~DQ." Duqu's purpose is to collect information from industrial infrastructure and system manufacturers in order to conduct a potential future attack [17]. Similar to Stuxnet, it targets only MS Windows-based systems. Further, it has a modular structure. It can also be reconfigured from a CCS. The malware is designed to self-delete after 36 days.

Duqu: Timeline of Major Activities

Table 18.5 shows the timeline of the major activities of Duqu-related events. The earliest time stamps observed on Duqu was in 2007 and the last seen Duqu activity was in 2011.

Duqu: Architecture and Organization

The following three groups of components are found to be present in the Duqu malware [17].

1. A Keylogger tool (keylogger.exe)
2. Jminet7 group (Jminet7.sys driver group)
3. CMI4432 group (CMI4432.sys driver group)

TABLE 18.5

Duqu's Timeline of Activities

Time	Activity
August 2007	This time stamp comes from the module which drops the Duqu driver in the system.
February 2008	A module was compiled. This is a driver similar to the one used in 2007 having MD5 and 20608 bytes in size.
June 2009	No activity of Duqu was seen, but earliest Stuxnet attacks observed.
November 2010	Most of the Duqu modules were compiled. They are drivers and pnf encrypted dll.
October 2011	Some more modules have been compiled. October 18, 2011, is the last seen date of Duqu activity.

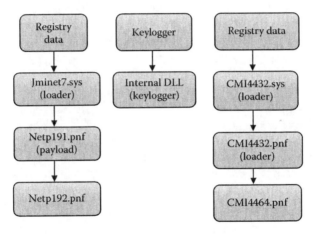

FIGURE 18.4
Modules in Duqu.

These three groups of components are depicted in Figure 18.4. A *Keylogger tool* is a standalone executable file that does not need prior installation. It contains a DLL which is encrypted internally and delivers keylogging functions. Main keylogger injects and controls keylogging processes.

A service is defined in the registry which loads *Jminet7.sys* driver during Windows bootup process. This driver loads the configuration data and injects netp191.pnf DLL payload into a system process. Configuration data is stored in netp192.pnf which is an encrypted configuration file. Similarly, a service is defined in the registry which loads *CMI4432.sys* driver during Windows bootup process. This driver injects CMI4432.pnf DLL payload into a system process and the configuration data is stored in CMI4464.pnf, the encrypted configuration file. Jminet7.sys and CMI4432.sys groups differ only in payload. CMI4432.sys contains a valid digital signature of the Taiwanese manufacturer C-Media [17].

Duqu: Installation Process

Duqu uses exploit in Microsoft Word document vulnerability. The Word document contains zero-day kernel exploit that allows attackers to install Duqu into the computer without the knowledge of the user. The installation process itself is very long and can be divided into two parts: (1) exploit shell code and (2) installer.

1. *Exploit shell code.* As soon as the affected Word document is opened, the exploit is triggered. This exploit contains kernel mode shell code which first checks if the computer is already compromised or not. If it has already been compromised, then the shell code exits. If the computer has not yet been infected, then the shell code decrypts two executable files from within the Word document: (a) driver file and (b) installer DLL. Then the execution is passed on to the driver file which then injects the code into services.exe. The code then executes installer DLL. Then the shell code wipes itself off from memory.

2. *Installer.* Installer decrypts three files from itself: (a) Duqu's main DLL, (b) a .sys driver file which is a load point that starts Duqu after reboot, and (c) an installer configuration file. The installer configuration file has two time stamps representing time for installation. It will terminate if executed outside this timeframe. The installer passes execution to main DLL by hooking to ntdll.dll in the same way as in Stuxnet. There are just three files left on the disk after the installation process: the driver, the encrypted main DLL, and its configuration file.

Duqu: Operational Scenario

The loader (Jminet7.sys) is responsible for injecting the main DLL (netp191 .pnf) into a specific process. The injected process is generally services.exe. Netp191.pnf file contains payload and an encrypted data block. It decrypts configuration data stored in netp192.pnf and checks the "lifetime" value in it. If it has been running for more than 36 days, then it cleans the routine. Otherwise it continues to function. Netp191.pnf then injects itself into one of the following four processes [15]:

1. Explorer.exe
2. IExplorer.exe
3. Firefox.exe
4. Pccntmon.exe

Duqu checks for antivirus processes running in the system, in a similar way as does Stuxnet. It further checks for the processes listed in Table 18.2.

If any one of them is found, then Duqu injects itself into that process. It carries on the same procedure as with Stuxnet including hooking to functions of ntdll.dll.

The rootkit provides functionality to install and start keylogger. Keylogger module is delivered via CCS to target after initial infection. The keylogger .exe contains an embedded jpeg file, the picture is used only for deception, and it contains the encrypted DLL. The DLL contains keylogger-related function calls.

The keylogger stores data in the %TEMP% directory of the target computer. It collects the following data [17]:

- Keystroke data
- Machine information
- Process list
- Network information
- List of shared folders
- List of machines on the same network
- Screenshots

It then sends the collected data to the remote CCSs using http and https protocols. The entire operational scenario of Duqu is schematically represented in Figure 18.5.

Duqu: Spreading Mechanisms

Duqu is a highly targeted cyber malware. Some of the spreading mechanisms known are (1) using zero-day Word exploit and (2) utilizing peer-to-peer command and control protocol.

1. *Zero-day Word exploit.* Duqu attacks Microsoft Windows system using zero-day Word vulnerability. The dropper uses Microsoft Word (.doc) that exploits Win32k TrueType font parsing engine and allows execution [18].
2. *Peer-to-peer command and control protocol.* The peer-to-peer command and control protocol use http over SMB. A newly infected computer will be typically configured to connect back to the infecting computer using a predefined SMB. The peer computer, which was previously the infecting computer then proxies the command and control traffic to external CCS.

Generally, most of the secure networks are configured to have a secure zone where internal servers are located. As Duqu spreads from a less secure

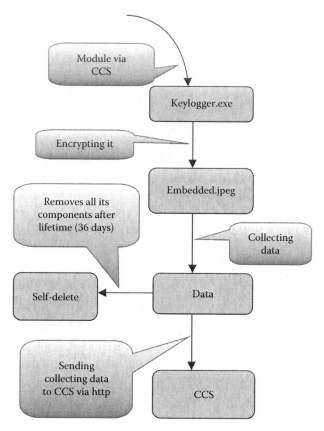

FIGURE 18.5
Installer activity diagram of Duqu.

zone to a more secure zone, it is able to retain a connection back to CCS by effectively building a private bridge between compromised computer leading back to the CCS [19].

Comparison of Stuxnet, Flame, and Duqu

An often-overlooked detail about Stuxnet, Duqu, and Flame is that the attacks targeted by them are only against Windows machines. For additional readings on Stuxnet and Duqu the reader may refer to Refs. [20,21]. Pirated software has been the major platform for these three malwares. A qualitative comparison of the three malwares under discussion in this chapter is given in Table 18.6.

TABLE 18.6

A Comparison between Stuxnet, Flame, and Duqu

Property	Stuxnet	Flame	Duqu
Size	500 KB	20 MB	300 KB
Information stealer	No	Yes	Yes
Type of attack	Targeted	Targeted	Targeted
Keylogger module	No	Yes	Yes
Kill command	No	Yes	Self-deletes after 36 days
Infection through local shares	Yes	Very likely	No
PLC functionality	Yes	No	No
Autorun.inf	Yes	Yes	No
Recording audio	No	Yes	No
Encryptions used	No	Yes	Yes
Languages used	Mostly C and C++	Lua scripting and C	Mainly in C++
RPC communication	Yes	Yes	Yes
Modular malware	Yes	Yes	Yes
Physical damage	Yes (by affecting PLCs)	No	No
Use of C&C servers	Yes	Yes	Yes
DLL injection into process	Yes	Yes	Yes
Records Skype conversations	No	Yes	No
Use of virtual machine	No	Yes (for Lua)	No

Conclusion

In this chapter, the description and comparison of three recently detected powerful malwares are presented. The cyber malwares compared include Stuxnet, Flame, and Duqu. Stuxnet is the first to exploit four zero-day vulnerabilities, compromise two digital certificates, inject code into ICSs, and hide the code from the operator. Flame is the most complicated malware ever discovered. Due to its huge size and use of encryption, it is even very difficult to analyze the malware. It took around five years to detect the malware and only limited information is known so far. Duqu, a Stuxnet-like malware, though it did not cause any physical damage, did target a few nations to get the secret information it needed. Stuxnet, Flame, and Duqu are the three most sophisticated malwares encountered in the recent past. We qualitatively compared the general features of the three malwares.

References

1. VirusBlokAda Ltd. http://www.anti-virus.by/en/index.shtml
2. R. Langner, *Langner communications*. http://www.langner.com/en/, October 2010.
3. R. Langner, "Stuxnet: Dissecting a Cyberwarfare Weapon," *IEEE Security and Privacy Magazine*, 9(3): 49–51, 2011.
4. N. Falliere, L. O. Murchu, and E. Chien, "W32.Stuxnet Dossier," Symantec, version 1.3 edition, November 2010.
5. D. P. Fidler, "Was Stuxnet an Act of War? Decoding a Cyber attack," *IEEE Security and Privacy Magazine*, 9(4): 56–59, 2011.
6. A. Matrosov, E. Rodionov, D. Harley, and J. Malcho, "Stuxnet Under the Microscope," ESET IT Security Company, 2011, http://www.go.eset.com/resources/white-papers/Stuxnet_Under_the_Microscope.pdf
7. A. Cardenas, S. Amin, Z.-S. Lin, Y.-L. Hua, and S. Sastry, "Attacks against Process Control Systems: Risk Assessment, Detection, and Response," *Proceedings of the 6th ACM Symposium on Information, Computer and Communications Security*, March, pp. 355–366, 2011.
8. Aleks, Kaspersky Lab expert, http://www.securelist.com/en/blog/208193522/The_Flame_Questions_and_Answers
9. "The Threat from flamer by European Network and Information Security Agency," http://www.enisa.europa.eu/media/news-items/The-threat-from-Flamer.pdf.
10. M. J. Schwartz, http://www.informationweek.com/security/attacks/flame-faq-11-facts-about-complex-malware/240001177
11. Aleks, Kaspersky Lab expert, http://www.securelist.com/en/blog/208193540/The_Roof_Is_on_Fire_Tackling_Flames_C_C_Servers
12. S. Yu, "Fast-flux Attack Network Identification Based on Agent Lifespan," *Proceedings of IEEE Conference on Wireless Communication, Network and Information Security (IEEE WCINS)*, June 2010.
13. Aleks, Kaspersky Lab expert, http://www.securelist.com/en/blog/208193538/Flame_Bunny_Frog_Munch_and_BeetleJuice
14. Microsoft Security Bulletin MS10-033. http://technet.microsoft.com/en-us/security/bulletin/MS10-033
15. http://nakedsecurity.sophos.com/2012/06/04/flame-malware-used-man-in-the-middle-attack-against-windows-update/.
16. Skywiper analysis team; CrySyS, "A Complex Malware for Targeted Attacks," http://www.crysys.hu/skywiper/skywiper.pdf
17. Boldizsar Bencsath, Gabor Pek, Levente Buttyan, and Mark Felegyhazi; CrySyS, "Duqu: A Stuxnet-Like Malware Found in the Wild," http://crysys.hu/~mfelegyhazi/publications/Bencsath2011duqu.pdf
18. IBM internet security systems. http://www.iss.net/threats/438.html
19. Pragya Jainand and Anjali Sardana, "Defending against Internet Worms Using Honeyfarm," *Proceedings of the CUBE International Information Technology Conference (CUBE)*, September 3–5, Pune, India: ACM, pp. 795–800, 2012.
20. T. M. Chen and S. Abu-Nimeh, "Lessons from Stuxnet," *IEEE Computer Magazine*, 44(4): 91–93, 2011.
21. E. Chien, L. OMurchu, and N. Falliere, "W32.Duqu: The Precursor to the Next Stuxnet," *Proceedings of the 5th USENIX Conference on Large-Scale Exploits and Emergent Threats (LEET)*, San Jose, CA, April 2012.

19

Trust in Cyberspace: New Information Security Paradigm

Roberto Uzal, Narayan C. Debnath, Daniel Riesco,
and German Montejano

CONTENTS

Introduction

Insecurity is an intellectual perception, accepted by an individual or a society as a clear example, model, or pattern of how things work in the cyberspace. Not all software and information operate the way people expect. Several recent examples have proven that cyber attacks may cause physical damage to premises, showing the need to reinforce security in data transmission channels. Therefore, a new paradigm for managing information security arises from the intention of obtaining people's trust in cyberspace. Definition of "paradigm" was used first by the US science fiction writer Thomas Kuhn [1] to refer to theoretical frameworks within which all scientific thinking and practices operate [2].

It is known that, in several nation-states, cyber war units from other "friendly nation-states" are placing a trapdoor in civilian networks, planting logic bombs in electric power grids, and in infrastructure for destruction [3]. It is evident that we are facing new and very important changes in the traditional information security paradigm. Paradigm shift means a fundamental change in an individual's or a society's view of how things work in the cyberspace, for example, the shift from the geocentric to the heliocentric paradigm, from "humors" to microbes as causes of disease and from heart to brain as the center of thinking and feeling [4]. Criminal hackers could detect some of those placed "military logic bombs" and use them for criminal purposes. This is not a theory. It is just a component of the new scenarios in current and actual information security.

This presentation is about the differences between traditional and new information security paradigms; the conceptual difference between "known computer viruses" and sophisticated cyber weapons; the existence of a cyber weapons "black market"; the differences between cyber war, cyber terrorism, and cyber crime; the new information security paradigm characteristics; and the author's conclusion about the new information security paradigm to be faced.

In this chapter, authors point to top cyber war and IT expert's opinions and trustworthy sources in order to obtain conclusions using the synergic effect of those expert's opinions.

"Traditional" Information Security Paradigm

The Internet's crucial role in modern life, commerce, and government underscores the need to study the security of the protocols and infrastructure that comprise it. For years, we've focused on endpoint security and ignored infrastructure weaknesses. Recent discoveries and initiatives highlight a simple fact: "the core is just as vulnerable as the edge" [5]. Facing those infrastructure weaknesses and overcoming them is a key issue in the context of the new information security paradigm.

Traditionally, it was generally admitted that information security concepts are important in creating security policies, procedures, and IT business decisions. This work examines the effective applicability of traditional information security just understood as confidentiality plus integrity plus availability plus authenticity. As an example, this chapter's contents show the importance of legal, political, and human resources management issues in the new information security paradigm.

Nowadays, the threats to information security have grown dramatically in scope and complexity. Sophisticated cyber weapons and cyber war strategies are present, and their actual impacts in corporations and other organizational environment have changed the "game rules" in the field of information security. This chapter shows the true state of government, corporations, and organizations in general terms of security environment, challenges, and readiness. We also present an outline of a needed new information security paradigm.

In advance, we can mention that, mainly for defense officials, cyberspace has become a new operational domain [6,7] as the traditional air–sea–land. Now, in actual terms, cyberspace includes virtually every use of a computer, including those built into weapons, vehicles, decision-support tools, and everyday life. Currently, most military missions, in some sense, occur in cyberspace. As examples, we can mention combat, including counterinsurgency; military support to national priorities (counterdrug, border security, etc.); disaster relief and humanitarian assistance; and conflicts between nation-states in the cyberspace [6–10].

From Computer Virus to Cyber Weapons

"A computer virus described as a cyber weapon and the most complex ever created, has been discovered in thousands of computers in the Middle East" [8]. It is known that this new kind of computer virus, called Flame, discovered by security experts at Kaspersky Labs, marks a new era in cyber warfare [10].

Flame (approximate size of 20 MB) is the third major cyber weapon detected/identified after Stuxnet (approximate size 2 MB), the worm that attacked Iran's uranium enrichment plant at Natanz in 2010, and Duqu, a data-stealing military-oriented malware.

Flame computing algorithm's complexity is over a hundred times the "usual" complexity of "standard/traditional" PC viruses. Flame is designed primarily as a spy, whose basic functions include stealing information, for example, the contact list of mobile phones close to infected computers, turning on microphones and Webcams also on infected PCs to listen to conversations, and to recognize persons. Flame acts also as a sort of "software bus" allocating different cyber weapons/computer viruses in its "software slot," according to the nature of different missions it must face. The complexity of Flame shows that, in its creation, the resources of one or several nation-states have been required. Flame is the most complex piece of malicious software discovered to date, said Kaspersky Lab senior researcher Roel Schouwenberg [11,12], whose company discovered the virus acting according to UN orientation. It is suspected that Flame was "working" undiscovered for 5 years. The authors of this chapter agree with Schouwenberg: "The only logical conclusion is that there are other operations ongoing that we don't know about."

Information security must now be mainly focused on new weapons and types of attacks we can describe using, as a reference, Stuxnet cyber weapon. Many security experts, including US officials, have said that it was likely that Stuxnet was made by the United States with Israel's assistance [9]. The Stuxnet worm scheme consists of the following modules: structural module, payload module, propulsion module, guide module, and communication module. This scheme looks like a missile conceptual architecture, as it is analogous to a missile conceptual design. Stuxnet could not be compared with "traditional" computer viruses or known worms and trojans. Stuxnet is also a sophisticated example of programming multiparadigms: imperative—algorithmic paradigm, object-oriented paradigm, and functional paradigm. The Stuxnet-like cyber weapons development era implies an important effort and knowledge level in information security.

Stuxnet is a reference of the new information security paradigm needed. Stuxnet is not the work of a small group of hackers because when its structure and algorithmic complexity is analyzed, it evidences several years of a high-level skilled programming team and very high development costs.

Cyber Weapons: From Cyber War to Cyber Terrorism and Cyber Crime

1. According to *The Guardian* [11,12], a 3-week wave of massive cyber attacks on the small Baltic country of Estonia, the first-known incidence of such an assault on a state, is causing alarm across the Western alliance, with NATO urgently examining the offensive and its implications. While Russia and Estonia are embroiled in their worst dispute since the collapse of the Soviet Union, a row that erupted at the end of last month over the Estonians' removal of the Bronze Soldier Soviet war memorial in central Tallinn, the country has been subjected to a barrage of cyber warfare, disabling the Web sites of government ministries, political parties, newspapers, banks, and companies. NATO has dispatched some of its top experts to Tallinn to investigate and to help the Estonians beef up their cyber defenses. "This is an operational security issue, something we're taking very seriously," said an official at NATO headquarters in Brussels.

2. Richard Clarke reported [6] that "Syria had spent billions of dollars on air defense systems. That 2007 September night, Syrian military

personnel were closely watching their radars. The skies over Syria seemed safe and largely empty as midnight rolled around. In fact, however, formations of Israel Eagles and Falcons had penetrated Syrian airspace through Turkey. Those aircraft, designed and first built in the 1970s were far from stealthy. Their steel and titanium airframes, their sharp edges and corners, and the bombs and missiles hanging on their wings should have lit up the Syrian radars like the Christmas tree illuminating New York's Rockefeller Plaza in December. But, they did not. What the Syrians slowly, reluctantly, and painfully concluded the next morning was that Israel had 'owned,' using a cyber weapon, Syrian air defense network the night before. The Syrian ground based controllers had seen no targets. This is how war would be fought in the information age, this was cyber war."

3. *USA Today* reported on July 9, 2009 [13] that, "U.S. authorities say they are eyeing North Korea as the origin of the cyber attack that overwhelmed government Web sites in the United States and South Korea. Targets of the most widespread cyber offensive of recent years also included the National Security Agency, Homeland Security Department and State Department, the NASDAQ stock market and *The Washington Post*, according to an early analysis of the malicious software used in the attacks."

4. Tony Capaccio and Jeff Bliss [14] reported that, "Computer hackers, possibly from the Chinese military, interfered with two U.S. government satellites four times in 2007 and 2008 through a ground station in Norway, according to a congressional commission. The intrusions on the satellites, used for earth climate and terrain observation, underscore the potential danger posed by hackers."

5. David E. Sanger and Eric Schmitt [9] reported that, "The top American military official responsible for defending the United States against cyber attacks said Thursday that there had been an important increment in computer attacks on American infrastructure between 2009 and 2011, initiated by criminal gangs, hackers and other nations."

6. *The Wall Street Journal* [15] reported in April 8, 2009 that Russian and Chinese spies had penetrated the US electric grid. Lawmakers are pushing at least three different proposals to boost cyber security in the electric sector, including measures that would give the federal government authority to issue regulations to combat imminent threats.

7. President Obama on May 29, 2009, admitted, "Cyber intruders have probed our electrical grids" (remarks by the President on securing our Nation's Cyber Infrastructure) [16].

8. Iftikhar Alam [17] from *The Nation* (Pakistan) on December 5, 2010, reported that the Friday night's cyber attack on the Web site of Central Bureau of Investigation (CBI)—the top civilian investigation agency of India—in response to the attack on 40 Pakistani Web sites has intensified the cyber war between India and Pakistan, which had started in 1998. According to the IT experts, there are hundreds of highly professional hackers operating in both the countries.

Could the cyber weapons, used in the aforementioned examples, be transferred from military environment to criminal hands? This question and its answer are closely related to the information security paradigm change introduced in this presentation.

A Cyber Weapons "Black Market"?

1. In his book *Confront and Conceal: Obama's Secret Wars and Surprising Use of American Power*, David E. Sanger [18] claims that the current US President Barack Obama has been involved in a secret cyber war with Iran, for his entire presidential term. Sanger reports that Obama has ramped up a cyber weapons operation originally started in the Bush regime, codenamed "Olympic Games." The "Olympic Games" cyber attacks were aimed against Iran's nuclear enrichment facility. The specific cyber weapon aimed at Iran's nuclear enrichment facility has been dubbed Stuxnet. Stuxnet is a virus used to sabotage the centrifuges for creating weapons-grade enriched uranium. It was used against Iran in 2010. The reports about this new Stuxnet virus came out just a few days after the United Nations warned the Middle East about a virus, called the Flame virus, now on the loose. Sanger's sources say that these cyber weapons are being created and used jointly by the United States and Israel. However, as is the nature of covert warfare, these allegations have been completely denied by the US and Israeli officials. The Obama regime has recently admitted to using cyber weapons, but only against Al-Qaeda terrorists. They have not admitted to using cyber weapons against the Iran government yet. Iran has found computers in their facilities that contain the Stuxnet and Flame viruses. The Flame virus is now on the loose and could be in anyone's hands. These cyber weapons could easily be sold on the black market to cyber terrorists. Now that the US/Israel's enemies have these cyber weapons in their hands, it is only a matter of time

before the coding of these viruses is reverse-engineered and used against us, as Sanger reported.

2. Sam Kiley [19] is the defense and security editor of *Sky News,* a 24-hour television news service operated by Sky Television, part of British Sky Broadcasting. He is an award-winning journalist with more than 20 years' experience, based at different times of his career in London, Los Angeles, Nairobi, Johannesburg, and Jerusalem. According to Sam Kiley, Stuxnet, the cyber weapon that was used to disrupt Iran's nuclear program, has been traded on the black market and could be used by terrorists.

Cyber War, Cyber Terrorism, and Cyber Crime Consequence: The Information Security Paradigm Change

The US government security expert, Richard A. Clarke, in his book *Cyber War* [6], defines "cyber warfare" as "actions by a nation-state to penetrate another nation's computers or networks for the purposes of causing damage or disruption."

In this sense, *The Economist* describes cyberspace as "the fifth domain of warfare," [6] and William J. Lynn [20], US Deputy Secretary of Defense, states that "as a doctrinal matter, the Pentagon has formally recognized cyberspace as a new domain in Warfare ... [which] has become just as critical to military operations as land, sea, air, and space."

In parallel, the Federal Bureau of Investigation (FBI) [21,22] defines terrorism as the unlawful use of force or violence against persons or property to intimidate or coerce a government, the civilian population, or any segment thereof, in furtherance of political or social objectives. Cyber terrorism could thus be defined as the use of computing resources to intimidate or coerce others. An example of cyber terrorism could be hacking into a hospital computer system and changing someone's medicine prescription to a lethal dosage as an act of revenge. It sounds theatrical, but these things can and do happen. "Cyber terrorism is a component of Information Warfare, but Information Warfare is not cyber terrorism."

Also the FBI recognizes four instances of cyber crime [21,22]: (1) cyber crimes against children (usually involving child pornography or child rape), (2) theft of intellectual property, (3) publication and intentional dissemination of malware, and (4) national and international Internet fraud.

Now we know that it is possible that sophisticated cyber weapons used in conflicts between nation-states can be traded in the black market and could be used by terrorists or criminals. It is not a theoretical speculation, but it is a real-world situation. Recently, the authors of this research met the engineer in charge of an effluent plant belonging to an important industrial complex in South America, who reported that a "Stuxnet-like" malware took the effluent plant control and changed the operation parameters causing polluting materials to be dumped, even when the effluent plant monitoring panel displayed that everything was in normal operation conditions.

New Information Security Paradigm Environment

Taking into account the new scenario description, information security needs an important paradigm shift in order to successfully protect information assets. Organizations must effectively change from an information-system-focused information security management to an organizational-focused information security management, requiring a well-established information security management system (ISMS) [23]. The current acceptable concept of ISMS must be improved and optimized regardless of the fact that this improvement and optimization requires higher risks, investment, knowledge, and skills. This improved and optimized ISMS must address all aspects in an organization that deals with creating and maintaining a secure information environment using a multidisciplinary and interdisciplinary approach. It also requires an intelligent mixture of aspects such as clear policies, a very intelligent use of standards, effective guidelines, technology support, legal support, political support, and a very specialized human resources management.

In this context, the two Koreas provide a very interesting paradigm contrast to be studied. South Korea has high-speed Internet access reaching 95% of its citizenry. This is the highest rate of access to the Internet in any nation today. With this national emphasis on connectivity, South Koreans typically store their medical, banking, and online shopping records digitally. This makes these networks more vulnerable to attacks as there are personal assets associated with these networks.

By contrast, North Korea has very little Internet connectivity, and is therefore not as vulnerable to external online attacks. Who would attack North Korea's Internet? By strongly restricting who has access to the Internet, North Korea can focus its limited resources on a few universities that may be the launching point for the recent cyber attacks, currently focused on their neighbor and rival South Korea, but which someday could be used on countries in the West. Generically, these are called asymmetric threats.

Obtaining *people's trust in cyberspace* is a new challenge which implies working on the improvement and reinforcement of information and communication security.

The mentioned new challenge was well focused by Christopher Painter, coordinator of cyber issues—US Department of State, during his lecture in the context of "The Third Worldwide Cyber Security Summit, New Delhi—2012" [24]. Mr Painter pointed thus: First, cyberspace is not a lawless space but a space where laws do apply and where there are constrains on state behavior. Second, there are practical things we can do to build better confidence, better transparency, better cooperation, and, ultimately, better stability. The point here is to avoid conflict and make sure that conflict does not benefit any state.

Our general idea about this matter is that people's trust in cyberspace must be founded on nation-state's trust in cyberspace. In this sense, it is important to take into account the US Secretary of State Hillary Clinton's remarks in the framework of the last Budapest Convention (October 2012): "That's why we are building an environment in which norms of responsible behavior guide states' actions, sustain partnerships, and support the rule of law in cyberspace" [25].

People's trust in cyberspace is not only a technological knowledge area. It has very important political, legal, social, and also psychological sides.

Why is this important? Cyberspace touches nearly every part of our daily lives. It is the broadband networks beneath us and the wireless signals around us, the local networks in the schools and hospitals and businesses, and the massive grids that power a nation. It is the classified military and intelligence networks that keep us safe, and the World Wide Web that has made us more interconnected than at any time in human history. We must secure our cyberspace to ensure that we can continue to grow the nation's economy and protect the way of life.

What must we do? A nation's cybersecurity strategy is twofold: (1) improve the resilience to cyber incidents and (2) reduce the cyber threat.

Improving the cyber resilience includes hardening our digital infrastructure to be more resistant to penetration and disruption; improving the ability to defend against sophisticated and agile cyber threats; and recovering quickly from cyber incidents—whether caused by malicious activity, accident, or natural disaster.

Where possible, we must also reduce cyber threats. We seek to reduce threats by working with allies on international norms of acceptable behavior in cyberspace, strengthening law enforcement capabilities against cybercrime, and deterring potential adversaries from taking advantage of our remaining vulnerabilities.

The last paragraphs are a synthesis of President Obama's way of thinking about the concept we are developing [16].

Nonfunctional Specification Changes

In general terms for both academia and practitioners [26], information systems nonfunctional specification follows the general scheme as shown in the following picture:

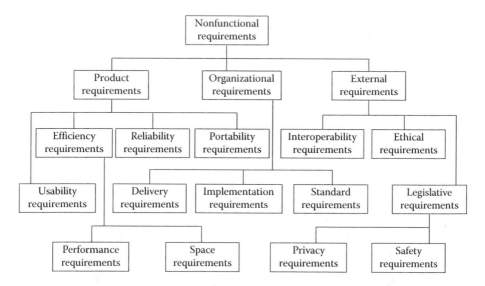

Taking into account this picture, it is important to remark that reliability engineering [27] is a field that deals with the study, evaluation, and life-cycle management of reliability: the ability of a system or component to perform its required functions under the stated conditions for a specified period of time. Reliability engineering is a subdiscipline within systems engineering. Reliability is often measured as probability of failure, frequency of failures, or in terms of availability, a probability derived from reliability and maintainability. Maintainability and maintenance are often important parts of reliability engineering.

Reliability engineering is closely related to safety engineering, in that they use common methods for their analysis and may require input from each other. Reliability engineering focuses on costs of failure caused by system downtime, cost of spares, repair equipment, personnel, and cost of warranty claims. The focus of safety engineering is normally not on cost, but on preserving life and nature, and therefore deals only with particular dangerous system failure modes.

Reliability engineering for complex systems requires a different, more elaborate systems approach than reliability for noncomplex software tools. Reliability analysis has important links with function analysis, requirements specification, systems design, hardware design, software design, manufacturing, testing, maintenance, transport, storage, spare parts, operations research, human factors, technical documentation, training, and more. Effective reliability engineering requires experience, broad engineering skills, and knowledge in many different fields of engineering.

Considering the scenario diagramed in this chapter, information systems reliability engineering includes the study of weaknesses and vulnerabilities because the preliminary stages of information system nonfunctional specification.

Additionally, subjects such as access control, application security, business continuity, and disaster recovery planning; cryptography; information security and risk management; legal, regulations, compliance, and investigations; operations security; physical (environmental) security; security architecture and design; and telecommunications and network security must be considered by project managers since the project feasibility analysis. Information security policies, standards, and procedures are key issues to face in a successful information system project.

Reliability engineering includes important aspects to be applied in a generic information system project. Reliability engineering is not linked to specific products.

The "classical" nonfunctional specification hierarchical diagram should be modified. The information system reliability specification should be allocated in the first priority level of nonfunctional specification. The following picture shows this point of view:

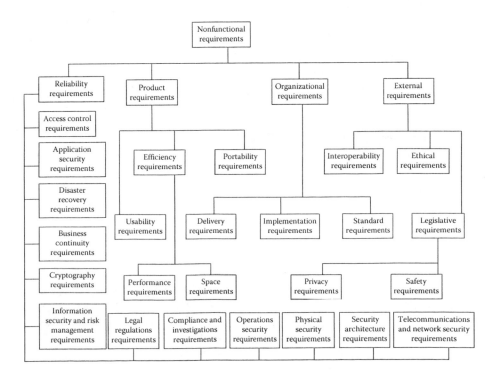

As we can see, reliability requirements are now in the priority first level. Reliability requirements are not depending on a specific software product. Additionally, reliability requirements must be considered taking into account a very detailed scheme.

New information security paradigm requires important changes in the field of software products nonfunctional specification.

Conclusions

1. The recently discovered cyber weapons can be easily described as one of the most complex IT threats ever discovered. They are big and incredibly sophisticated. They pretty much redefine the notion of information security.

2. Considering the existence of a sort of cyber weapon black market, very sophisticated malware in terrorist and criminal hands changes the information security scenario and the information security paradigm.

3. When a developed country government decided, for example, to bomb the nuclear installations belonging to a developing country or a terrorist's training area, the developing country may not be able to respond using conventional military forces and/or conventional weapons. It may respond by developing cyber war capabilities, destroying, for example, an important part of the international financial system in which the country has very little share. There are evidences that this kind of cyber weapons can easily be transferred from government area to criminal and terrorist groups.

4. The use of "Flame-like" and/or "Stuxnet-like" worms/trojans, in the context of dirty competition between huge corporations, could possibly start (if not already has).

5. Cyber warfare is seen as a technology problem by technologists, a policy problem by politicians, and a profit problem by businessmen. This confluence of concerns is likely due to the prevailing nature of technology in our daily lives. Additionally, the confluence of concerns implies a multidisciplinary approach we need in the context of a new information security paradigm.

6. The most effective protection against "cyber warfare-like" attacks is securing information and networks. Important investment in security should be applied to all systems, including those "not critical," because any vulnerable system can be coopted and used to carry out attacks. Measures to mitigate the potential damage of a "cyber war-like" attack

include comprehensive disaster recovery planning that includes provisions for extended devastations.

7. At the level of nation-states, cyber defense units and at its related agencies, most intelligent and highly skilled human resources must develop plans and capabilities to achieve "dominance in cyberspace" to maintain the nation-state in the condition of a safe environment for government institutions, corporations, and all kind of organizations.

8. In the short-term, the defense agenda must include the implantation of a sort of "defensive triad": (a) capability of stopping sophisticated malware on the Internet at the backbone of Internet service provider level; (b) prioritizing and strengthening the controls of the electric grid; and (c) increasing the security of the defense area networks including the IT support for the top-level decision-making process.

9. It must be established that, information system specifications must have a new balance between functional and nonfunctional specifications. Currently, and in actual terms, functional specifications have priority over the nonfunctional ones. Project budget and other resources must be allocated to obtain goals related to concepts such as confidentiality, integrity, availability, and authenticity in the context of nonfunctional specifications. In the new information security paradigm, security must be the key issue in an information system conceptual design, development, implantation, use, and maintenance.

References

1. Kuhn, T. S., *The Structure of Scientific Revolution*, Chicago, IL: University of Chicago Press, 1962.
2. Business Dictionary, http://www.businessdictionary.com/definition/paradigm.html
3. Clarke, R. A. and Knake, R. K., *Cyber War: The Next Threat to National Security and What to Do About It*, New York: HarperCollins, 2012.
4. Business Dictionary, http://www.businessdictionary.com/definition/paradigm-shift.html
5. Nazario, J. and Kristoff, J., "Internet Infrastructure Security," *IEEE Computer Society*, 10(4):24–25, 2012.
6. *The Economist*, "Cyberwar: War in the Fifth Domain," July 1, 2010, http://www.economist.com/node/16478792
7. Lynn, W. J. III, "Defending a New Domain: The Pentagon's Cyber Strategy," *Foreign Affairs*, September/October 2010, pp. 97–108.
8. Waugh, R., *Daily Mail*, May 28, 2012, http://www.dailymail.co.uk/sciencetech/article-2151199/A-new-era-cyber-warfare-Virus-weapon-lurked-inside-thousands-computers-Middle-East-years.html

9. Sanger, D. E. and Schmitt, E., "Rise is seen in cyberattacks targeting US infrastructure," *New York Times*, July 26, 2012, http://www.nytimes.com/2012/07/27/us/cyberattacks-are-up-national-security-chief-says.html?_r=0

10. McElroy, D. and Williams, C., "Flame: world's most complex computer virus exposed," *The Telegraph*, May 28, 2012, http://www.telegraph.co.uk/news/worldnews/middleeast/iran/9295938/Flame-worlds-most-complex-computer-virus-exposed.html

11. Traynor, I., "Russia accused of unleashing cyberwar to disable Estonia," *The Guardian*, May 17, 2007, http://www.theguardian.com/world/2007/may/17/topstories3.russia

12. Spiegel on line international, "Old wars and new: Estonians accuse kremlin of cyberwarfare," May 17, 2007, http://www.spiegel.de/international/world/old-wars-and-new-estonians-accuse-kremlin-of-cyberwarfare-a-483394.html

13. *USA Today*, "US officials eye N. Korea in cyber attack," July 9, 2009, http://usatoday30.usatoday.com/news/washington/2009-07-08-hacking-washington-nkorea_N.htm

14. Capaccio, T. and Bliss, J., *Business Week*, October 27, 2011, http://www.dailymail.co.uk/sciencetech/article-2151199/A-new-era-cyber-warfare-Virus-weapon-lurked-inside-thousands-computers-Middle-East-years.html

15. Gorman, S., "Electricity industry to scan grid for spies," June 18, 2009, http://online.wsj.com/article/SB124528065956425189.html

16. President Barack Obama remarks, May 29, 2009, http://www.whitehouse.gov/issues/foreign-policy/cybersecurity

17. Alam, I., "Pakistan-India cyber war begins," December 5, 2010, http://www.nation.com.pk/pakistan-news-newspaper-daily-english-online/politics/05-Dec-2010/PakistanIndia-cyber-war-begins

18. Sanger, D. E., *Confront and Conceal*, New York: Crown Publishers, 2012.

19. Kiley, S., "Super virus a target for cyber terrorists," November 25, 2010, http://news.sky.com/story/820902/super-virus-a-target-for-cyber-terrorists

20. Parrish, K., "Lynn: Cyber Strategy's Thrust Is Defensive," *American Forces News Service*, July 14, 2011.

21. Tafoya, W. L., "Cyber terror," November 2011, http://www.fbi.gov/stats-services/publications/law-enforcement-bulletin/november-2011/cyber-terror

22. Federal Bureau of Investigations, "Cyber Investigations," http://www.fbi.gov/about-us/investigate/cyber/cyber

23. ISO STANDARD (International Standard Organization), "ISO 27001 & information security," ISO standards, 2013, http://www.itgovernance.co.uk/iso27001.aspx

24. Painter, C., "The Third Worldwide Cyber security summit," New Delhi, India, 2012, http://cis-india.org/news/india-conference-on-cyber-security-and-cyber-governance

25. Clinton, H. R., "Video remarks for the Budapest cyber conference," Budapest, October 5, 2012, http://www.state.gov/secretary/rm/2012/10/198672.htm

26. Sommerville, I., *Software Engineering*, Boston, MA: Pearson, 9th Edition, 2011.

27. Aven, T. and Krohn, B. S., "A new perspective on how to understand, assess and manage risk and the unforeseen," 121:1–10, 2013.

20

A Comprehensive Survey of Antiforensics for Network Security

Rahul Chandran and Wei Q. Yan

CONTENTS

Introduction

The increasing scale of cyber attacks and computer crime has led investigators to utilize the latest technology to discover new ways of investigative methodologies for forensic process. But on the other hand, the attackers and lawbreakers tend to invent new ways of attacks and ways to hide their source of attack and identity thereby hindering investigation. This mode of operation is called *antiforensics*. Current forensics deals with two types of evidence analysis such as live analysis and offline analysis. Live analysis mainly monitors and gathers evidence from live networks and systems [1,2]. Offline analysis deals with evidence processing after physical or logical imaging of the entire system.

Computer forensics can be defined as the investigative and analytical techniques to identify, collect, examine, and preserve electronic information and data that can be potentially used as evidence in a court. It always appears to have legal issues with the acceptance of evidence, and questions are raised on the integrity of evidence [3]. Frameworks, policies, and methodologies are implemented for better forensic investigation [4,5]. Computer forensics is classified into two main categories—traditional computer system forensics [6] that deals with the investigation regarding the hard disks, personal computers, USBs [7], and network forensics that deals with computer networks. A classical computer forensics process can be classified into four main phases such as collection of evidence, evidence processing, analysis, and reporting. All these phases are accompanied by the preservation of evidence and documentation. The main drawback of the forensic process is to discover whether the evidence has been modified prior to collection by the investigators [8]. Two ways of obstructing computer forensics is to destroy the evidence or to hide it [9]. One of the main problems that hinders digital investigation is that the investigators fail to evaluate whether the evidence they gathered is adequate to prove the events of the crime, detect any antiforensics attacks, and lighten these effects on the compromised evidence [10].

The major concerns in forensics today are the rapid advances of the technology toward wireless technology [11–13], peer-to-peer networks [14–16], and increasing influence of the social networks. Most of the tools and techniques of forensics are available as open source that helps the hackers locate vulnerability and there are vast number of tutorials on the Internet regarding hacking and penetration testing. Another field which threats forensic investigation is the antiforensic tools and techniques which make the forensic process even harder and taunt its reliability and integrity [17]. An example is the antiforensic techniques used for compression of JPEG images [18–20].

Antiforensics can be defined as [21] *The methods used to prevent (or act against) the application of science to those criminal and civil laws that are enforced by police agencies in a criminal justice system.* The main aim of antiforensics is to destroy the logical evidence gathered by the investigators so the evidence proves to be insufficient and incapable of confirming that the crime occurred. For example, once the attacker had succeeded in compromising a system, the first step carried out is to delete traces of the events that occurred. Antiforensics techniques are carried out to reduce the qualitative and quantitative substantial evidence [22] on the compromised systems [10,23,24]. As technology advances, new antiforensic tools and techniques are discovered and implemented.

Antiforensics deals with deletion of evidence from network logs and deletion of files from compromised system such as sudden death in mobile phones when a forensic procedure is detected and android antiforensics which deletes log files from android phones [25–27]. Another way is to avoid detection by spoofing, zombie attacks, misinformation, disabling logs, and traditional ways such as encryption and steganography. The main antiforensic tool is Metasploit antiforensic framework [28] which is an open source collaborative, investigating into the limitation of the computer forensics tools and helping them improve digital forensic process and evidence validation. Metasploit antiforensic investigation arsenal (MAFIA) [29–31] had provided a suite of programs such as SAM Juicer, Slacker, Transmogrify, and Timestomp [29] that reveal the ways by which forensic examinations can get confused. Detailed explanation on this will be given later. A new victim of antiforensics is network security field that is one of the crucial components of any network infrastructure.

Network security has always been an issue at present as the entire IT world revolves around it. The possibilities of intrusion and data theft are growing as companies move from their LAN network to the public domain and the Internet. Corporate security mainly deals with securing the key assets of the company in which the data, valuable information, and knowledge of how the information can be used [32]. In order to employ and maintain a robust security of the computer networks, the network security professionals spend hours in implementing the monitoring tools, intrusion detection, and prevention systems. The security measures should be one step ahead of the current attack strategies of the hackers. The leading forms of attacks by the hackers are by IP spoofing, port scanning, packet sniffing, denial of service (DoS) attacks. According to the Open Web Application Security Project (OWASP), top Web security risks are cross-site scripting, cross-site request forgery, injection [33], security misconfiguration, and broken authentication and session management [34]. The investigation of attacks in networks is what we call network forensics. It can be defined as techniques used to collect or capture, analyze and identify, and record network traffic. In network forensics, network logs and packets are retrieved using network security software, and is analyzed and investigated to trace back the attack methodology and even the source of the attack and attacker [35].

Network forensics can be divided into two main streams such as static analysis and dynamic analysis. The static analysis is the process of identifying the conduct of the attacks or crime without executing it. The phases include the analysis of the system file, log files, firewall logs, network logs, checking the presence of malware and virus, and reverse engineering. The dynamic analysis on the other hand deals with live network analysis, analysis of network traffic, network packet capture, file system monitoring for changes, and registry file analysis [36]. One of the main approaches of dynamic analysis is the use of what is called "Honeypots" [37,38]. Collection of honeypots, known as honeynets, which can be isolated from the rest of the network, can be used for network traffic analysis and prevent unwanted traffic on public networks [39,40]. The success of network forensics lies in identification of the source, approach, and techniques of the attack. This can be achieved by reverse engineering of the network attacks. One of the algorithms that help in tracing out the path of the attack is the network attack graph. These network attack graphs are used to analyze the path of attacks from the known vulnerabilities of the system. Detailed explanation on this will be given later.

This chapter mainly focuses on implementation of antiforensic techniques in computer networks and its analysis so as to contribute to enhancement of network security. The rest of this chapter is systematized as follows: A detailed review of the antiforensic technique is presented in the section "Fundamentals of Antiforensic Techniques." Network infrastructures are explained in the section "State of the Art." Network attacks, attack graphs, tools, and network forensic process are explained in the section "Approaches, Methodologies, and Techniques." The section "Evaluations and Comparisons of Tools and Techniques" covers various network security and monitoring tools (NMT), network forensic analysis tools (NFATs), and antiforensic tools. In the section "Trends," discussion of the research problem is introduced. The last section concludes with the trends and future work. The antiforensic tools and techniques will be studied in detail and this chapter intends to explore the new possibility of constructive deployment of antiforensics for improvement of network security.

Fundamentals of Antiforensic Techniques

Digital evidence can be easily altered, removed, hidden, and prevented from re-creating the source without any trace. To cope with these, investigators must be familiar with the antiforensic techniques. The main antiforensic goals are (1) avoiding detection of the attack, (2) disruption and prevention of collection of valid evidence, (3) increasing the time duration for collection and analysis of evidence, (4) subverting forensic tools from gathering the right evidence, and (5) leaving no trace of the antiforensic tool or technique deployed [41]. In order to achieve these goals, various tools and techniques are introduced [8,10,21,29,42,43]. The subsections below detail the techniques used for antiforensics.

Data Destruction

The basic antiforensic technique is data destruction which causes the investigation to stop. It can also be called as secure deletion. It can be either logical or physical destruction of the data. Logical destruction is accomplished through frequent overwriting. Physical destruction can be carried out with the help of magnetic tapes by degaussing the media [8]. Data destruction is used to remove the residue of the deleted files, erase the logs, timestamps, and registries of the system activities, thereby securing the identity of the crime. CCleaner [44] is a software that supports the permanent deletion and removal of all temporary files and unnecessary files from the system. Necrofile [2] rewrites the selected partition or portion of the hard disk with mock data destroying the evidence completely. Active Eraser is another data destruction tool which is used for secure erasure of data. In networks, data packet destruction (DPD) using NS2 (network simulator) and random packet destruction (RPD) using DDoS attacks are the main data destruction techniques [45].

Data Hiding

Data hiding is one of the most traditional and successful antiforensic techniques. Inserting the data in different places where it should not be or in metadata files is called data hiding. The data or the information can be stored in slack spaces, scattered all over the memory and empty spaces in the disk sectors [46,47]. But it largely relies on the forensic tools used and the lack of the ability of the investigator to find hidden data which is outside the normal visibility as the main factor. Renaming the file is one example of incriminating the data by hiding. Encryption, watermarking, covert channels [10], and steganography are the main techniques used to obscure network traffic and data. It should be ensured that the data or information should not be lost while encrypting and using covert channels. This technique can be exploited for both constructive and destructive purpose.

The data are hidden in portions of the medium which are outside the specific format of that medium such as slack space at the end of the partition and fake bad sectors. Data hidden in these areas are hard to discover as that needs special tools. But it is very difficult to hide from normal analysis. Another way is to hide the data inside the specific format of the medium and the data should not be of any format other than the medium. It mainly relies on security through obscurity as it is easy to discover once the method is known. Virus hiding within the exe code section and steganography—hidden texts in documents—are other forms of data hiding [48]. It is very hard to detect without specific tools and have complex algorithms especially in steganography. Hiding information in empty headers of network layer and transport layer packets is hard to be traced [33].

Slack Space

The areas in the hard disk that have limited access are considered to be the slack space. The unused space of the sector in a RAM that cannot be addressed by an OS is known as RAM slack space [46]. Because the files in a hard disk do not end within the last sector of the block which leads to slack space in the consecutive sector, a volume of slack space can be defined as the unused space between the end of the file system and the end of the partition where the file resides. Faked bad clusters can also be used for data hiding [49]. The NTFS file system identifies bad clusters ($BadClus) that have defects in it using the master file table (MAT). Once some clusters are marked as bad clusters, they can be used to hide data of unlimited size [50]. The tools such as bmap and Slacker from Metasploit can be used for data hiding in slack spaces.

Encryption

The evidence files can be found by search methods. The detected evidence cannot be accessed if it is encrypted [10]. Thus, encryption becomes another kind of antiforensic technique. Antiforensic tool using encryption methodology has been developed [51]. There are lot of encryption algorithms such as XOR, Blowfish, AES, and RSA. Strong and good encryption algorithms are easy to be misused and make the forensic analysis hard due to the key management. The encryption can be used in network communications which make the network analysis harder. For example, encrypted packets are difficult to analyze by network forensic tools.

Steganography

Steganography can be defined as hiding information in messages, images, and files. The art of steganography dates centuries back in which messages are sent hidden in pictures [52]. There are several methods and algorithms for hiding data in various files. In steganography, only the sender and the receiver are aware of the information hidden in the files [3]. The detection of steganographic files is a challenge for the investigators.

Steganography in network communications makes use of covert channels to hide secret data into the user's normal data transmission that cannot be seen by third parties. Steganography not only provides security but also anonymity and privacy. As Internet has provided covert channel communications, network steganography is currently rising and is a threat to network security. Network steganography utilizes communication protocol control elements which make it harder to detect and destroy. It can use more protocols in the OSI layer such as HTTP Header manipulation in application layer and the LSB of voice sample modification for VoIP.

Network steganography can also be classified according to the modification of the protocol data unit such as modification of service data units (SDUs), modification of protocol control information (PCI), and time relation between PDUs.

HIdden Communication system for CorrUPted networkS (HICCUPS) [53] is another steganography technique for wireless communications especially for voice data.

Data/Trail Obfuscation

The main function of this technique is to divert digital forensic process. It can be successfully achieved by modification of metadata, anonymity techniques such as IP spoofing, MAC spoofing, VPNs, proxies, and covering traces of evidence. Digital forensic investigators can be misled by the attacker through false e-mail header generation, log alteration, and SMTP proxies [10]. Timestamp alteration and modification of headers is another form of trail obfuscation. The major techniques that implement the data/trail obfuscation are as follows [29]:

1. Log Cleaners
2. Spoofing
3. Misinformation
4. Zombie accounts
5. Trojan commands

Traffic content obfuscation is successfully implemented using virtual private networks (VPNs) and SSH tunneling [53].

Attack against Forensic Tools

The attacker introduces modification to the target machine of the investigator so that they provide wrong evidence. This includes rootkit attacks, file signature alteration, and exploitation of the vulnerabilities in the hash algorithms to create hash collisions [10,54]. The time and cost of analysis and digital investigation are the key features for an organization. If the attacker is able to control these constraints, then the investigators will be forced to stop the forensic procedures. The use of an intermediate system by the attacker makes the investigation difficult as it requires cooperation of different system administrators is an example of this antiforensic attack technique. Development of disk-avoiding tools prevents the forensic tools from detecting the attacker's activities by direct access to the memory [55].

State of the Art

Network infrastructures in any organization demand 100% security so that their assets are secure from threats. Thus, network security becomes a crucial component of the corporate environment. Today's technology provides a wide variety of security features such as intrusion detection system and intrusion prevention system (IDS/IPS), firewalls, antivirus guards, honeypots [56], and computer forensic tools [55,12]. Even if these tools impart a sufficient defensive mechanism, attackers are able to penetrate the networks. It has become difficult to investigate network attacks as the attackers utilize recently developed robust antiforensic tools and techniques to hide their identity and attack paths [57]. IP spoofing, trace obstruction, covert channels [2,10], tunneling, antihoneypot technology [37], and network steganography are some of the techniques used by the attackers for the defense strategy.

In the past 3 to 5 years, antiforensic techniques had been utilized by the attackers for data destruction, data hiding, and data obfuscation in traditional computer systems and storage devices. Advanced technology has helped them to extend the application of antiforensic techniques to computer networks and network infrastructure [58]. This makes the investigative process which includes evidence collection, evidence processing, and analysis very challenging.

In order to prevent various threats and attacks, various network security and monitoring tools can be implemented on different nodes of a network. Similarly, network forensic tools support in investigation and analysis of attacks and help to discover the origin of the attack, analysis of the evidence, and present evidence report. The various network forensic frameworks suggested by the Digital Forensics Research Workshop (DFRWS) and other researchers such as framework for distributed forensic, soft computing-based frameworks [59], honeypot-based framework [60], and attack graphs provide ample proof of research in this area [61–66].

The key objective is an inclusive survey of the tools and techniques utilized for antiforensics, network forensics, and network monitoring and security tools. This survey will help to study the wide range of tools used for forensics in computer networks and antiforensics. Understanding the techniques and algorithms used by the attackers assists in better and proper network security framework [67]. For successful implementation of a robust defensive infrastructure, this can be an effective measure.

The survey is the basis for practical experimentation of tools and techniques. The most common antiforensic framework used is Metasploit framework tools such as Timestomp and Slacker [29]. Antiforensic techniques such as data hiding, encryption, destruction, obfuscation, and data wiping can be tested in networks. The main platforms used for the implementation of the above techniques are Windows 7 and Linux Back Track 5 R2. Most of the tools can be run on multiple platforms even in the latest Windows 8. The analysis of the techniques is carried out with the aid of forensic tools such as Encase, Access-Data Forensic Tool Kit (FTK), and Internet Evidence Finder. Identification of antiforensic techniques and their effect in network evidence is the key part of the experimental analysis, thereby deducing effective counter measures to improve network security.

This chapter discusses the main methodologies, algorithms, and techniques used by digital forensic investigators. Taking into account the antiforensic techniques categorized in Table 20.1, various counter approaches for identifying the main source of the attack are also discussed.

TABLE 20.1

Antiforensics Tools and Techniques

Technique	Tools	Functions	Features
Data Destruction			
Physical	Magnetic fields	Degaussing the medium such as hard disks and other storage devices.	• Wipes entire drive • Cleans and restores • Supports SATA, USB, and SCSI
Logical	Drive Scrubber 3	Permanently and securely deletes data form drives. It also wipes free spaces.	• Securely overwrites and destroys all data on physical drive or logical partition • Supports IDE/ATA/SCSI hard disk, HDD/Floppies/Zip/FlashMedia drives disk eraser software • Supports large (more than 128 GB) size drives
	Active Eraser/ Active Kill Disk v6.0	Destroys all the data securely. It erases partitions, unused spaces, and logical drives. Supports all formats such as FAT and NTFS. Erases Internet activities (temporary internet files, cookies, history, etc.). Wipes out drive's free space out of previously deleted data.	• Data verification could be performed after erasing is completed • Scans drives and previews files on FAT, FAT32, and NTFS before erasing • Can be placed and run from USB Disk

(Continued)

TABLE 20.1
Continued

Technique	Tools	Functions	Features
	Disk Wipe2.3.1	Secure file wiping application which uses quick format before disk wiping for better performance and replaces the 0's and 1's with all zeros or ones new data.	• S-ATA (SATA), IDE, SCSI, USB, and FIREWIRE interfaces are supported
Data Hiding Slack space	Slacker FragFS Rootkit	All these applications hide the data in slack space, bad cluster of NTFS file system, and rootkit.	
Encryption	TrueCrypt	Tools are used to encrypt the drives for protection and inaccessibility. Algorithms such as AES, RSA, and Blowfish are mainly used. Encrypts an entire partition or storage device such as USB flash drive or hard drive. Encrypts a partition or drive where MS Windows is installed (pre-boot authentication).	• Creates a virtual encrypted disk within a file and mounts it as a real disk • Encryption is automatic, real-time (on-the-fly), and transparent • Parallelization and pipelining allow data to be read and written as fast as if the drive was not encrypted • Encryption can be hardware-accelerated on modern processors • Provides plausible deniability, in case an adversary forces you to reveal the password

Steganography	Steghide Stegdetect	Hides data files in images, audio, and video files, especially pictures. Stegdetect can detect hidden content in the file.

- Compression of embedded data
- Encryption of embedded data
- Embedding of a checksum to verify the integrity of the extracted data
- Support for JPEG, BMP, WAV, and AU files
- Only reports images that are likely to have steganographic content
- -h Only calculates the DCT histogram. Use the -d option to display the values
- -n Enables checking of JPEG header information to suppress false-positives. If enabled, all JPEG images that contain common fields will be treated as negatives. Out guess checking will be disabled if the JFIF marker does not match version 1.1

Network Steganography Tools	Stegtunnel	Using covert channels for communication using HTTP GET request and TCP connection.

- It can hide the data underneath real TCP connections, using real, unmodified clients and servers to provide the TCP conversation. In this way, detection of odd-looking sessions is avoided

	Hcovert	

- It provides covert channels in the sequence numbers and IPIDs of TCP connections. Latest version added a reliable file transfer mode using.
- Hamming-style error correction, and removes the requirement for a proxy IP address on some operating systems

(Continued)

TABLE 20.1
Continued

Technique	Tools	Functions	Features
	Socat	Socat is a command line tool which creates packets for IP6, IP4, TCP, and UDP protocols.	• It supports broadcasts and multicasts, abstract Unix sockets, Linux tun/tap, GNU readline, and PTYs • It provides forking, logging, and dumping, and different modes for interprocess communication • Many options are available for tuning socat and its channels. Socat can be used, for example, as a TCP relay (one-shot or daemon), as a daemon-based socksifier, as a shell interface to Unix sockets, as an IP6 relay, or for redirecting TCP-oriented programs to a serial line
	OpenPuff	OpenPuff supports many carrier formats such as images, audio, and video.	• Lets users hide data in more than a single carrier file. When hidden data are split among a set of carrier files you get a carrier chain, with no enforced hidden data theoretical size limit (256 MB, 512 MB, etc., depending only on the implementation) • Implements three layers of hidden data obfuscation (cryptography, whitening, and encoding) • Extends deniable cryptography into deniable steganography

Hide IP		
A4 Proxy	Anonymity 4 Proxy is mainly used for active hiding of IP address while surfing. Generates fake IP address, blocks cookies, and modifies HTTP variables. Also used for sharing Internet connection with other users over a LAN.	• Confuses the Web sites further by sending them a fake IP address along with your requests • Downloads files with programs like GetRight and other download managers staying anonymous to the sites from which you download • Learns more about the inside of the Internet and how it works • Thoroughly checks the anonymity status of proxy servers and their performance • Chooses to use only those proxies that meet particular anonymity requirements • A4Proxy supports HTTP (Web sites), secure HTTP (HTTPS, SSL secure Web sites), and FTP protocols • Uses a different anonymous proxy server for each request • Blocks cookies, and selectively modify any information sent out by your browser • Finds the anonymous proxy server which is the fastest for your location or the fastest for a particular URL. (FTP server or Web site) • Simulates ordinary requests, as if they are made not through a proxy but directly • Simulates non-anonymous requests from proxy servers with randomly selected IP addresses • Redirects and modifies • HTTP requests to anonymous proxy servers according to the rules defined by yourself • Uses stop-lists for sites and network clients • Associates each computer in your LAN with its own anonymous proxy server

(Continued)

TABLE 20.1

Continued

Technique	Tools	Functions	Features
Data/Trail Obfuscation			
IP spoofing, MAC Spoofing, SMTP Proxies, Log Cleaners, Others	Obfuscate payload	Obfuscate pay load to bypass the SNORT IDS	
	Back Track 5 R2—OS	The latest version of this Linux OS contains more than 150 antiforensic tools. This is one of the best OS to carry out the experiments with the tools and techniques.	
	SamJuicer	Acquires the hashes from the NT security access manager (SAM) files without changing the data on the hard disk.	
	Slacker	Hides files within the slack space of the NT file system (NTFS)	
	Timestomp	Alters all four NTFS file times: modified, access, creation, and file entry update.	
	Evidence Eliminator	In-depth wiping of data from storage devices. Deletes all the files including plug-in modules, slack space. It deletes and modifies the date and time of all files including the windows registry and log files.	• This software deletes files so effectively that they cannot be recovered by any of the current commercial or government recovery methods • Removes the traces of files and your internet history so that they cannot be recovered • Removal tools for both your online activity and offline

Approaches, Methodologies, and Techniques

Network attacks have always been a threat to Internet technology. Recent studies provide evidence that most of the antiforensic tools and techniques were applied in the normal attacks, in order to conceal the identity and source of the attacker. These techniques have been previously deployed against the traditional forensics [6]. As technology advances, new ways of attacks are discovered with the help of antiforensics techniques such as data hiding, obfuscation, and destruction.

Network attacks have always been a challenge for the security field and digital forensic investigators. The network attack process is divided into five stages in Howard taxonomy of computer and network attacks. They are the attackers, tools which are used by attackers, access using vulnerabilities and unauthorized users, results of the attacks, and the objectives [68]. Another approach mentioned is Lough's taxonomy called the validation exposure randomness deal-location improper condition taxonomy (VERDICT) which is based on the characteristics of the attacks [69]. A dimensional classification with sublevels of the different attacks gives a good overview of the attack paths and attack scenarios.

Most of the attacks use the vulnerabilities of the network infrastructure, system, or software. Common vulnerabilities and exposures (CVEs), vulnerability database (VDB) from security focus, open source vulnerability database (OSVDB) [70], and National Vulnerability Database (NVD) [71] are vulnerability repositories which provide a good range of vulnerability descriptions used for investigative purpose. The open source vulnerability and assessment language (OVAL) and common vulnerability scoring system (CVSS) [72] are two standardized frameworks for rating vulnerabilities in IT industries [70].

Traditional antiforensics deals with hiding data in the disk and slack space and destruction of data and data obfuscation through MACE alteration. Antiforensic techniques have been extended to network infrastructure such as hiding IP through proxy [71], encrypted packets, deleting logs, steganography, and covert tunneling. Due to integration of antiforensic techniques in network attacks, the attack path identified from forensic analysis will be different from the original and will be a strenuous effort to acquire the latter. The first step is to differentiate between an antiforensic attack and a normal attack. Normal attacks can be easily identified as there will not be any ambiguity in the process of analysis of evidence and attack paths. There are many methodologies [73–75] and approaches suggested in various studies to identify an antiforensic attack.

Network Attack Graphs

To investigate these kinds of attacks is challenging and new approaches such as attack graphs should be deployed. Attack graphs can be defined as an instrument to compute the hierarchical steps of an attack scenario with the help of known vulnerabilities and configuration. They are used by system administrators and investigators to analyze the type of attacks, different ways of attacks, and precautionary and preventive measures applied to counter those attacks [70]. IP trace back process is not a straightforward process because of IP spoofing and compromised intermediate host [12,76]. Automated forensic analysis of network attacks utilizing attack graphs focuses on better analysis of evidence to detect attacks [77]. Incorporation of antiforensic nodes onto the attack graphs [78] may provide sufficient information regarding the attacker's intention of reducing the generation of evidence and gives two possibilities of trace path, one with normal attack nodes and the other with antiforensic nodes.

Three different approaches are provided [79]. The first is the attack alert aggregation that utilizes leader follower similarity-based alert correlation [80], the second for building evidence graph [79] and the third for expansion of the attack graph to gather hidden members of the attack group. Automated analysis of evidence graph is employed using fuzzy cognitive map (FCM). Minimization of attack graphs using various algorithms provides identification of precise path of attacks [81].

Using the attack evidence graphs, investigators can determine the existence of antiforensic attacks and identify the tools and techniques used by the attacker. Thus, they can reconstruct the attack scenario with the minimum evidence they have.

Tools for Generating Attack Graphs

1. *Topological analysis of network attack vulnerability (TVA)*. It generates attack graphs using a graph search algorithm. It utilizes dependency graphs to create pre- and postconditions [71].

2. *Network security planning architecture (NETSPA)*. It is a framework for generation of network models using known vulnerabilities and firewall rules. This acts as a source for generation of attack graphs to identify the potential attacks and trace out the paths [69].

3. *Multihost, multistage vulnerability analysis (MULVAL)* [82]. It is a framework for integration of vulnerabilities and network configurations which uses Datalog as its language. It consists of a scanner and an analyzer. The reasoning engine which has datalog rules captures system behavior.

The integration of attack graph workflow with the IDS management using VDBs and attack graph generation tool is an effective forensic measure [83]. The attack intention analysis algorithm [84] provides a new method for network forensics that helps in identifying similar attacks for evidence analysis using alert correlation and distance-based similarity measure to identify the relationship strength between attack evidences. It can be suggested that integration of attack intention analysis [85,86] with the IDS may provide precise attack alerts and identify accurate attack paths.

Network Forensic Process

The network forensic process can be divided into three main phases.

Phase 1: Network Data/Traffic Capture via Network Monitoring

This can be articulated as a collection of evidence from the network for analysis. Evidence acquisition can be carried out either offline or online/live. There are a large number of tools and systems that can be used for monitoring and capturing the packets. TCPDump and WireShark are two of the most common tools used for monitoring.

The detection of network attacks is the base objective behind network monitoring. It is very challenging task in today's Internet technology. It has been very difficult to ensure that the attack is a true positive one, as lots of attacks are carried out in disguise with the help of antiforensic tools and techniques. Considerable amount of work has been seen in the network attack detection area. The recent work of autonomous network security for detection of network attacks is an attempt to implement the independent system that identifies intrusions automatically without statistical learning using clustering method for unsupervised anomaly detection. Most of the IDSs use data-mining algorithms, neural network, support vector machine (SVM), genetic algorithm, and fuzzy logic for behavioral and anomaly-based detection methodologies [87]. These algorithms help in detecting failed attacks and false positives [88,89].

In order to secure a network from outside attacks, it is necessary to understand the network traffic flow and content of the network packets. Content- and context-based monitoring [68] is another effective approach for network monitoring and detection of attacks which incorporates data-mining and database-auditing techniques [68]. The data-mining techniques utilized in IDS help in pattern analysis [90], sequence analysis, and identify attacks in an effective manner [91]. The output from various network monitoring tools is the network traffic packets such as .pcap extension files that can be analyzed using network forensic tools.

Phase 2: Network Forensics and Analysis

The evidence consists of network packets, firewall logs, IDS logs, system logs, router logs, and audit logs. The gathered information should be documented using techniques such as OpenSVN subversion [34]. Once the packets are captured, they can be analyzed using various network forensic tools such as WireShark, Encase, Network Miner, and Net Detector. The forensic tools also incorporate the IDS and IPS. Network intrusion detection system

(NIDS) such as SNORT uses pattern-matching algorithms and techniques [90] for network packet analysis and attack detection.

Time-based network traffic visualizer (TNV) is another tool used for analysis of network traffic over a time period. Filtering mechanisms and ID analysis help in identifying anomalous behavior [68]. The main evidences scrutinized in network forensics are authentication logs, operating system logs, application logs, and network device logs which constitute date and time stamps of IP address and error boots. As we mentioned earlier, with the help of antiforensic techniques the attacker may delete the evidence logs. The main network forensic tools are described in Table 20.2.

Phase 3: Developing the Attack Graphs

In order to investigate network attacks, to find the source of the attack and the attackers, one has to trace back the entire path of the attack. The attack path can be resolved using reverse engineering of the attack from the destination with the help of attack graph technique. Using vulnerability and system configuration as input, attack graphs can be created using various tools.

Due to the current scenario of high probability of application of antiforensic techniques, the normal construction of attack graphs with system configuration and vulnerability information cannot aid in gathering sufficient evidence and tracing back the path. This is because the graph created without antiforensic technique nodes may obfuscate the investigator. Apart from these attributes, integration of a new attribute such as antiforensic database which includes all techniques may prove helpful in successful generation of precise and accurate attack paths. Antiforensic technique such as trace path obstruction technique hinders the development of attack graphs [79].

All the phases have to be properly documented and should report the motive, strategy, preventive and precautionary measure for the attacks.

Evaluations and Comparisons of Tools and Techniques

Network Forensic Analysis Tools

Network forensic analysis tools (NFATs) help in the analysis phase of the evidence collected as a part of the forensic procedure [12,33]. These tools are used for both offline and online analysis which includes real-time acquisition and event viewer. Most of the tools are network traffic capturing ones except for NetDetector that has inbuilt IDS, and Kismet is mainly used for wireless networks. Table 20.2 shows the comparison of the features and functions of various network forensic tools used for network evidence analysis.

TABLE 20.2

Network Forensic Analysis Tools

Network Forensic Tools	Functions	Features
NetDetector	Signature analyzing IDS is incorporated which detect known and unknown threats, analyzes network packets, provides e-mail traffic monitoring, untrusted URL activity, and helps to resolve sophisticated cyber security attacks, real-time alerting on security, and performance-related events.	• Signature analysis tool • Event viewer • Application reconstruction tool • Uses a Flash-based Web interface
Network Miner v1.0	Packet Capturing tool which collects data regarding operating systems and open ports. It is a passive sniffer and the files are extracted using parsing PCAP file.	• Offline analysis • Supported protocols are FTO, HTTP, SMB, and TFTP
Iris v5.1.065	Analyzes the network traffic and reassembles in its own format and reconstructs the session and packets. Also used for electronic discovery.	• Service-oriented architecture for packet capture • Statistical measurement for packet size and protocol distribution • Reconstruction of e-mail messages, Web browsing sessions, and instant message sessions
Xplico v1.0.0	Network traffic capturing and is a protocol analyzer which has multithreading, TCP reassembling, and Reverse DNS look-up option for better analysis, and the result is presented in a visual form.	• Data capture • Real-time acquisition • Reverse DNS look-up
Silent Runner	Network Packet capturing, analyzing, host detection, and anomaly detection are the main functions. Reverse engineering of events, actual network traffic, and security incidents in the proper sequence are the main features.	• Real-time data capture • Incident response • Graphical visualization of result
Kismet	For 802.11 layer 2 wireless network capturing, analysis, and IDS. Detects hidden networks, passive collection of network packets (TCP, ARP, DHCP, and UDP).	• 802.11b, 802.11g, 802.11a, 802.11n sniffing • Multicard and channel hopping support • Runtime WEP decoding • Tap virtual network interface drivers for real-time export of packets • Hidden SSID decloaking • Distributed remote sniffing with Kismet drones and XML logging

(Continued)

TABLE 20.2

Continued

Network Forensic Tools	Functions	Features
Solera Network DS Series Applications	Mainly used for Packet Capture, Network Forensics, Security Intelligence, and Analytics. The DeepSee forensic suite reconstructs network attributes such as Web pages, PDF files, and images.	• High-speed data capture application for network traffic • Reconstruction and sequencing

Network Security and Monitoring Tools

Network monitoring tools are the key tools utilized for observing anomalies in the network traffic and detecting threats and attacks. The key features of the tools are protocol analysis and packet analysis. Table 20.3 shows the comparisons of functions and features of the main network security and monitoring tools used for the collection and analysis of evidence [12,92,93].

Overview of Antiforensics Tools and Techniques

As explained earlier, there are wide varieties of antiforensic tools built on utilizing antiforensic techniques to aid the attackers. Most of the antiforensic tools are now open-source ones which can be downloaded for free and can be modified according to the situation and requirement. The Back Track 5 R2 operating system provides a wide range of both forensic and antiforensic tools. Table 20.1 shows the comparison of the features and functions of different antiforensic tools. Most of the tools are used for evidence source destruction and obfuscation.

Trends

Network security has always been a crucial issue in the current world of technology as the entire corporate environment relies on the Internet, and the corporate assets have to be secured for the successful management of their organizations. Its advanced technology aims to deploy better secured network to prevent threats and attacks. Advanced intrusion detection and prevention system (IDS/IPS) [94] with inbuilt data mining, intentional analysis, and neural fuzzy logic helps in alerting about attacks and threats with the least possibility of false-positive alerts. Numerous researches have been done in various fields of network security and forensics to discover better defensive measures against attacks and threats especially in the field of wireless technology. Various researches should be carried out to gather effective evidence from wireless network [95–97]. Network forensics is one of the sensitive areas in digital forensics as it contributes evidence to identify

TABLE 20.3

Network Security and Monitoring Tools

Network Monitoring Tools	Functions	Features
TCPDump	Packet sniffer for protocol debugging and acquisition of data. Used for trouble shooting network activity and diagnosis of DoS attacks and has the "Berkley Packet Filter" (BPF).	• Command line tool • A portable C/C++ library for network traffic
TCPFlow	Investigation and management of network traffic and data flow in TCP/IP network. Captured file stored separately and reconstructs the data stream.	• Protocol analysis • Packet capture
Nmap	Network mapper used for security auditing. The GUI module is Zenmap. Raw IP packets are used for various functionalities.	• Port scanning • OS detection
TCPDStat	Reads TCPDump files with the aid of the pCap library and finds the trace. Gives a vague idea of content of the trace. Output may include protocol breakdowns, source and destination address, and number of packets.	• Protocol breakdown • PCAP library • High-level traffic pattern monitoring
TCP Trace	Reads packets captured form various packet capturing tools such as TCPDump and gives the output such as round-trip time, hops, time elapsed, and amount of data sent and received.	• TCPDump file analysis
WireShark	Protocol analyzer that provides in-depth inspection of protocol, live capture, and VoIP analysis.	• Rich display filter • Can run on multiple platforms such as Windows, LINUX, and Solaris • Supports more than 100 protocols
Ethereal	Open source packet analyzer that has filter capabilities and works in both promiscuous and non-promiscuous mode.	• Reconstructs TCP session • Captures data from ethernet, token ring, and 802.11 wireless
Snort v2.9.3.1	It is an Open Source IPS/IDS which incorporates signature, anomaly-based and protocol inspection. Protocol analysis and content searching are the main function.	• Supports Unix and Windows platforms
Bro	Network analysis framework with IDS. In-depth analysis of protocols and can be used in high performance networks, focus on application level.	• Protocol analysis • Semantic analysis and thorough activity logging

the identity and source of the attacker. The forensic investigation process has changed from traditional system forensics to live forensics and has incorporated various methodologies to defend antiforensic techniques and/or to reduce their effect in the collected digital evidence. The difficulty level of investigation of network attacks has risen in the recent years. As technology advances, new tools are portable and handy [98], and relevant techniques are developed for digital forensic investigation. But on the other side, criminals exploit the technology and find new ways to thwart the forensic process.

As mentioned earlier, due to advancement in antiforensic techniques, collection and analysis of evidence from computer networks, which have been vulnerable to attacks, have been very challenging. Antiforensics is not completely about tools which assist to cover up the trace but is a combination of techniques, tactics, and strategy. The current trend in antiforensics shows that the application of techniques has moved from conventional areas of data hiding and deletion of trace of evidence and logs in systems to computer network. Techniques such as network steganography, covert tunneling, trace obstruction, and hiding IP are now frequently used by the attackers for the defense mechanism. Apart from this, law-breakers try to obfuscate the forensic investigators by providing fake evidence, attacking forensic tools with compromised systems.

The main effect of antiforensic techniques is on the integrity [99] and dependence of the evidence collected. The tools and techniques are robust enough to alter (modify, delete, and hide) the evidence source and the evidence itself. The validation of evidence thus becomes a vital factor during the forensic investigative process. It becomes necessary to validate the evidence in each and every step of the process by detecting the presence of antiforensic techniques or use of antiforensic tools, especially during live forensics. Robust methodologies and frameworks [29,100] will be developed for this purpose. Integration of intelligent analysis such as fuzzy logic and neural networks [101], antiforensic detection algorithms and frameworks in forensic tools, network security, and monitoring systems such as IDS/IPS will prove to be effective countermeasures.

The current security features in computer networks have various flaws and vulnerabilities that are exploited by the attackers. Improper system configuration, IDS/IPS configuration, and firewall rules are another weakness utilized to gain access to systems and network infrastructure. Another key point is the compatibility of network infrastructure with the current forensic tools. The latest version of forensic tools such as Encase and Access Data FTK tries to cope up with the advancement of technology [74]. These tools incorporate techniques for network forensics and Internet forensics [95] such as Web analysis [102,103], blog analysis [104], and e-mail forensics [105,106]. In order to trace out path of the attack, to find the source and identity of the attacker, several methods such as attack graph theory, packet analysis, and Metasploit forensic frameworks can be handy. To conclude, key areas where advancement has to be carried out are tools for developing network evidence graphs [105] and attack graphs, detection of antiforensic attack tools and reduce their effect [21] in evidence so integrity is not lost completely.

Conclusions

Antiforensics was confined only to storage devices and computer systems for the past few years. Network forensics is one of the main challenging fields of digital forensics in this current era of latest technology. As new and robust attack techniques are discovered, it has become almost impossible to find the exact source of the attack. When antiforensics combines with these network attacks, it will be for a more robust and intense way of attacking and even more difficult to gather evidence, analyze, and find the trace route and source. One of the latest forensic process to identify the trace route [107] is reverse engineering of the attacks using network attack graphs.

In this chapter, an in-depth survey on the antiforensics techniques has been conducted. The survey describes the main antiforensic tools which are classified with relevance to the techniques and algorithms they exploit. As the survey deals with how antiforensics could be combined with network attacks, a review of the common network attacks have also been mentioned. The evidence collection and analysis of the network attacks are carried out using network security and monitoring tools (NSMs) and network forensics analysis tools (NFATs). A detailed review of network tools has been carried out in the survey.

The key issue for forensic investigators during the forensic process is the validation of evidence [108]. The integrity of the collected evidence has to be questioned at each stage of analysis. Hash analysis and signature analysis are helpful to a certain extend. Sometimes hash collision techniques obfuscate the investigators. Another aspect is that the forensic investigation process itself will be under attack using rootkits [109], compromised hosts, and attack on forensic tools [110].

The challenges and issues in various tools and techniques have to be studied so that the vulnerabilities can be discovered. Antiforensics techniques will focus on the vulnerabilities of the digital forensic software by obfuscation and misinformation. In order to defend such kind of attacks against the forensic tools, antiforensics techniques and network antiforensics have to be investigated further in depth.

References

1. Nikkel, B. J. (2006) "Improving Evidence Acquisition from Live Network Sources," *Digital Investigation*, 3(2): 89–96.
2. Hartley, W. M. (2007) "Current and Future Threats to Digital Forensics," *ISSA Journal*, 12–14.
3. Losavio, M., Nasraoui, O., Thacker, V., Marean, J., Miles, N., Yampolskiy, R., & Imam, I. (2009) "Assessing the Legal Risks in Network Forensic Probing," in *Advances in Digital Forensics V*, Berlin/Heidelberg: Springer, pp. 255–266.

4. Endicott-Popovsky, B., & Frincke, D. (2007) "Embedding Hercule Poirot in Networks: Addressing Inefficiencies in Digital Forensic Investigations," in *Foundations of Augmented Cognition*, Berlin/Heidelberg: Springer, pp. 364–372.

5. Manzano, Y., & Yasinsac, A. (2001) "Policies to Enhance Computer and Network Forensics," in *Proceedings of the 2nd Annual IEEE Systems, Man and Cybernetics Information Assurance Workshop*, United States Military Academy, West Point, NY, June 5–6, pp. 289–295.

6. Benjamin, T., & Jill, S. (2007) "Wireless Forensic Analysis Tools for Use in the Electronic Evidence Collection Process," in *40th Annual Hawaii International Conference on System Sciences*, Waikoloa, HI, January, pp. 267–267.

7. Bosschert, T. (2007) "Battling Anti-Forensics: Beating the U3 Stick," *Journal of Digital Forensic Practice*, 1(4): 265–273.

8. Caloyannides, M. A. (2009) "Forensics Is So 'Yesterday'," *Security & Privacy, IEEE*, 7(2): 18–25.

9. Johansson, C. (2002) "Forensic and Anti-Forensic Computing," Thesis, Blekinge Institute of Technology, Sweden.

10. Rekhis, S., & Boudriga, N. (2010) "Formal Digital Investigation of Anti-forensic Attacks," in *5th IEEE International Workshop on Systematic Approaches to Digital Forensic Engineering (SADFE)*, Oakland, CA, May, pp. 1–8.

11. Qureshi, A. (2009) "802.11 Network Forensic Analysis" [White Paper], SANS Institute InfoSec Reading Room, pp. 3–48.

12. Meghanathan, N., Allam, S. R., & Moore, L. A. (2009) "Tools and Techniques for Network Forensics," *International Journal of Network Security & Its Applications*, 1(1): 14–25.

13. Gorodetski, V., & Kotenko, I. (2002) "Attacks against Computer Network: Formal Grammar-Based Framework and Simulation Tool," in *Recent Advances in Intrusion Detection*, Gorodetski, V., & Kotenko, I. (eds.), *LNCS* 2516. Berlin: Springer, pp. 219–238.

14. Ieong, R., Lai, P., Chow, K. P., Law, F., Kwan, M., & Tse, K. (2009) "A Model for Foxy Peer-to-Peer Network Investigations," in *Advances in Digital Forensics V*, pp. 175–186.

15. Piper, S., Davis, M., Manes, G., & Shenoi, S. (2005) "Detecting Hidden Data in Ext2/Ext3 File Systems," in *Advances in Digital Forensics*, 194, pp. 245–256.

16. Meng, K. et al. (2009) "Building a Wireless Capturing Tool for WiFi," *Security and Communication Networks*, 2(6): 654–668.

17. Dahbur, K., & Mohammad, B. (2011) "Toward Understanding the Challenges and Countermeasures in Computer Anti-Forensics," *International Journal of Cloud Applications and Computing*, 1(3): 22–35.

18. Stamm, M. C., Tjoa, S. K., Lin, W. S., & Liu, K. J. R. (2010) "Anti-Forensics of JPEG Compression," in *IEEE International Conference on Acoustics Speech and Signal Processing (ICASSP)*, Dallas, TX, March 14–19, pp. 1694–1697.

19. Cao, G. et al. (2010) "Anti-Forensics of Contrast Enhancement in Digital Images," in *Proceedings of the 12th ACM Workshop on Multimedia and Security*, Rome, Italy, pp. 25–34.

20. Stamm, M. C., Tjoa, S. K., Lin, W. S., & Liu, K. J. R. (2010) "Undetectable Image Tampering through JPEG Compression Anti-Forensics," in *IEEE International Conference on Image Process*, Hong Kong, September, pp. 2109–2112.

21. Harris, R. (2006) "Arriving at an Anti-Forensics Consensus: Examining How to Define and Control the Anti-Forensics Problem," *Digital Investigation*, 3: 44–49.

22. Takahashi, D., & Xiao, Y. (2008) "Retrieving Knowledge from Auditing Log-Files for Computer and Network Forensics and Accountability," *Security and Communication Networks*, 1(2): 147–160.
23. Behr, D. J. (2008) "Anti-Forensics: What It Is? What It Does and Why You Need to Know?" *New Jersey Layer Magazine*, 255: 4–9.
24. Hilley, S. (2007) "Anti-Forensics with a Small Army of Exploits," *Digital Investigation*, 4(1): 13–15.
25. Azadegan, S., Yu, W., Liu, H., Sistani, M., & Acharya, S. (2012) "Novel Anti-Forensics Approaches for Smart Phones," in *45th Hawaii International Conference on System Science (HICSS)*, Maui, HI, January, pp. 5424–5431.
26. Albano, P. et al. (2011) "A Novel Anti-forensics Technique for the Android OS," in *International Conference on Broadband and Wireless Computing, Communication and Applications*, Barcelona, Spain, October, pp. 380–385.
27. Distefano, A. et al. (2010) "Android Anti-Forensics through a Local Paradigm," *Digital Investigation*, 7: 83–94.
28. Kessler, G. C. (2007) "Anti-Forensics and the Digital Investigator," in *5th Australian Digital Forensics Conference*, Perth, Western Australia, pp. 1–7.
29. Shanmugam, K. et al. (2011) "An Approach for Validation of Digital Anti-Forensic Evidence," *Information Security Journal: A Global Perspective*, 20(4–5): 219–230.
30. Schlicher, B. (2008) "Emergence of Cyber Anti-Forensics Impacting Cyber Security," in *Proceedings of the 4th Annual Workshop on Cyber Security and Information Intelligence Research: Developing Strategies to Meet the Cyber Security and Information Intelligence Challenges Ahead*, Oak Ridge, TN, pp. 1–12.
31. Simmons, C. B. et al. (2011) "A Framework and Demo for Preventing Anti-Computer Forensics," *Information Systems*, 12(1): 366–372.
32. Khan, S. A., & Uddin, N. (2012) "A Survey on Network Attacks and Defence-in-Depth Mechanism by Intrusion Detection System," Bachelor Thesis, BRAC University, Dhaka, Bangladesh.
33. Samalekas, K. (2010) "Network Forensics: Following the Digital Trail in a Virtual Environment," Master Thesis, University of Gothenburg, Gothenburg.
34. Chen, A. H. C., Chu, H. C., & Su, F. (2012) "The Study on Performing Network Forensics—A Case of Web Application Attack," *Business and Information*, pp. 839–357.
35. Garfinkel, S., Beverly, R., & Cardwell, G. (2011) "Forensic Carving of Network Packets and Associated Data Structures," Masters Thesis, Naval Postgraduate School, Monterey, CA.
36. Nikkel, B. J. (2004) "Domain Name Forensics: A Systematic Approach to Investigating an Internet Presence," *Digital Investigation*, 1(4): 247–255.
37. Krawetz, N. (2004) "Anti-Honeypot Technology," *IEEE Security & Privacy*, 2(1): 76–79.
38. Boran, S. (1999) "An Overview of Corporate Information Security," http://boran.com/security/sp/security_space.html
39. Annis, J. (2009) "Zombie Networks: An Investigation into the Use of Anti-Forensic Techniques Employed by Botnets," Master Thesis, The Open University, Leeds, UK.
40. Eggendorfer, T. (2008) "Methods to Identify Spammers," in *1st International Conference on Forensic Applications and Techniques in Telecommunications, Information, and Multimedia and Workshop*, Adelaide, South Australia, pp. 1–7.

41. Rekhis, S., & Boudriga, N. (2012) "A System for Formal Digital Forensic Investigation Aware of Anti-Forensic Attacks," *IEEE Transactions on Information Forensics and Security*, 7(2): 635–650.
42. Forte, D., & Power, R. (2007) "A Tour through the Realm of Anti-Forensics," *Computer Fraud & Security*, 2007(6): 18–20.
43. Sartin, B. (2006) "Anti-Forensics—Distorting the Evidence," *Computer Fraud & Security*, 2006(5): 4–6.
44. Velupillai, H., & Mokhonoana, P. (2008) "Evaluation of Registry Data Removal by Shredder Programs," in *Advances in Digital Forensics IV*, 285, pp. 51–58.
45. Zhou, W., Marshall, A., Zhou, W., & Kun, Y. (2008) "A Random Packet Destruction DoS Attack for Wireless Networks," in *IEEE International Conference on Communications (ICC)*, pp. 1658–1662.
46. Berghel, H. (2007) "Hiding Data, Forensics, and Anti-Forensics," *Communications of the ACM*, 50(4): 15–20.
47. Dahbur, K., & Mohammad, B. (2011) "The Anti-Forensics Challenge," in *ACM International Conference on Intelligent Semantic Web-Services and Applications*, Amman, Jordan, pp. 1–7.
48. Daniels, J. (2008) "Forensic and Anti-Forensic Techniques for OLE2-Formatted Documents," MS Thesis, Utah State University, Logan, UT.
49. Chan, E., Venkataraman, S., Tkach, N., Larson, K., Gutierrez, A., & Campbell, R. H. (2011) "Characterizing Data Structures for Volatile Forensics," in *IEEE Symposium on Systematic Approaches to Digital Forensic Engineering (SADFE)*, Oakland, CA, May, pp. 1–9.
50. Pollitt, M. (2008) "Applying Traditional Forensic Taxonomy to Digital Forensics," in *Advances in Digital Forensics IV*, 285, pp. 17–26.
51. Sang Su, L. et al. (2007) "A New Anti-Forensic Tool Based on a Simple Data Encryption Scheme," in *Future Generation Communication and Networking (FGCN)*, Jeju Island, Korea, December, pp. 114–118.
52. Suhyung, J., & Dowon, H. (2008) "Defense Technology of Anti-Forensic," in *International Conference on Control, Automation and Systems (ICCAS)*, Singapore, October, pp. 884–887.
53. Szczypiorski, K. (2004) "HICCUPS: Hidden Communication System for Coruppted Networks," in *10th International Multi-Conference on Advanced Computer Systems (ACS)*, Marseilles, France, October, pp. 31–40.
54. Baier, H., & Breitinger, F. (2011) "Security Aspects of Piecewise Hashing in Computer Forensics," in *6th International Conference on IT Security Incident Management and IT Forensics (IMF)*, Fraunhofer Institutszentrum Stuttgart, Germany, May, pp. 21–36.
55. Smith, A. (2007) "Describing and Categorizing Disk-Avoiding Anti-Forensics Tools," *Journal of Digital Forensic Practice*, 1(4): 309–313.
56. Fairbanks, K. D., Lee, C. P., Xia, Y. H., & Owen, H. L. (2007) "TimeKeeper: A Metadata Archiving Method for Honeypot Forensics," in *IEEE Information Assurance and Security Workshop (IAW)*, West Point, NY, June, pp. 114–118.
57. Arnold, T., & Yang, T. A. (2011) "Rootkit Attacks and Protection: A Case Study of Teaching Network Security," *Journal of Computing Sciences in Colleges*, 26(5): 122–129.
58. Hunt, R., & Slay, J. (2010) "Achieving Critical Infrastructure Protection through the Interaction of Computer Security and Network Forensics," in *8th Annual International Conference on Privacy Security and Trust (PST)*, Ottawa, ON, IEEE, August, pp. 23–30.

59. Endicott-Popovsky, B., Frincke, D. A., & Taylor, C. A. (2007) "A Theoretical Framework for Organizational Network Forensic Readiness," *Journal of Computers*, 2(3): 1–11.
60. Seifert, C., Endicott-Popovsky, B., Frincke, D., Komisarczuk, P., Muschevici, R., & Welch, I. (2008) "Justifying the Need for Forensically Ready Protocols: A Case Study of Identifying Malicious Web Servers Using Client Honeypots," in *4th Annual IFIP WG 11.9 International Conference on Digital Forensics*, Montreal, QC, October, pp. 1–14.
61. Nilsson, D. K., & Larson, U. E. (2008) "Conducting Forensic Investigations of Cyber-Attacks on Automobile In-Vehicle Networks," in *1st International Conference on Forensic Applications and Techniques in Telecommunications, Information, and Multimedia and Workshop*, Adelaide, South Australia, pp. 1–6.
62. Harshbarger, B. (2010) "Social Networking Websites as a Tool for Investigators," *Journal of Network Forensics*, 2(1): 25–33.
63. Pilli, E. S., Joshi, R., & Niyogi, R. (2010) "Network Forensic Frameworks: Survey and Research Challenges," *Digital Investigation*, 7(1): 14–27.
64. Jiang, D., & Shuai, G. (2011) "Research on the Clients of Network Forensics," in *3rd International Conference on the Computer Research and Development (ICCRD)*, Guangzhou, China, March, pp. 466–468.
65. Lewthwaite, J., & Smith, V. (2008) "Limewire Examinations," *Digital Investigation*, 5: 96–104.
66. Barford, P. et al. (2002) "A Signal Analysis of Network Traffic Anomalies," in *2nd ACM SIGCOMM Workshop on Internet Measurement*. Marseille, France, pp. 71 82.
67. Ding, X., & Zou, H. (2011) "Time Based Data Forensic and Cross-Reference Analysis," in *ACM Symposium on Applied Computing*, TaiChung, Taiwan, pp. 185–190.
68. Goodall, J. R., Lutters, W. G., Rheingans, P., & Komlodi, A. (2006) "Focusing on Context in Ntraffic Analysis," *IEEE Computer Graphics and Applications*, 26(2): 72–80.
69. Zheng, W., Yang, O., & Yujun, L. (2011) "A Taxonomy of Network and Computer Attacks Based on Responses," in *International Conference on Information Technology, Computer Engineering and Management Sciences (ICM)*, pp. 26–29.
70. Sheyner, O., & Wing, J. (2004) "Tools for Generating and Analyzing Attack Graphs," in *Workshop on Formal Methods for Components and Objects*, January, Berlin/Heidelberg: Springer, pp. 344–371.
71. Chandankhede, P. H., & Nimbhorkar, S. U. (2012) "Autonomous Network Security for Detection of Network Attacks," *International Journal of Scientific and Research Publications*, 2(1): 109.
72. Ou, X., & Singhal, A. (2011) *Quantitative Security Risk Assessment of Enterprise Networks*. New York: Springer, pp. 1–28.
73. Weihan, G., Peng Chor, L., & Chai Kiat, Y. (2009) "A Trusted Platform Module Based Anti-Forensics System," in *International Conference on Network and Service Security (N2S)*, Paris, France, June, pp. 1–5.
74. Pajek, P., & Pimenidis, E. (2009) "Computer Anti-Forensics Methods and Their Impact on Computer Forensic Investigation," in *Global Security, Safety, and Sustainability*, Tenreiro de Magalhães, S., Jahankhani, H., & Hessami, A. G. (eds.), Berlin: Springer, 45, pp. 145–155.
75. Erasani, S. (2010) "Implementation of Anti-Forensic Mechanisms and Testing with Forensic Methods," Graduate Project Report, Texas A&M University-Corpus Christi, Corpus Christi, TX.

76. Nikkel, B. J. (2007) "An Introduction to Investigating IPv6 Networks," *Digital Investigation*, 4(2): 59–67.
77. Wang, W., & Daniels, T. E. (2008) "A Graph Based Approach toward Network Forensics Analysis," *ACM Transactions on Information and System Security*, 12(1): 1–33.
78. Changwei, L., Singhal, A., & Wijesekera, D. (2012) "Using Attack Graphs in Forensic Examinations," in *7th International Conference on Availability, Reliability and Security (ARES)*, pp. 596–603.
79. Wei, W., & Daniels, T. E. (2005) "Building Evidence Graphs for Network Forensics Analysis," in *Computer Security Applications Conference*, pp. 266.
80. Ingols, K., Chu, M., Lippmann, R., Webster, S., & Boyer, S. (2009) "Modeling Modern Network Attacks and Countermeasures Using Attack Graphs," in *Annual Computer Security Applications Conference*, pp. 117–126.
81. Jian, B. et al. (2010) "Research on Network Security of Defense Based on Honeypot," in *International Conference on Computer Application and System Modeling (ICCASM)*, pp. 299–302.
82. Ou, X., Govindavajhala, S., & Appel, A. W. (2005) "MulVAL: A Logic-Based Network Security Analyzer," in *14th Conference on USENIX Security Symposium*, Baltimore, MD, p. 8.
83. Roschke, S. et al. (2010) "Using Vulnerability Information and Attack Graphs for Intrusion Detection," in *6th International Conference on Information Assurance and Security (IAS)*, pp. 68–73.
84. Jantan, A. et al. (2012) "A Similarity Model to Estimate Attack Strategy Based on Intentions Analysis for Network Forensics," in *Recent Trends in Computer Networks and Distributed Systems Security*, Thampi, S. M., Zomaya, A. Y., Strufe, T., Alcaraz Calero, J. M., & Thomas, T. (eds.), Berlin: Springer, 335, pp. 336–346.
85. Rasmi, M., & Jantan, A. (2011) "Attack Intention Analysis Model for Network Forensics," in *Software Engineering and Computer Systems*, pp. 403–411.
86. Sy, B. K. (2009) "Integrating Intrusion Alert Information to Aid Forensic Explanation: An Analytical Intrusion Detection Framework for Distributive IDS," *Information Fusion*, 10(4): 325–341.
87. Ammann, P. et al. (2002) "Scalable, Graph-Based Network Vulnerability Analysis," in *Proceedings of the 9th ACM Conference on Computer and Communications Security*, Washington, DC, pp. 217–224.
88. Almulhem, A. (2009) "Network Forensics: Notions and Challenges," in *IEEE International Symposium on the Signal Processing and Information Technology (ISSPIT)*, pp. 463–466.
89. Jha, S., Sheyner, O., & Wing, J. (2002) "Two Formal Analysis of Attack Graphs," in *15th IEEE Workshop on Computer Security Foundations*, pp. 49.
90. Nehinbe, J. O. (2011) "Emerging Threats, Risks and Mitigation Strategies in Network Forensics," in *24th Canadian Conference on Electrical and Computer Engineering (CCECE)*, pp. 1228–1232.
91. Catania, C. A., & Garino, C. G. (2012) "Automatic Network Intrusion Detection: Current Techniques and Open Issues," *Computers & Electrical Engineering*, 38(5): 1062–1072.
92. Nomad, S. (2003) "Covering Your Tracks: ncrypt and ncovert," Research Report, Black Hat. http://www.blackhat.com/html/bh-media-archives/bh-archives-2003.html

93. Seewald, A. K., & Gansterer, W. N. (2010) "On the Detection and Identification of Botnets," *Computers & Security*, 29(1): 45–58.
94. Vasiliadis, G., Antonatos, S., Polychronakis, M., Markatos, E., & Ioannidis, S. (2008) "Gnort: High Performance Network Intrusion Detection Using Graphics," in *Proceedings of Recent Advances in Intrusion Detection, LNCS* 5230, pp. 116–134.
95. Berghel, H. (2003) "The Discipline of Internet Forensics," *Communications of the ACM*, 46(8): 15–20.
96. Taylor, M. et al. (2011) "Digital Evidence from Peer-to-Peer Networks," *Computer Law & Security Review*, 27(6): 647–652.
97. Ngobeni, S. J., & Venter, H. S. (2009) "Design of a Wireless Forensic Readiness Model (WFRM)," in *Information Security South Africa (ISSA) Conference*, University of Johannesburg, South Africa, pp. 1–17.
98. Nikkel, B. J. (2006) "A Portable Network Forensic Evidence Collector," *Digital Investigation*, 3(3): 127–135.
99. Peron, C. S. J., & Legary, M. (1995) "Digital Anti-Forensics: Emerging Trends in Data Transformation Techniques," in *E-Crime and Computer Evidence Conference*. http://www.ide.bth.se/~andersc/kurser/DVC013/PDFs/Seccuris-Antiforensics.pdf
100. Mansfield-Devine, S. (2010) "Fighting Forensics," *Computer Fraud & Security*, 2010(1): 17–20.
101. Saad, S., & Traore, I. (2010) "Method Ontology for Intelligent Network Forensics Analysis," in *8th Annual International Conference on Privacy Security and Trust (PST)*, pp. 7–14.
102. Kiley, M., Dankner, S., & Rogers, M. (2008) "Forensic Analysis of Volatile Instant Messaging," in *Advances in Digital Forensics IV*, Ray, I., & Shenoi, S. (eds.), *LNCS* 285. Berlin: Springer, pp. 129–138.
103. Mishra, S., Jing, D., & Han, R. (2004) "Intrusion Tolerance and Anti-Traffic Analysis Strategies for Wireless Sensor Networks," in *International Conference on Dependable Systems and Networks*, pp. 637–646.
104. Dardick, G. S., & Roche, C. L. (2007) "Blogs: Anti-Forensics and Counter Anti-Forensics," in *5th Australian Digital Forensics Conference*, pp. 199.
105. Cohen, F. (2009) "Bulk Email Forensics," in *Advances in Digital Forensics V*, pp. 51–67.
106. Ehlert, S. et al. (2010) "Survey of Network Security Systems to Counter SIP-Based Denial-of-Service Attacks," *Computers & Security*, 29(2): 225–243.
107. Heydari, S., Martin, M. V., Rjaibi, W., & Lin, X. (2010) "Emerging Trends in Network Forensics," Research Report, IBM Corporation, pp. 389–390.
108. Chan, E., Chaugule, A., Larson, K., & Campbell, R. (2010) "Performing Live Forensics on Insider Attacks," in *The CAE Workshop on Insider Threat*, pp. 1–9.
109. Levi, A., & Güder, C. B. (2009) "Understanding the Limitations of S/MIME Digital Signatures for E-mails: A GUI Based Approach," *Computers & Security*, 28(3–4): 105–120.
110. Forte, D. (2008) "Dealing with Forensic Software Vulnerabilities: Is Anti-Forensics a Real Danger?" *Network Security*, 2008(12): 18–20.

21

Password Security and Protection

Emery Moodley, Gary Huo, Mindy Hsieh, Stephen Cai, and Wei Q. Yan

CONTENTS

Introduction

The implementation of passwords as a form of electronic security has been common for many years. Users rely heavily on passwords to safeguard their data on various devices. This is done to prevent personal and sensitive information from falling into the hands of those who would use it for exploitation.

In recent years, it has become apparent that the traditional method of entering an alphanumeric username and password to verify the identity of an individual, or authenticate them, has become inadequate and in some cases obsolete [1]. Subsequently, new methods of authentication must now be developed in order to cope with the growing demand for security and the growing number of attacks which traditional password systems have become susceptible to Ref. [2].

As we become more reliant on technology and the information, we accumulate and share via this technology, the act of securing this data becomes paramount. With organizations utilizing multiple devices, which are constantly connected to the world around us through the Internet, making sure that each of these devices is secure becomes difficult [3]. If a user's data are not properly protected, particularly where large-scale organizations are concerned, loss of revenue becomes a serious hazard for the said organization [2]. Therefore, numerous alternatives to the traditional password system have been developed; these alternatives aim to provide a more holistic approach to authentication [4]. However, it must be said that no system is foolproof; this is especially true if users do not have an understanding of why proper password practices and etiquette are required. Educating those who will use an authentication system is vital to maintaining its security [5]. A chain is only as strong as its weakest link. Subsequently, we will endeavor to describe how to increase the recall rate among users of an authentication system, while maintaining convenience, and most importantly, security [6].

Section "Common Password Authentication Methods" details the various methods of authentication that are currently available, describing each of their strengths and weaknesses. Section "Security Analysis" details the various attacks that are commonly used today to bypass authentication systems and illegally retrieve data. Section "Password Encryption" explains the use of encryption methods that are used to secure authentication systems. Section "Comparison of Algorithms" explains algorithms, and the comparisons of these methods are detailed which will help to understand the use of encryption in next section. Section "Password Storage" details how and where a password is stored and securely protected in different systems. Section "Balancing Security, Convenience, and Education" details the need to educate users, and give them the tools they need to create more secure passwords and maintain the integrity of a system. Section "Multiple-Layer Authentication" will introduce the concept of multiple authentications, which is a layered approach to security.

Common Password Authentication Methods

Password authentication is facilitated in four main forms: (1) traditional passwords, (2) dynamic passwords, (3) cognitive authentication, and (4) graphical passwords. Each of these systems has its own strengths and weaknesses [7–10].

Traditional alphanumeric passwords are the ones that are most commonly used and understood. Users have been utilizing these for decades and subsequently feel very comfortable using them [11]. They are also the most cost effective and convenient to implement because of their general lack of complexity. However, they are also the most vulnerable, because they have been employed for so long, various methods of bypassing them have been developed for nefarious purposes [12]. These passwords form three main subcategories: (1) *word-only passwords*, where only letters can be used; (2) *mixed passwords*, where letters, numbers, and symbols can be employed; and (3) *symbol-based passwords*, where only special characters are permitted. All of these rely on the well-known system of utilizing a username, password, and verification of a user's identity [13].

Dynamic passwords also employ an alphanumeric sequence and username to facilitate authentication; however, they utilize an additional layer of security to prevent attacks. Typically, a user will authenticate him- or herself using a password and username via a secure channel and will then be issued with a second dynamic password. This password can only be used once, and is usually only valid for a fixed length of time or until the user logs off. Dynamic passwords will change each time a user attempts to be authenticated [9]. In this way, repeat attacks are prevented, as the same passphrase cannot be used twice. This technique is often coupled with a token of some kind, a small device detached from a user's PC, which can display the dynamic password to the user discreetly [14].

The channel that the user attains the dynamic password from is usually separated completely from the secure system that the password is used to access, often on internal network. This layered approach reduces the chances of a successful attack [15]. While dynamic passwords are more secure than traditional passwords, they are more costly to implement and less convenient. If a user loses a token or forgets his or her login details to attain the dynamic password, the process of authentication could be time-consuming, which in turn may result in a loss of revenue when the system is deployed across an entire organization [9].

Graphical passwords are the newest of the four authentication methods; they rely on visual aids and cues to authenticate a user. The general premise behind graphical passwords is that it is easier for a user to recall an image or a series of images, than it is to remember a long alphanumeric passphrase, which has no significance to the user personally. Graphical passwords come in two main forms: (1) recall based and (2) recognition based. Recall-based authentication relies on the user being able to recall a series of images in the correct order, whereas recognition-based authentication entails the user selecting a specific picture from a grid or list [7,16]. While these are the two main forms, the implications for graphical passwords are numerous; some methods even employ the use of a 3D environment where objects can be interacted with. While this may seem excessive, it brings to light the fact that the strength of graphical authentication lies in its variability [6]. Having multiple systems that rely on the same cognitive processes, recall and recognition, but are seemingly different in terms of the way they function, technically will make it difficult to develop attacks which can be used for more than one specific authentication system. Graphical passwords are however more costly to implement and being a relatively new approach to authentication, users can sometimes struggle to use them. However, this is something that can be corrected over time as users move away from traditional passphrases and become more familiar with more inventive methods of authentication. They are also more susceptible to shoulder-surfing attacks, but they are conversely immune to dictionary attacks and key logging [8].

Cognitive authentication is often used as a secondary form of user verification. It involves the user answering predetermined questions about him- or herself. These questions are usually of personal significance to the user, and may involve details about their place of birth, country of origin, or mother's maiden name, for example. Cognitive authentication has a very high recall rate as users identify with the answers personally and subsequently are more likely to remember them; however, cognitive attacks are extremely susceptible to guess attacks, especially when the attacker knows the user personally. This is why they are generally only employed as a second form of authentication [15,17].

Security Analysis

Authentication by way of passwords is the most common tool to verify a user and subsequently allow them to access data and sensitive information [18]; however, it must be said that this process is not always entirely secure [4,6]. There are numerous ways to protect a system, but creating an authentication process which is highly secure can be extremely time-consuming and expensive [19,20]. This is made increasingly difficult by the growing number of attacks and vulnerabilities that authentication systems are susceptible to; this is especially true of traditional alphanumeric password systems [19,21]. In order to secure an authentication system, regardless of its type, we must first have an understanding of the common type of attacks and which type of system they are designed to exploit [22–24]. The ideal password system should be able to deter hackers from cracking passwords and prevent them from stealing or tampering with sensitive data [6,25–29].

Major Attacks with Potential Risks

Dictionary Attack

Dictionary attacks involve the use of a dictionary that contains encrypted alphanumeric phrases. The attacker cross-checks this encrypted dictionary with the encrypted file containing a user's password and enabling the password to be revealed if it is in alphanumeric form [12,30]. However, the attacker must have knowledge of which encryption and algorithm is used, and more specifically, which operating system. The password will be cracked if the encrypted password is matching the encrypted value from the password file. Further, the attackers must have the copy of the password file and do the cracking on another machine because the attacked server's CPU is likely to be under heavy strain to its workload which will cause the cracking process to fail [31]. These attacks are particularly successful against weak password systems where constraints are not specified as to the length or format of the passwords [30,32,33].

Brute-Force Attack

This is a cryptographic attack that generates various key combinations; the attacker attempts to force entry into the system by attempting all the possible combinations for an alphanumeric or mixed-symbol password. This attack is said to be computationally intensive, which takes a longer time to compete than the dictionary attack [19]; however, the brute-force attack will only work if the system has no limitation on login attempts, and if the network and system administrator does not check the log regularly, and therefore, does not notice the alarming rate of failed login attempts [23].

Shoulder-Surfing Attack

Shoulder surfing is to watch if someone enters the password without permission, or try to pick up the messages or hints that could possibly be related to the real password [23]. Shoulder surfing is the most successful attack that can obtain graphical passwords, if the action of logging in is performed in a secure environment; this attack is not possible where there is no direct line of sight to the screen or keyboard. However, if authentication is carried out in a public place where an attacker can clearly see the screen, keyboard, or other input device, there is a chance that the attacker may obtain your login details [34,35]. Research shows that graphical passwords are more susceptible to shoulder-surfing attacks, particularly where mobile devices are concerned. This is because it is easier to identify login information in a graphical form, than it is to identify alphanumeric sequences [14,36]. In some cases, passwords can even be saved in unencrypted plaintext in Web browsers, further increasing the chance of shoulder-surfing attacks, allowing hackers to gain access to the system with minimal effort [37].

Key Logging

Key logging is the process of recording user's keystrokes, thereby making it possible to record keyboard input in password and username fields [38].

This is of course done without the user's consent and can be implemented through the use of software or hardware. The hardware application of this attack however involves physically installing a device on a user's keyboard and is therefore less common [39]. Conversely, due to the growing popularity of the Internet and subsequently the growing number of ways, users can be tricked into downloading malicious files; the software variant of this attack is far more popular [40–42]. However, both applications are equally dangerous once in place, as they will enable an attacker to remote retrieve authentication information as well as any other sensitive information that the user types [43–45].

Preventions of Attacks

Dictionary Attack

Most traditional authentication systems are now backed by password policy and etiquette documentation. Such policy requires passwords to be of a certain length; this is now hardcoded into the systems, forcing users to choose passwords of a particular length [46]. At present, eight digits seem to be the preferred number required to facilitate adequate security. This stifles dictionary attacks as the longer the password, the more difficult it is for the attacker to correlate phrases in the dictionary with the user's actual passwords. This is bolstered by the security policy that makes it compulsory for digits and symbols to be included in passwords. This is most effective when these characters are inserted into the phrase as opposed to being appended at the beginning or the end [33]. For an example, an ideal password would contain numbers, letters, and digits, but not in any specific order, the letters need not form words and the digits need not have any personal significance to the user, such as their date of birth or age. Using such personal details may make the password easier to crack [47]. Use of upper and lower case letters further increases security. However, capitalized letters should not always feature at the beginning of a password or phrase, in order to further stifle dictionary attacks [37]. In addition to this, using multiple authentications, featuring a graphical or cognitive passwords, removes the risk of key logging alone that allows intruders to gain access to a system [8,13,48].

Brute-Force Attack

Implementation of a limit on login attempts is the main way in which brute-force attacks are prevented. As these attacks rely on multiple logins and essentially try to guess every possible combination for a password [33,49], limiting the number of failed logins and then forcing a user to seek approval or help from an administrator before continuing is essential. For systems where high security is required, accounts could also be disabled when too many failed login attempts are made. The recommended number of attempts should be set to three; this gives the user ample chance to correctly input their login details while compensating for typos and other user errors, while negating the effects of a brute-force attack [48,50]. If disabling accounts permanently leads to too many legitimate interruptions that prevent users from carrying out their work, a delay time can be set between failed login attempts, or three failed logins could result in a 15-minute lockout rather than a permanent lockout until an administrator allows the user access. This will severely hinder any brute-force attack attempts [45]. Regular auditing of network traffic and system logs by network administrators will also reveal if any alarming number of failed logins have been attempted on one account. Automating a notification process to alert the network administrator of any such suspicious activity will allow him/her to look into the situation and stop a brute-force attack before it is successful [31]. However, it must be mentioned again that the use of either a password expiry system through dynamic authentication, or the use of graphical passwords significantly negates the effects of brute-force attacks. An intruder may still be able to gain access to a dynamic authentication system, but he/she would not be given access indefinitely, or until caught. Graphical passwords are immune to brute-force attacks currently as they do not utilize alphanumeric passwords [48,51,52].

Shoulder Surfing

Shoulder surfing is difficult to prevent, particularly where graphical authentication systems are concerned [42]. The only foolproof method is to ensure that a user is alone when authenticating themselves [34]. As this is not practical, it is suggested that graphical passwords should be implemented in conjunction with traditional alphanumeric passwords to reduce the risk of a shoulder-surfing attack [48]. The concept of "Cued Click Points" (CCPs) must also be considered for graphical authentication systems; this system reduces the chance of shoulder surfing by prompting users to click specific points on an image or set of images repeatedly in order to be authenticated. This would make it more difficult for a shoulder surfer to attain login information as opposed to the more common concept of only clicking a picture or object once [14,50].

Key Logging

Auditing of network traffic and system logs is paramount in preventing key logging attacks. As these can be either hardware or software based, it is important for system administrators to regularly audit their hardware and software, and conduct regular scans to attain if anything is not as it should be [53–55]. In our case, a multiple authentication could prevent key logging because we will be using clicking graphical password as one approach to prevent the key logging attack [44,56].

Password Encryption

Encryption is the process of encoding a plaintext message into an unreadable cipher text by using algorithms. There are two types of encryption schemes: (1) symmetric key and (2) public key. However, it is not entirely secure; there are methods to decrypt cipher text depending on the knowledge of the key, time, and monetary resources [57].

Hash function is an encryption algorithm that converts arbitrary data messages into fixed-size bit string messages called message digests or hash values. The ideal cryptographic hash function has three fundamental properties: (1) it must be able to easily convert digital information into a fixed-length hash value; (2) it must be computationally infeasible to derive any information about the input message from just the hash or message digest; and (3) it must be computationally infeasible to find two files that have the same hash [58,59]. Hash functions need to be able to resist cryptanalytic attacks, so it needs to have properties such as preimage resistance, second preimage resistance, and collision resistance. The ways to achieve these are to use one-way hash function.

The most commonly used hash algorithms are message-digest algorithm 5 (MD5) and secure hash algorithm 1 (SHA-1). The creation of MD5 is attributed to the cryptographer Ron Rivest, who believed it to be a more secure alternate to its predecessor MD4 which first came into use during 1992. The algorithm involves breaking down a plaintext file into separate 512-bit blocks; these blocks are then used as input for the algorithm that returns a 128-bit hash value which is unique to that file [58,60]. The SHA-1 is a collection of hash algorithms developed by National Institute of Standards and Technology (NIST) [61]. Recently, there have been significant advances in the understanding of hash functions; the research conducted shows that none of the algorithms currently used are 100% secure. However, in many cases it is infeasible to reverse these algorithms, in a process that is often extremely expensive and time-consuming [62]. This is true of both MD5 and SHA-1, which are widely implemented. These algorithms are susceptible to attacks known as collision attacks, preimage attacks, and second image attacks. Other prominent attacks include rainbow tables, brute-force attacks, and the birthday attack, which is based on the birthday paradox theory [61].

A group of scientists led by Dr. Wang published a paper in the year 2004, detailing research they had conducted, which lead them to create the same message digest using two different hashes. This contradicts one of the fundamentals of hashing, which states that it should be infeasible to find two files with the same hash. This was accomplished on average in 1 hour. Theoretically, these scientists have cracked MD5 [63]. Many scientists are attempting to build on the work of Dr. Wang, by finding ways to reduce the time taken to attack MD5 and SHA-1 using differing algebraic techniques. This leads to believe that MD5 and SHA-1 are no longer secure, even though they are widely used in industry [64].

To improve security of hash functions, new algorithms are being developed. SHA-2 was first published in 2001; it has four hash functions with digests that are 224, 256, 384, and 512 bits long. However, it is not perfect, in 2008 attacks were developed which overcame SHA-2's preimage resistance. It is also now possible to use collision attacks in order to overcome the first 24 out of 64 steps of SHA-256 and 46 out of 80 steps for SHA-512 [65]. However, there has never been a successful attack demonstrated on SHA-2. It is not as widely used as SHA-1 even if it is more secure, and this may be due to lack of support on Microsoft Windows or there is no urgency to implement SHA-2. This lack of urgency can be attributed to the fact that the attacks SHA-1 is susceptible to very time-consuming to implement, and subsequently it is not a widely used method of cracking passwords or intercepting sensitive data [62]. The next successor in the SHA family has been developed, but is currently still being tested and making it too new to use for serious security applications [66].

Comparison of Algorithms

SHA-1 and MD5 are both designed to be as fast and efficient as possible; a good computer can hash millions of password in a second. Because the hashed passwords are saved in database, brute-force attacks are a concern, if an attacker is able to retrieve the information in database. Hackers reverse calculation of equal speed can be performed by the hacker to obtain the hashed passwords. To solve this type of problems, several enhanced algorithms, such as bcrypt, PBKDF2, and scrypt, are developed to slow down brute-force attacks [67].

Compared to the MD5 and SHA-1 functions, MD5 will generate the 128 bits of output value with various length of input message, while SHA-1 processes the messages into 160 bits output. Both algorithms broke the message into pieces of 512-bit blocks, and each piece is 16 32-bit words. The message block process contains four stages called rounds. MD5 goes 64 rounds SHA-1 and does 80 rounds in the calculation. However, for SHA-2, the process only has 64 rounds, but because the internal state size increases from 160 to 512, compared to SHA-1, it has more combination probabilities; therefore, it is more secure than SHA-1 and MD5 functions against brute-force attacks. These figures are even higher for SHA-512/384, the internal state size is double the SHA-256/224 and the block size is also twice as much, process rounds are 80 steps [67].

SHA-1 function uses a sequence of logical functions from f_0 to f_{79}; every function f_t operates on three 32-bit words: x, y, and z, and then generates a 32-bit word as output. Each function also includes Ch(x, y, z) and Maj(x, y, z) functions. The exclusive-OR operation ($\oplus$) in these functions may be replaced by a bitwise OR operation ($\vee$) and may produce identical results. SHA-1 function of $f_t(x, y, z)$ is defined as follows:

$$
f_t(x, y, z) = \begin{cases}
\text{Ch}(x, y, z) = (x \wedge y) \oplus (x \wedge z) & 0 \le t \le 19 \\
\text{Parity}(x, y, z) = x \oplus y \oplus z & 20 \le t \le 39 \\
\text{Maj}(x, y, z) = (x \wedge y) \oplus (x \wedge z) \oplus (y \wedge z) & 40 \le t \le 59 \\
\text{Parity}(x, y, z) = x \oplus y \oplus z & 60 \le t \le 79
\end{cases} \quad (21.1)
$$

Equation 21.1 is the SHA-1 function of $f_t(x, y, z)$ [67]. The functions of SHA-2 family are using the familiar. They are divided into sections, SHA-224/SHA-256, and SHA-384/512. SHA-224 and 256 use six logical functions, and each function operates 32-bit words, and output of each function is a new 32-bit word.

$$
\text{Ch}(x, y, z) = (x \wedge y) \oplus (x \wedge z)
$$

$$
\text{Maj}(x, y, z) = (x \wedge y) \oplus (x \wedge z) \oplus (y \wedge z)
$$

$$
\sum_{0}^{\{256\}} (x) = \text{ROTR}^2(x) \oplus \text{ROTR}^{13}(x) \oplus \text{ROTR}^{22}(x)
$$

$$
\sum_{1}^{\{256\}} (x) = \text{ROTR}^6(x) \oplus \text{ROTR}^{11}(x) \oplus \text{ROTR}^{22}(x) \qquad (21.2)
$$

$$
\overset{\{256\}}{\underset{0}{\sigma}} (x) = \text{ROTR}^7(x) \oplus \text{ROTR}^{18}(x) \oplus \text{SHR}^3(x)
$$

$$
\overset{\{256\}}{\underset{1}{\sigma}} (x) = \text{ROTR}^{17}(x) \oplus \text{ROTR}^{19}(x) \oplus \text{SHR}^{10}(x)
$$

Equation 21.2 is the official function of SHA-224 and SHA-256. SHA-384 and SHA-512 also use six logical functions, but each function operates 64-bit words; the output result is a new 64-bit word.

$$\mathrm{Ch}(x, y, z) = (x \wedge y) \oplus (x \wedge z)$$

$$\mathrm{Maj}(x, y, z) = (x \wedge y) \oplus (x \wedge z) \oplus (y \wedge z)$$

$$\sum_{0}^{\{512\}}(x) = \mathrm{ROTR}^{28}(x) \oplus \mathrm{ROTR}^{34}(x) \oplus \mathrm{ROTR}^{39}(x)$$

$$\sum_{1}^{\{512\}}(x) = \mathrm{ROTR}^{14}(x) \oplus \mathrm{ROTR}^{18}(x) \oplus \mathrm{ROTR}^{41}(x)$$

$$\overset{\{512\}}{\underset{0}{\sigma}}(x) = \mathrm{ROTR}^{1}(x) \oplus \mathrm{ROTR}^{8}(x) \oplus \mathrm{SHR}^{7}(x)$$

$$\overset{\{512\}}{\underset{1}{\sigma}}(x) = \mathrm{ROTR}^{19}(x) \oplus \mathrm{ROTR}^{61}(x) \oplus \mathrm{SHR}^{6}(x)$$

(21.3)

Equation 21.3 is the official functions of SHA-384 and SHA-512 [68].

Password Storage

Regardless of the device, passwords should never be stored in the form of plaintext for security reasons. Instead, they must be first encrypted before being stored. Hashing algorithms, such as the ones previously discussed, facilitate this encryption. The process of authentication usually involves the user inputting their login details; their password is then hashed. This hash value or message digest is then compared with the hash value that was created when the user first stipulated his or her password. If the hash values are identical, the user is authenticated and granted access. As one of the fundamentals of hashing states that it should be infeasible to find two files with the same hash, security is facilitated in this manner.

Automatic Teller Machine

Automatic teller machines (ATMs) also adopt a layered approach to security; this approach is becoming more and more popular in the security industry as security becomes more important to users and customers. ATMs utilize a token and knowledge-based password in conjunction to form a two-layered authentication process [69]. The token is ATM card, and the knowledge key is four-digit personal identification number (PIN). Upon a user inserting their card, the machine reads the magnetic strip or chip embedded in the card; this strip or chip contains information about user and an identification number which can be used by the ATM to retrieve the account information.

When the user inputs their PIN, this four-digit number is hashed. Using the information stored on the card, the ATM is able to communicate with a server and retrieve a hashed version of your PIN [70].

The ATM compares the digest stored on the server with the one that it has just created; both digests are created using the same algorithm and subsequently should be identical if the user has entered the correct PIN. If the two digests match, the user is authenticated and granted access to their bank accounts [71].

While this approach may seem secure in theory, it is extremely susceptible to shoulder-surfing attacks and skimming attacks, which involve the attack storing additional hardware on the ATM. Recently, the incorporation of radio-frequency identification (RFID) chips in cards has become prominent, and this allows users to make contactless transactions by simply waving their card near a scanner; while this is extremely convenient, purchases below a certain value often do not require entering a PIN number, which obviously raises security issues, especially in the case of lost or misplaced cards [72]. Subsequently, numerous advances have been made with regard to incorporating biometric scanners into ATMs. Companies have begun testing ATMs equipped with fingerprint scanners as well as retina scanners, removing the need for users to use a card at all, or fortifying security when used in conjunction with a card. Research has also been conducted into developing a card which has a finger print scanner embedded into itself, thereby making it impossible for anybody but the owner to use it. While these measures may seem extreme and unnecessary at this point, as traditional approaches become less secure, the need for more innovative authentication methods such as those mentioned earlier becomes apparent. This technology may not be implemented immediately but will find its place in society in the coming years.

Cloud Computing

Cloud computing is becoming increasingly popular, not just for home users but for organizations as well. Passwords are stored in the cloud, in much the same way as they are stored on local machines. The user creates a password, and this is then hashed and used for comparison each subsequent time the user attempts to be authenticated. The main difference is that the hash is not stored on the user's machine, but rather in the cloud on a database server. When a user attempts authentication, their password input is hashed and compared to a hash stored on the cloud server. While cloud computing presents many benefits and has many modern-day applications, it is not also entirely secure. Numerous companies have fallen prey to online attackers who gained access to numerous passwords by accessing cloud servers, such as LinkedIn and DropBox [73,74].

Cloud computing can be used for malicious purposes too; hackers employ "dark clouds" or botnets to conduct illegal activities. Botnets are large groups of computers, which unbeknownst to the owner are being used to spread malware and crack passwords. If a hacker was to infiltrate a cloud password database, the hacker would be presented with a series of hash values; these values can be susceptible to a brute-force attack in order to retrieve passwords. This would be an extremely long process that would not be feasibly on a single computer, but using botnets hackers can "farm out" sections of the process to computers all over the world, subsequently achieving in hours what would usually take years. This has been demonstrated by a German researcher that utilized Amazon's cloud computing facilities to crack a six-character password in 49 minutes, at a cost of US$2.10. This illustrates how dark cloud computing can be used in a cost- and time-effective manner for negative purposes [75].

Unix/Linux System

Passwords in Unix or Linux are stored in the directory/etc/passwd, in an encoded rather than encrypted format. This is because a traditional algorithm is not used; subsequently, the digest is encoded and not encrypted. Before the password is encoded, a random key known as the "salt" is chosen. This key is a numeric value between 1 and 4096. Appending the "salt" to the password, and then performing a one-way hash, provides the digest that is then stored. When the user attempts to be authenticated, the salt is once again added to their input before the hashing procedure is carried out and the two hashes compared. If both hashes are identical, access is granted. Despite utilizing this method, an attacker would still be able to conduct a dictionary attack to retrieve a password if they had access to the/etc/passwd file. To combat this, a technique known as shadow passwording was developed for Linux. This technique creates a second file which is only available to the system administrator; this shadow password file contains all the hashes of the passwords, while the other/etc/passwd file is left either empty or with false data to throw off an attacker. This is an effective way of deterring attackers from attempting to conduct time-consuming dictionary attacks on Linux systems, as the passwords they may attain from the public/etc/passwd file may be completely obsolete [42].

Microsoft Windows

In the Microsoft Windows NT/XP/Vista/Windows 7 environment, Security Account Manager (SAM) is the database where system passwords are stored, using either LM/NTLM hashing [76]. In Microsoft Windows, access to this database is not granted and modification is not possible while Windows is running and even while Windows is shutdown or the "Blue Screen of Death" occurs, the lock of SAM files will still be effective. It is said that an in-memory copy of SAM file could be dumped by a technique, which would then make brute-force attacks possible. For added security, SAM can implement SYSKEY, which could prevent offline password cracking even if an attacker had a copy of it. LM hashing is the primary hash function of the Microsoft LAN Manager and Microsoft Windows [76]. Moreover, NTLM is more secure because of its implementation of Unicode support [77]. SAM is the part of the registry which can be found on the hard disk. However, LM and NTLM are no longer implemented as they have been cracked, by default NTLM2 is now the standard implemented by Microsoft for use with the SAM database in conjunction with the Kereberos authentication method.

Mobile

The most common iOS devices such as the iPad and iPhone are both mobile devices, which could potentially store extremely sensitive data. Also they are gateway devices that can be synced with one's home computer, e-mail, and other various cloud accounts; therefore, if a device was to be compromised, other accounts could also be attacked. iOS systems implement authentication systems in the form of a four-digit numeric PIN by default. If this PIN is entered incorrectly five times, the device gets locked for a minute to prevent further guess attempts. By way of configuration, the user can specify a permanent lockout if a number of login attempts are unsuccessful forcing the device to be connected to the computer it is synced with in order for it to be unlocked once again. For increased security, the user has the option of turning off the simple passcode feature in the Settings tab; this will enable alphanumeric passwords to be used as opposed to standard numeric passwords. The required password feature can also be utilized to prompt the user to reenter a password after a fixed period of time. iOS devices employ the Advanced Encryption Standard (AES) algorithm or AES-256 [12,53]. This encryption method was developed by the US government in 2001 [12]. In theory, AES-128 has already been cracked, by a group of Microsoft researchers, however they found that it would take a trillion computers, two billion years to crack one AES-128 key. By extrapolation, this makes AES-256 virtually impenetrable at current [12]. It has been reported that jailbreaking iPhones or iPads, in order to be able to install third-party apps, will in fact make it susceptible to attacks which could steal passwords. Conversely, if a phone were to be stolen, an attacker could gain access to data [12].

Balancing Security, Convenience, and Education

For an authentication system to be successful, it must be both secure to prevent unwanted intrusion and convenient so the user is not hindered by it. A user must feel as though security is an integral part of the work that they are carrying out, rather than perceiving it to be an unrelated task or set of guidelines that they are forced to follow. The user therefore must see the benefits of maintaining security, to them personally and to the organization as a whole [78].

When passwords systems become too complex, or stipulate too many constraints, convenience for the user is decreased, this results in poor recall rates of passwords [79]. This ultimately is inefficient as more time is spent trying to authenticate users, as the help desk staff must get involved when passwords are forgotten, resulting in increased expenditure. In contrast, if a password is too simple, security issues become more apparent, although authentication may now be more efficient due to higher recall rates, the chances of unwanted intruders accessing the system is also increased. Subsequently, the need to balance these two variables, security and convenience become apparent. Both are equally important and neglecting either will have adverse effects on an organization as a whole [14,15,80].

It has been suggested that educating users on the need for security and ultimately the need to protect sensitive data will bolster security and convenience. Teaching users good password etiquette and practices while equipping them with skills to remember their passwords facilitates this [6]. Forty-seven percent of British employees admitted to using personal details in their passwords, such as their name or date of birth; this is an example of bad password etiquette [81]. Likewise, using the same password for multiple accounts should also be avoided, because if an account that is not as well protected is compromised, an attacker may then have access to other more private information on another system. Users must also refrain from writing down passwords in an effort to remember them; while this may seem like common sense, studies show that many users still engage in this practice [79,80]. It has been suggested that implementing proper security policies within larger organizations can help overcome issues such as this. Such policies can be stipulated in IT code of conduct manuals/agreements or IT policy documents, and reinforced through training programs, but can also be built into security systems. For example, removing the ability for employees to utilize their user names as part of their passwords, or requiring a certain number of characters to be used for a password with a number of those characters being symbols or digits [5,82].

IT policy and code of conduct requirements may however be overlooked by staff in favor of convenience. If they feel that requirements are too stringent, they may perceive security as a process external to their work, as opposed to something that they can directly benefit from, subsequently it is important to educate staff through seminars or meetings on the value of maintaining the organizations IT security [79,82]. Convenience for staff can also be improved by teaching them memory recall techniques for memorizing passwords or helping them create and remember stronger passwords through various password creation techniques [83]. For example, the base password creation technique is especially useful for remembering passwords for multiple accounts; employees are encouraged to use a base suffix or prefix for each password such as "e$r8," which then becomes the suffix or prefix for each password making them easier to remember. This technique is strengthened if employees are taught password formation strategies, which encourage them to use letters, numbers, and symbols. As opposed to using a word followed by a series of numbers, which is quite common, an employee could for example use a set number of random characters, followed by a number, a symbol, and then three more numbers. This formula can then be used not only to create stronger passwords but also to recall said passwords [13].

Multiple-Layer Authentication

As each method of authentication has its own flaws and strengths, the optimum way to ensure that both security and convenience are balanced is to employ a layered approach in the form of a multiple authentication system. Such a system would employ a series of different authentication methods in order to verify the identity of the user. Each layer adds additional security, while having multiple layers promotes convenience, for example, users can rely on a graphical password in conjunction with a passphrase as opposed to a sole passphrase, which is extremely long, in order to facilitate security. While different authentication methods are susceptible to different attacks on their own, when combined with other methods, they can produce a more holistic security system. Layered security mechanisms will deter intruders, by making it far harder to access the system illegally. Also if an attack were to be developed to infiltrate one layered security system, it would not necessarily be successful against another, as different combinations of layers would be able to prevent different forms of attacks [14,15,55,84].

For example, a three-layer authentication system could utilize the traditional username and password method as the initial layer; while this layer on its own may not facilitate security adequately, it is convenient as most users have had experience using this type of authentication technique [15]. Further, the second and third layers will strengthen the system. The second layer will be the most secure, as it is the most innovative, it involves a graphical system where the user must select familiar pictures from a grid in the correct sequence. However, an additional constraint is added, each picture must be clicked a specific number of times [16,85,86]. The pictures that are to be clicked, the order in which they must be clicked, and the number of times each must be clicked are all stipulated by the user during account creation. This is a means of promoting recall rates, as constraints are not forced upon users, but determined themselves within a framework [79]. The pictures themselves however will be selected from a random set, removing the risk of the user uploading personal pictures, which would make the system partially susceptible to guess attacks. Such a technique will work well when coupled with an education program, which shows users the value of the system and how to implement it properly. While many users may be unfamiliar with this kind of authentication method, the concept is easily enough to grasp, and with proper education users would grow to become comfortable with it [87]. The third layer involves a cognitive question and answer system where the user must answer a random question from a set of predetermined questions. This is a very user-friendly method of authentication as users are more likely to remember personal details. Again, this is not the most secure method, but when coupled with the other layers, it produces a more holistic system, which is as user-friendly as it is secure, provided the users are educated adequately [88].

Conclusion

The need for authentication systems to safeguard electronic data is a constant; as information technology advances, it becomes apparent that the technology and mechanisms in place to safeguard our data must evolve too. Traditional systems of authentication, although widely implemented, often fail to meet security needs [89,90]; this is in part due to the fact that the number of ways these systems can be attacked is steadily growing. Therefore, the implementation of new systems that use innovative methods of authentication in conjunction with traditional methods to produce multiple-layered authentication systems becomes necessary.

Conversely, while users are familiar with many of these systems and utilize them on a day-to-day basis, they are often not very user-friendly or convenient. This leads to bad password practice or etiquette, where users resort to using the same password for multiple accounts or even writing them down. This is an affront to the security of an authentication system, but must be attributed to the poor design of the system in the first place. Equilibrium must be achieved between convenience and security. More specifically, the system must be simple to use, but also secure [62]. This can be achieved through educating users, implementing security policies, and making them aware of why such policies are necessary. Giving users the tools to create secure passwords, remembering those passwords, and helping them understand why security is a priority are paramount. If this technique is coupled with the system that is built with security, as well as convenience in mind, a holistic approach to user authentication and security in general can be achieved.

References

1. Skaff, G. (2007). An alternative to passwords? *Biometric Technology Today*, 15(5), pp. 10–11.
2. Matthews, T. (2012). Passwords are not enough. *Computer Fraud & Security*, 2012(5), p. 18.
3. Renaud, K. & Goucher, W. (2012). Email passwords: Pushing on a latched door. *Computer Fraud & Security*, 2012(9), p. 16.
4. Schaffer, K. (2011). Are password requirements too difficult? *Computer*, 44(12), pp. 90–92.
5. Charoen, D., Raman, M., & Olfman, L. (2008). Improving end user behaviour in password utilization: An action research initiative. *Systemic Practice and Action Research*, 21(1), pp. 55–72.
6. Zhang, J., Luo, X., Akkaladevi, S., & Ziegelmayer, J. (2009). Improving multiple-password recall: An empirical study. *European Journal of Information Systems*, 18(2), pp. 165–176.
7. Biddle, R., Chiasson, S., & Van Oorschot, P. C. (2012). Graphical passwords: Learning from the first twelve years. *ACM Computing Surveys*, 44(4), pp. 1–41.
8. Eljetlawi, A. M. & Ithnin, N. B. (2009). Graphical password: Usable graphical password prototype. *Journal of International Commercial Law and Technology*, 4(4), pp. 299–310.
9. Huang, C., Ma, S., & Chen, K. (2011). Using one-time passwords to prevent password phishing attacks. *Journal of Network and Computer Applications*, 34(4), pp. 1292–1301.
10. Dell'Amico, M., Michiardi, P., & Roudier, Y. (2010). Password strength: An empirical analysis, in *InfoCom*, San Diego, CA, March 14–19, pp. 1–9.
11. Singh, C., Singh, L., & Marks, E. (2011). Investigating the combination of text and graphical passwords for a more secure and usable experience. *International Journal of Network Security & Its Applications*, 3(2), pp. 78–95.
12. Neumann, P. G. (1994). Risks of passwords. Association for computing machinery. *Communications of the ACM*, 37(4), p. 126.
13. Helkala, K. (2011). Password education based on guidelines tailored to different password categories. *Journal of Computers*, 6(5), pp. 969–975.
14. Gyorffy, J. C., Tappenden, A. F., & Miller, J. (2011). Token-based graphical password authentication. *International Journal of Information Security*, 10(6), pp. 321–336.
15. Kennedy, D. (2007). Power passwords. *ABA Journal*, December, p. 59.
16. Van Oorschot, P. C. & Biddle, R. (2007). *Graphical Password Authentication Using Cued Click Points*. Berlin: Springer, pp. 359–374.
17. Haga, W. J. & Zviran, M. (1991). Question-and-answer passwords: An empirical evaluation. *Information Systems*, 16(3), pp. 335–343.
18. Shim, K. (2006). Security flaws of remote user access over insecure networks. *Computer Communications*, 30(1), pp. 117–121.
19. Weber, J. E., Guster, D., Safonov, P., & Schmidt, M. B. (2008). Weak password security: An empirical study. *Information Security Journal*, 17(1), p. 45.
20. Thilmany, J. (2004). Password protected plus. *Mechanical Engineering*, 126(4), p. 14.
21. Das, M. L., Saxena, A., & Gulati, V. P. (2004). A dynamic ID-based remote user authentication scheme. *IEEE Transactions on Consumer Electronics*, 50(2), pp. 629–631.

22. Chakrabarti, S. & Singbal, M. (2007). Password-based authentication: Preventing dictionary attacks. *Computer*, 40(6), pp. 68–74.

23. Walters, M. & Matulich, E. (2011). Assessing password threats: Implications for formulating university password policies. *Journal of Technology Research*, 2, p. 1.

24. Arutyunov, V. V. & Natkin, N. S. (2010). Comparative analysis of biometric systems for information protection. *Scientific and Technical Information Processing*, 37(2), pp. 87–93.

25. Kirda, E. & Kruegel, C. (2006). Protecting users against phishing attacks. *The Computer Journal*, 49(5), p. 554.

26. Katz, J., Ostrovsky, R., & Yung, M. (2010). Efficient and secure authenticated key exchange using weak passwords. *Journal of the Association for Computing Machinery*, 57(1), p. 3.

27. Marechal, S. (2008). Advances in password cracking. *Journal in Computer Virology*, 4(1), pp. 73–81.

28. Sood, S. K. (2011). Cookie-based virtual password authentication protocol. *Information Security Journal*, 20(2), p. 100.

29. Sun, H., Chen, Y., & Lin, Y. (2012). oPass: A user authentication protocol resistant to password stealing and password reuse attacks. *IEEE Transactions on Information Forensics and Security*, 7(2), pp. 651–663.

30. Zhang, L. & McDowell, W. C. (2009). Am I really at risk? Determinants of online users' intentions to use strong passwords. *Journal of Internet Commerce*, 8(3–4), pp. 180–197.

31. Magruder, S., Lewis Jr., S. X., & Burks, E. J. (2007). Technical report: More secure passwords. *Journal of International Technology and Information Management*, 16(1), p. 87.

32. Connolly, P. J. (2011). Passwords done poorly. *eWeek*, 28(5), p. 42.

33. Vu, K. L., Proctor, R. W., Bhargav-Spantzel, A., Tai, B., Cook, J., & Eugene Schultz, E. (2007). Improving password security and memorability to protect personal and organizational information. *International Journal of Human-Computer Studies*, 65(8), pp. 744–757.

34. Alsulaiman, F. A. & El Saddik, A. (2008). Three-dimensional password for more secure authentication. *IEEE Transactions on Instrumentation and Measurement*, 57(9), pp. 1929–1938.

35. Seng, L. K., Ithnin, N., & Mammi, H. K. (2011). User affinity of choice-features of mobile device graphical password scheme anti-shoulder surfing mechanism. *International Journal of Computer Science Issues*, 8(4), pp. 255–261.

36. Kumar, R., Kumar, N., & Vats, N. (2011). Protecting the content—Using digest authentication system and other techniques. *International Journal of Computer Science and Information Security*, 9(8), pp. 76–85.

37. Antonopoulos, A. (2009). New secure password rules. *Network World*, 26(30), p. 17.

38. Nam, J., Paik, J., Kim, U. M., & Won, D. (2008). Security enhancement to a password-authenticated group key exchange protocol for mobile ad-hoc networks. *IEEE Communications Letters*, 12(2), pp. 127–129.

39. Chang, T., Tsai, C., & Lin, J. (2012). A graphical-based password keystroke dynamic authentication system for touch screen handheld mobile devices. *The Journal of Systems & Software*, 85(5), pp. 1157–1165.

40. Golinsky, E. & Ginsberg, S. (2009). Be careful with your online passwords. *Work & Family Life*, 23(9), p. 3.
41. Furnell, S. (2007). An assessment of website password practices. *Computers & Security*, 26(7), pp. 445–451.
42. Yang, S. S. & Choi, H. (2010). Vulnerability analysis and the practical implications of a server-challenge-based one-time password system. *Information Management & Computer Security*, 18(2), pp. 86–100.
43. Inman, M. (2008). Malicious hardware steals passwords. *New Scientist*, 198(2654), p. 26.
44. Jang, W., Cho, S., & Lee, H. (2011). *User-Oriented Pseudo Biometric Image Based One-Time Password Mechanism on Smart Phone*. Berlin: Springer, pp. 395–404.
45. Juang, W.-S., Chen, S.-T., & Liaw, H.-T. (2008). Robust and efficient password-authenticated key agreement using smart cards. *IEEE Transactions on Industrial Electronics*, 55(6), pp. 2551–2556.
46. Wood, C. C. (1997). A secure password storage policy. *Information Management & Computer Security*, 5(2), pp. 79–80.
47. Hancock, B. (2001). Risk management experts think passwords are too weak. *Computer & Security*, 20(8), pp. 651–652.
48. Malempati, S. & Mogalla, S. (2011). User authentication using native language passwords. *International Journal of Network Security & Its Applications*, 3(6), pp. 149–160.
49. Anonymous. (2000). Securing user passwords. Association for Computing Machinery. *Communications of the ACM*, 43(4), p. 11.
50. Singh, S. & Agarwal, G. (2011). Integration of sound signature in graphical password authentication system. *International Journal of Computer Applications*, 12(9), pp. 11–13.
51. Lovelace, H. W. (2005). Password complexity puts security at risk. *InformationWeek*, 1027, p. 68.
52. Lin, C. & Hwang, T. (2003). A password authentication scheme with secure password updating. *Computers & Security*, 22(1), pp. 68–72.
53. Shim, K. (2005). Off-line password-guessing attacks on the generalized key agreement and password authentication protocol. *Applied Mathematics and Computation*, 169(1), pp. 511–515.
54. Shifrin, T. (2005). Password risks grow. *Computer Weekly*, October, p. 10.
55. Teh, P. S., Teoh, A. B. J., Tee, C., & Ong, T. S. (2010). Keystroke dynamics in password authentication enhancement. *Expert Systems with Applications*, 37(12), pp. 8618–8627.
56. Oreku, G. S. & Li, J. (2009). End user authentication (EUA) model and password for security. *Journal of Organizational and End User Computing*, 21(2), pp. 28–43.
57. Basar, M. (2011). Summarizing data for secure transaction: A hash algorithm. *African Journal of Business Management*, 5(34), pp. 13211–13216.
58. Kasgar, A. K., Agrawal, J., & Shahu, S. (2012). New modified 256-bit MD5 algorithm with SHA compression function. *International Journal of Computer Applications*, 42(12), pp. 47–51.
59. Thompson, E. (2005). MD5 collisions and the impact on computer forensics. *Digital Investigation*, 2(1), pp. 36–40.
60. Matusiewicz, K. & Pieprzyk, J. (2006). Finding good differential patterns for attacks on SHA-1, in *International Workshop on Coding and Cryptography*.

61. Lee, C.-Y., Lin, C.-H., Chen, C.-J., Yeh, Y.-S. (2012). Generalized secure hash algorithm: SHA-X. *International Journal of Advancements in Computing Technology*, 4(7).
62. Tam, L., Glassman, M., & Vandenwauver, M. (2010). The psychology of password management: A tradeoff between security and convenience. *Behaviour & Information Technology*, 29(3), pp. 233–244.
63. Liang, J. & Lai, X. (2007). Improved collision attack on hash function MD5. *Journal of Computer Science and Technology*, 22(1), pp. 79–87.
64. Stevens, M. (2005). Fast collision attack on MD5, *Cryptology ePrint Archive, Report No. 2006/104*.
65. Sasaki, Y., Wang, L., & Aoki, K. (2009). Preimage attacks on 41-step SHA-256 and 46-step SHA-512, eprint.iacr.org/2009/479.pdf
66. Bertoni, G., Daeman, J., Peeters, M., & Van Assche, G. (2011). The Keccak SHA-3 submission, keccak.noekeon.org.
67. Gueron, S., Johnson, S., & Walker, J. (2010). SHA-512/256. *IACR Cryptology ePrint Archive*, 2010, p. 548.
68. Federal Information Processing Standards Publication. (2002). *Secure Hash Standard*. Gaithersburg, MD: National Institute of Standards and Technology.
69. Moncur, W. & Leplâtre, G. (2007). Pictures at the ATM: Exploring the usability of multiple graphical passwords, pp. 887–894 , http://dl.acm.org/citation.cfm?id=1240758
70. ATMs pin customers down to single translations (to control misuse of ATM cards). (2011). *India Business Insight*, January 6.
71. Trombley, P. (2008). ATM pin security. *Credit Union Management*, 31(7), p. 38.
72. Heydt-Benjamin, T. S., Bailey, D. V., Fu, K., Juels, A., & O'Hare, T. (2009). Vulnerabilities in first-generation RFID-enabled credit cards. *Economic Perspectives*, 33(1), p. 50.
73. Drop box password breach highlights cloud security weaknesses. (August 3, 2012). *eWeek*.
74. LinkedIn password theft underscores cloud security dangers. (June 18, 2012). *Baseline*.
75. Antonopoulos, A. (2010). Password cracking in the cloud. *Network World*, 27(22), p. 17.
76. Goel, C. K. & Arya, G. (2012). Hacking of passwords in windows environment. *International Journal of Computer Science and Communication Networks*, 2(3), pp. 430–435.
77. Munro, K. (2006). FT.com site: Security matters: Passwords. FT.com, p. 1.
78. Duggan, G. B., Johnson, H., & Grawemeyer, B. (2012). Rational security: Modeling everyday password use. *International Journal of Human-Computer Studies*, 70(6), pp. 415–431.
79. Campbell, J., Ma, W., & Kleeman, D. (2011). Impact of restrictive composition policy on user password choices. *Behavior & Information Technology*, 30(3), pp. 379–388.
80. Fuglerud, K. & Dale, O. (2011). Secure and inclusive authentication with a talking mobile one-time-password client. *IEEE Security & Privacy Magazine*, 9(2), pp. 27–34.
81. Brits are poor password pickers. (2001). *Computer Fraud & Security*, 2001(7), p. 3.
82. Farrell, S. & Farrell, S. (2008). Password policy purgatory. *IEEE Internet Computing*, 12(5), pp. 84–87.

83. El Emam, K., Moreau, K., & Jonker, E. (2011). How strong are passwords used to protect personal health information in clinical trials? *Journal of Medical Internet Research*, 13(1), p. 18.

84. Jacobs, D. L. (2011). Password protection. *Forbes*, 187(6), p. 62.

85. Khan, W. Z., Aalsalem, M. Y., & Xiang, Y. (2011). A graphical password based system for small mobile devices. *International Journal of Computer Science Issues*, 10(5), pp. 145–154.

86. Roach, T. J. (2008). Password pose a problem. *Rock Products*, 111(6), p. 4.

87. Wiedenbeck, S., Waters, J., Birget, J., Brodskiy, A., & Memon, N. (2005). PassPoints: Design and longitudinal evaluation of a graphical password system. *International Journal of Human-Computer Studies*, 63(1), pp. 102–127.

88. Lazar, L., Tikolsky, O., Glezer, C., & Zviran, M. (2011). Personalized cognitive passwords: An exploratory assessment. *Information Management & Computer Security*, 19(1), pp. 25–41.

89. Hayes, F. (2009). Password fail. *Computerworld*, 43(2), p. 36.

90. Schwartz, M. J. (March 2012). Security fail: Apple iOS password managers. *InformationWeek*, http://www.informationweek.com/security/encryption/security-fail-apple-ios-password-manager/232602738

91. Thangavel, T. S. & Krishnan, A. (2010). Provable secured hash password authentication. *International Journal of Computer Applications*, 1(19), pp. 38–45.

Index